Lial Video Library Workbook

Christine Verity

Beginning and Intermediate Algebra

Sixth Edition

Margaret L. Lial
American River College

John Hornsby
University of New Orleans

Terry McGinnis

PEARSON

Boston   Columbus   Indianapolis   New York   San Francisco
Amsterdam   Cape Town   Dubai   London   Madrid   Milan   Munich   Paris   Montreal   Toronto
Delhi   Mexico City   São Paulo   Sydney   Hong Kong   Seoul   Singapore   Taipei   Tokyo

The author and publisher of this book have used their best efforts in preparing this book. These efforts include the development, research, and testing of the theories and programs to determine their effectiveness. The author and publisher make no warranty of any kind, expressed or implied, with regard to these programs or the documentation contained in this book. The author and publisher shall not be liable in any event for incidental or consequential damages in connection with, or arising out of, the furnishing, performance, or use of these programs.

Reproduced by Pearson from electronic files supplied by the author.

Copyright © 2016, 2012, 2008 by Pearson Education, Inc. or its affiliates. All Rights Reserved. Printed in the United States of America. This publication is protected by copyright, and permission should be obtained from the publisher prior to any prohibited reproduction, storage in a retrieval system, or transmission in any form or by any means, electronic, mechanical, photocopying, recording, or otherwise. For information regarding permissions, request forms and the appropriate contacts within the Pearson Education Global Rights and Permissions Department, please visit www.pearsoned.com/permissions.

ISBN-13: 978-0-321-96972-9
ISBN-10: 0-321-96972-3

1 2 3 4 5 6 RRD-H 19 18 17 16 15

www.pearsonhighered.com

PEARSON

CONTENTS

CHAPTER R	PREALGEBRA REVIEW	1
CHAPTER 1	THE REAL NUMBER SYSTEM	19
CHAPTER 2	LINEAR EQUATIONS AND INEQUALITIES IN ONE VARIABLE	57
CHAPTER 3	LINEAR EQUATIONS AND INEQUALITIES IN TWO VARIABLES	103
CHAPTER 4	EXPONENTS AND POLYNOMIALS	141
CHAPTER 5	FACTORING AND APPLICATIONS	181
CHAPTER 6	RATIONAL EXPRESSIONS AND APPLICATIONS	213
CHAPTER 7	GRAPHS, LINEAR EQUATIONS, AND SYSTEMS	251
CHAPTER 8	INEQUALITIES AND ABSOLUTE VALUE	313
CHAPTER 9	RELATIONS AND FUNCTIONS	337
CHAPTER 10	ROOTS, RADICALS, AND ROOT FUNCTIONS	359
CHAPTER 11	QUADRATIC EQUATIONS, INEQUALITIES AND FUNCTIONS	403
CHAPTER 12	INVERSE, EXPONENTIAL, AND LOGARITHMIC FUNCTIONS	459
ANSWERS		501

Name: Date:
Instructor: Section:

Chapter R PREALGEBRA REVIEW

R.1 Fractions

Learning Objectives
1. Learn the definition of *factor*.
2. Write fractions in lowest terms.
3. Convert between improper fractions and mixed numbers.
4. Multiply and divide fractions.
5. Add and subtract fractions.
6. Solve applied problems that involve fractions.
7. Interpret data from a circle graph.

Key Terms

Use the vocabulary terms listed below to complete each statement in exercises 1–9.

numerator denominator proper fraction

improper fraction equivalent fractions lowest terms

prime number composite number prime factorization

1. Two fractions are _____ when they represent the same portion of a whole.

2. A fraction whose numerator is larger than its denominator is called an _____.

3. In the fraction $\frac{2}{9}$, the 2 is the _____.

4. A fraction whose denominator is larger than its numerator is called a _____.

5. The _____ of a fraction shows the number of equal parts in a whole.

6. A _____ has at least one factor other than itself and 1.

7. In a _____ every factor is a prime number.

8. The factors of a _____ are itself and 1.

9. A fraction is written in _____ when its numerator and denominator have no common factor other than 1.

Name: Date:
Instructor: Section:

Objective 1 Learn the definition of *factor*.

Video Examples

Review this example for Objective 1:
1. Write the number in prime factored form.

    48

    We use a factor tree, as shown below. The prime factors are boxed.

    Divide by the least prime factor of 54, which is 2. $54 = 2 \cdot 27$

    Divide 27 by 3 to find two factors of 27. $54 = 2 \cdot 3 \cdot 9$

    Now factor 9 as $3 \cdot 3$. $54 = 2 \cdot 3 \cdot 3 \cdot 3$

    $$\begin{array}{c} 54 \\ / \ \backslash \\ \boxed{2} \cdot 27 \\ / \ \backslash \\ \boxed{3} \cdot 9 \\ / \ \backslash \\ \boxed{3} \cdot \boxed{3} \end{array}$$

Now Try:
1. Write the number in prime factored form.
   210

   _____

Objective 1 Practice Exercises

For extra help, see Example 1 on page 2 of your text.

Write each number in prime factored form.

1. 98

2. 256

3. 546

1. _____

2. _____

3. _____

Objective 2 Write fractions in lowest terms.

Video Examples

Review this example for Objective 2:
2. Write the fraction in lowest terms.

    $\dfrac{15}{25}$

    $\dfrac{15}{25} = \dfrac{3 \cdot 5}{5 \cdot 5} = \dfrac{3}{5} \cdot \dfrac{5}{5} = \dfrac{3}{5} \cdot 1 = \dfrac{3}{5}$

Now Try:
2. Write the fraction in lowest terms.

   $\dfrac{9}{15}$

   _____

Name: Date:
Instructor: Section:

Objective 2 Practice Exercises

For extra help, see Example 2 on page 3 of your text.

Write each fraction in lowest terms.

4. $\dfrac{42}{150}$ 4. _____

5. $\dfrac{180}{216}$ 5. _____

6. $\dfrac{132}{292}$ 6. _____

Objective 3   Convert between improper fractions and mixed numbers.

Video Examples

Review these examples for Objective 3:	**Now Try:**
3. Write $\dfrac{53}{6}$ as a mixed number. We divide the numerator of the improper fraction by the denominator. $\begin{array}{r} 8 \\ 6\overline{)53} \\ \underline{48} \\ 5 \end{array}$ $\dfrac{53}{6} = 8\dfrac{5}{6}$	3. Write $\dfrac{74}{5}$ as a mixed number. _____
4. Write $5\dfrac{3}{8}$ as an improper fraction. We multiply the denominator of the fraction by the whole number and add the numerator to get the numerator of the improper fraction. $8 \cdot 5 + 3 = 40 + 3 = 43$ The denominator of the improper fraction is the same as the denominator in the mixed number, which is 8 here. Thus, $5\dfrac{3}{8} = \dfrac{43}{8}$	4. Write $12\dfrac{2}{7}$ as an improper fraction. _____

Name: Date:
Instructor: Section:

Objective 3 Practice Exercises

For extra help, see Examples 3–4 on page 4 of your text.

Write the improper fraction as a mixed number.

7. $\dfrac{321}{15}$ 7. _____

Write each mixed number as an improper fraction.

8. $13\dfrac{5}{9}$ 8. _____

9. $22\dfrac{2}{11}$ 9. _____

Objective 4 Multiply and divide fractions.

Video Examples

Review these examples for Objective 4:

5. Find the product, and write it in lowest terms.

   $\dfrac{5}{12} \cdot \dfrac{3}{10}$

   $\dfrac{5}{12} \cdot \dfrac{3}{10} = \dfrac{5 \cdot 3}{12 \cdot 10}$ Multiply numerators.
   Multiply denominators.

   $= \dfrac{5 \cdot 3}{4 \cdot 3 \cdot 2 \cdot 5}$ Factor the denominator.

   $= \dfrac{1}{4 \cdot 2}$ $\dfrac{3}{3} = 1$ and $\dfrac{5}{5} = 1$

   $= \dfrac{1}{8}$ Write in lowest terms.

6. Find the quotient, and write it in lowest terms.

   $\dfrac{2}{5} \div \dfrac{8}{7}$

   $\dfrac{2}{5} \div \dfrac{8}{7} = \dfrac{2}{5} \cdot \dfrac{7}{8}$ Multiply by the reciprocal.

   $= \dfrac{2 \cdot 7}{5 \cdot 4 \cdot 2}$ Multiply and factor.

   $= \dfrac{7}{20}$

Now Try:

5. Find the product, and write it in lowest terms.

   $\dfrac{7}{15} \cdot \dfrac{3}{14}$

   _____

6. Find the quotient, and write it in lowest terms.

   $\dfrac{6}{7} \div \dfrac{9}{8}$

   _____

Name: Date:
Instructor: Section:

Objective 4 Practice Exercises

For extra help, see Examples 5–6 on pages 4–6 of your text.

Find each product or quotient, and write it in lowest terms.

10. $\dfrac{25}{11} \cdot \dfrac{33}{10}$ 10. _____

11. $\dfrac{5}{4} \div \dfrac{25}{28}$ 11. _____

12. $4\dfrac{3}{8} \cdot 2\dfrac{4}{7}$ 12. _____

Objective 5   **Add and subtract fractions.**

Video Examples

Review these examples for Objective 5:

7. Add. Write the sum in lowest terms.

$$\dfrac{5}{24} + \dfrac{7}{24}$$

Add numerators. Keep the same denominator.
$$\dfrac{5}{24} + \dfrac{7}{24} = \dfrac{5+7}{24} = \dfrac{12}{24}, \text{ or } \dfrac{1}{2}$$
Write in lowest terms.

8. Add. Write the sum in lowest terms.

$$\dfrac{5}{21} + \dfrac{3}{14}$$

Step 1 To find the LCD, factor the denominators to prime factored form.
    $21 = 3 \cdot 7$ and $14 = 2 \cdot 7$
7 is a factor of both denominators.

$$\begin{array}{cc} 21 & 14 \\ \wedge & \wedge \end{array}$$

Step 2   LCD $= 3 \cdot 7 \cdot 2 = 42$
In this example, the LCD needs one factor of 3, one factor of 7 and one factor of 2.

Now Try:

7. Add. Write the sum in lowest terms.

$$\dfrac{5}{16} + \dfrac{7}{16}$$

_____

8. Add. Write the sum in lowest terms.

$$\dfrac{7}{12} + \dfrac{3}{8}$$

_____

Name: Date:
Instructor: Section:

Step 3 Now we can use the second property of 1 to write each fraction with 42 as the denominator.

$$\frac{5}{21} = \frac{5}{21} \cdot \frac{2}{2} = \frac{10}{42} \quad \text{and} \quad \frac{3}{14} = \frac{3}{14} \cdot \frac{3}{3} = \frac{9}{42}$$

Now add the two equivalent fractions to get the sum.

$$\frac{5}{21} + \frac{3}{14} = \frac{10}{42} + \frac{9}{42}$$
$$= \frac{19}{42}$$

9. Subtract. Write difference in lowest terms.

$$\frac{26}{9} - \frac{5}{9}$$

Subtract numerators. Keep the same denominator.

$$\frac{26}{9} - \frac{5}{9} = \frac{26-5}{9}$$
$$= \frac{21}{9}$$
$$= \frac{7}{3}, \text{ or } 2\frac{1}{3}$$

9. Subtract. Write difference in lowest terms.

$$\frac{11}{18} - \frac{7}{18}$$

_____

Objective 5 Practice Exercises

For extra help, see Example 7–9 on pages 7–10 of your text.

Find each sum or difference, and write it in lowest terms.

13. $\frac{23}{45} + \frac{47}{75}$ 13. _____

14. $2\frac{3}{4} + 7\frac{2}{3}$ 14. _____

15. $12\frac{5}{6} - 7\frac{7}{8}$ 15. _____

Name: Date:
Instructor: Section:

Objective 6 Solve applied problems that involve fractions.

Video Examples

Review this example for Objective 6:

10. Pauline and her two children picked cherries. Pauline picked $2\frac{3}{4}$ quarts, Jennie picked $1\frac{2}{3}$ quarts, and Dan picked $1\frac{1}{2}$ quarts. How many quarts of cherries did they pick?

Use Method 2.

$$2\frac{3}{4} = 2\frac{9}{12}$$
$$1\frac{2}{3} = 1\frac{8}{12}$$
$$+1\frac{1}{2} = 1\frac{6}{12}$$
$$\overline{\phantom{+1\frac{1}{2}=}4\frac{23}{12}}$$

Since $\frac{23}{12} = 1\frac{11}{12}$, $4\frac{23}{12} = 4 + 1\frac{11}{12} = 5\frac{11}{12}$ quarts.

Now Try:

10. A punch is made with $3\frac{1}{3}$ cups of ginger ale, $1\frac{1}{2}$ cups of orange juice, $1\frac{2}{3}$ cups of lemonade, and $2\frac{1}{4}$ cups of pineapple juice. Find the total number of cups in the punch.

_____

Objective 6 Practice Exercises

For extra help, see Example 10 on page 10 of your text.

Solve each applied problem. Write each answer in lowest terms.

16. Arnette worked $24\frac{1}{2}$ hours and earned $9 per hour. How much did she earn?

16. _____

17. Debbie made a shirt with $3\frac{1}{8}$ yards of material, a dress with $4\frac{7}{8}$ yards, and a jacket with $3\frac{3}{4}$ yards. How many yards of material did she use?

17. _____

18. Three sides of a parking lot are $35\frac{1}{4}$ yards, $42\frac{7}{8}$ yards, and $32\frac{3}{4}$ yards. If the total distance around the lot is $145\frac{1}{2}$ yards, find the length of the fourth side.

18. _____

Name: Date:
Instructor: Section:

Objective 7   Interpret data from a circle graph.

Video Examples

In August 2014, 1300 workers were surveyed on where they eat during their lunch time. The circle graph shows the approximate fractions of locations.

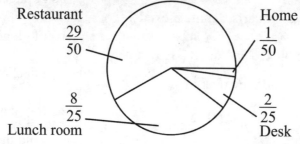

Review these examples for Objective 7:

11.

a. Which location has the largest share of workers? What was the share?

In the circle graph, the sector for Restaurant is the largest, so Restaurant had the largest share of workers, $\frac{29}{50}$.

b. How many actual workers ate in the Lunch Room?

Multiply the actual fraction from the graph of the Lunch Room by the number of workers surveyed.

$$\frac{8}{25} \cdot 1300 = \frac{8}{25} \cdot \frac{1300}{1}$$
$$= \frac{10,400}{25}$$
$$= 416$$

Thus, 416 workers ate in the Lunch Room.

Now Try:

11.

a. Which location had the smallest share of workers? What was the share?

_____

b. How many actual workers ate at their Desk?

_____

8   Copyright © 2016 Pearson Education, Inc.

Name: Date:
Instructor: Section:

Objective 7 Practice Exercises

For extra help, see Example 11 on page 11 of your text.

In August 2014, 1300 workers were surveyed on where they eat during their lunch time. The circle graph shows the approximate fractions of locations.

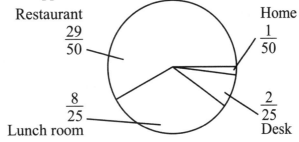

19. What location had the second-largest number of workers?

19. _____

20. Estimate the number of workers who eat at a Restaurant.

20. _____

21. How many actual workers ate at a Restaurant?

21. _____

Name: Date:
Instructor: Section:

Chapter R PREALGEBRA REVIEW

R.2 **Decimals and Percents**

Learning Objectives
1 Write decimals as fractions.
2 Add and subtract decimals.
3 Multiply and divide decimals.
4 Write fractions as decimals.
5 Write percents as decimals and decimals as percents.
6 Write percents as fractions and fractions as percents.
7 Solve applied problems that involve percents.

Key Terms

Use the vocabulary terms listed below to complete each statement in exercises 1–3.

decimals place value percent

1. We use _____ to show parts of a whole.

2. A _____ is assigned to each place to the left or right of the decimal point.

3. _____ means per one hundred.

Objective 1 Write decimals as fractions.

Video Examples

Review these examples for Objective 1:
1. Write each decimal as a fraction. Do not write in lowest terms.

   a. 0.87

   We read 0.87 as "eighty-seven hundredths," so the fraction form is $\frac{87}{100}$. Using the shortcut method, since there are two places to the right of the decimal point, there will be two zeros in the denominator.

   $0.87 = \frac{87}{100}$
   2 places   2 zeros

   b. 0.043

   We read 0.043 as "forty-three thousandths."

   $0.043 = \frac{43}{1000}$
   3 places   3 zeros

Now Try:
1. Write each decimal as a fraction. Do not write in lowest terms.

   a. 0.72

   _____

   b. 0.053

   _____

Name: Date:
Instructor: Section:

c. 5.3084

Here we have 4 places.
$5.3084 = 5 + 0.3084$
$= \frac{50,000}{10,000} + \frac{3084}{10,000}$ The LCD is 10,000.
$= \frac{53,084}{10,000}$
4 zeros

c. 3.7058

_____

Objective 1 Practice Exercises

For extra help, see Example 1 on page 17 of your text.

Write each decimal as a fraction. Do not write in lowest terms.

1. 0.007

2. 18.03

3. 30.0005

1. _____

2. _____

3. _____

Objective 2   Add and subtract decimals.

Video Examples

Review these examples for Objective 2:
2. Add or subtract as indicated.

   a. $7.34 + 15.6 + 2.419$

   Place the digits of the numbers in columns, so that tenths are in one column, hundredths in another column, and so on.

   ```
     7.34
    15.6
   + 2.419
   ───────
    25.359
   ```
   Align decimal points.

   To avoid errors, attach zeros to make all the numbers the same length.

   ```
     7.34              7.340
    15.6    becomes   15.600
   + 2.419           + 2.419
   ──────            ───────
                      25.359
   ```

Now Try:
2. Add or subtract as indicated.

   a. $5.23 + 28.9 + 3.741$

   _____

Name: Date:
Instructor: Section:

b. 56.8 − 42.307

$$\begin{array}{r} 56.8 \\ -\ 42.307 \end{array}$$ becomes $$\begin{array}{r} 56.800 \\ -\ 42.307 \\ \hline 14.493 \end{array}$$

b. 64.5 − 37.218

_____

Objective 2 Practice Exercises

For extra help, see Example 2 on pages 17–18 of your text.

Add or subtract as indicated.

4. 45.83 + 20.923 + 5.7   4. _____

5. 768.5 − 13.402   5. _____

6. 689 − 79.832   6. _____

Objective 3   Multiply and divide decimals.

Video Examples

Review these examples for Objective 3:

3. Multiply.

   a. 37.4×5.26

   There is 1 decimal place in the first number and 2 decimal places in the second number.
   Therefore, there are 1 + 2 = 3 decimal places in the answer.

$$\begin{array}{r} 37.4 \\ \times\ 5.26 \\ \hline 2244 \\ 748 \\ 1870 \\ \hline 196.724 \end{array}$$

   b. 0.08×0.6

   Here $8 \times 6 = 48$. There are 2 decimal places in the first number and 1 decimal place in the second number. Therefore, there are 2 + 1 = 3 decimal places in the answer.
   $0.08 \times 0.6 = 0.048$

Now Try:

3. Multiply.

   a. 26.8×9.37

   _____

   b. 0.04×0.7

   _____

Name: Date:
Instructor: Section:

4. Divide.

   a. $1191.45 \div 23.5$

   Write the problem as follows.

   $23.5 \overline{)1191.45}$

   To change 23.5 into a whole number, move the decimal point one place to the right. Move the decimal point in 1191.45 the same number of places to the right, to get 11,914.5.

   $235 \overline{)11,914.5}$

   Move the decimal point straight up and divide as with whole numbers.

   ```
         50.7
   235)11,914.5
       1175
       1645
       1645
          0
   ```

   b. $9.581 \div 5.78$ (Round the answer to two decimal places.)

   Move the decimal point two places to the right in 5.78, to get 578. Do the same thing with 9.581 to get 958.1.

   $578 \overline{)958.1}$

   Move the decimal point straight up and divide as with whole numbers.

   ```
        1.657
   578)958.100
       578
       3801
       3468
        3330
        2890
         4400
         4046
          354
   ```

   We carried out the division to three decimal places so that we could round to two decimal places, obtaining the quotient 1.66.

4. Divide.

   a. $2472.12 \div 76.3$

   _____

   b. $7.648 \div 5.36$ (Round the answer to two decimal places.)

   _____

Name: Date:
Instructor: Section:

Objective 3 Practice Exercises

For extra help, see Examples 3–5 on pages 18–20 of your text.

Multiply or divide as indicated.

7. 14.64×0.16

7. _____

8. $498.624 \div 21.2$

8. _____

9. $429.2 \div 1000$

9. _____

Objective 4 Write fractions as decimals.

Video Examples

Review these examples for Objective 4:

6. Write each fraction as a decimal.

   a. $\dfrac{27}{8}$

   Divide 27 by 8. Add a decimal point and as many 0s as necessary.

$$\begin{array}{r} 3.375 \\ 8\overline{)27.000} \\ \underline{24} \\ 30 \\ \underline{24} \\ 60 \\ \underline{56} \\ 40 \\ \underline{40} \\ 0 \end{array}$$

   $\dfrac{27}{8} = 3.375$

Now Try:

6. Write each fraction as a decimal.

   a. $\dfrac{7}{20}$

   _____

Name: Date:
Instructor: Section:

b. $\dfrac{26}{9}$ **b.** $\dfrac{32}{9}$

$$\begin{array}{r} 2.888... \\ 9\overline{)26.000...} \\ \underline{18} \\ 80 \\ \underline{72} \\ 80 \\ \underline{72} \\ 80 \\ \underline{72} \\ 8 \end{array}$$

$\dfrac{26}{9} = 2.888...$

The remainder is never 0. Because 8 is always left after the subtraction, this quotient is a repeating decimal. A convenient notation for a repeating decimal is a bar over the digit (or digits) that repeats.

$\dfrac{26}{9} = 2.888...$ or $2.\overline{8}$

Objective 4 Practice Exercises

For extra help, see Example 6 on page 20 of your text.

Write each fraction as a decimal. For repeating decimals, write the answer two ways: using the bar notation and rounding to the nearest thousandth.

10. $\dfrac{3}{7}$ 10. _____

11. $\dfrac{4}{9}$ 11. _____

12. $\dfrac{151}{200}$ 12. _____

Name: Date:
Instructor: Section:

Objective 5   Write percents as decimals and decimals as percents.

Video Examples

Review these examples for Objective 5:

9. Convert each percent to a decimal and each decimal to a percent.

    a. 54%

    54% = 0.54

    b. 3%

    3% = 0.03

    c. 0.29

    0.29 = 29%

    d. 4.6

    4.6 = 460%

Now Try:

9. Convert each percent to a decimal and each decimal to a percent.

    a. 91%

    _____

    b. 6%

    _____

    c. 0.43

    _____

    d. 5.2

    _____

Objective 5 Practice Exercises

For extra help, see Examples 7–9 on pages 21–22 of your text.

Convert each percent to a decimal and each decimal to a percent.

13.   362%   13. _____

14.   0.4%   14. _____

15.   0.084   15. _____

Objective 6   Write percents as fractions and fractions as percents.

Video Examples

Review these examples for Objective 6:

10. Write each percent as a fraction. Give answers in lowest terms.

    a. 12%

    Recall writing 12% as a decimal.
       $12\% = 12 \div 100 = 0.12$
    Because 0.12 means 12 hundredths,
       $0.12 = \dfrac{12}{100} = \dfrac{12 \div 4}{100 \div 4} = \dfrac{3}{25}$

Now Try:

10. Write each percent as a fraction. Give answers in lowest terms.

    a. 30%

    _____

Name: Date:
Instructor: Section:

b. 350%

$$350\% = \frac{350}{100} = \frac{350 \div 50}{100 \div 50} = \frac{7}{2} = 3\frac{1}{2}$$

11. Write each fraction as a percent. Round to the nearest tenth as necessary.

a. $\frac{3}{5}$

$$\frac{3}{5} = \left(\frac{3}{5}\right)(100\%) = \left(\frac{3}{5}\right)\left(\frac{100}{1}\%\right) = \left(\frac{3}{5}\right)\left(\frac{20 \cdot 5}{1}\%\right)$$
$$= \frac{60}{1}\% = 60\%$$

b. $\frac{1}{15}$

$$\frac{1}{15} = \left(\frac{1}{15}\right)(100\%) = \left(\frac{1}{15}\right)\left(\frac{100}{1}\%\right)$$
$$= \left(\frac{1}{3 \cdot 5}\right)\left(\frac{5 \cdot 20}{1}\%\right)$$
$$= \frac{20}{3}\% = 6\frac{2}{3}\%$$

$6\frac{2}{3}\%$ rounds to 6.7%.

b. 125%

_____

11. Write each fraction as a percent. Round to the nearest tenth as necessary.

a. $\frac{3}{20}$

_____

b. $\frac{1}{18}$

_____

Objective 6 Practice Exercises

For extra help, see Examples 10–11 on pages 22–23 of your text.

Convert each percent to a fraction and each fraction to a percent.

16. 140%   16. _____

17. 55.6%   17. _____

18. $\frac{11}{40}$ 18. _____

Name: Date:
Instructor: Section:

Objective 7   Solve applied problems that involve percents.

Video Examples

Review this example for Objective 7:

12. A calendar with a regular price of $12 is on sale at 46% off. Find the amount of the discount and the sale price of the calendar.

   The discount is 46% of 12.

   46%   of   12
   ↓   ↓   ↓
   0.46 ·   12
   = 5.52

   The discount is $5.52. The sale price is found by subtracting.

   Original price – discount = sale price
   $12 – $5.52 = $6.48

Now Try:

12. Olympic t-shirts are on sale at 56% off. The regular price is $19. Find the amount of the discount and the sale price.

Objective 7 Practice Exercises

For extra help, see Example 12 on page 23 of your text.

Solve each problem.

19. Ranee bought a pair of shoes with a regular price of $70, on sale at 20% off. Find the amount of the discount and the sale price.

19. _____

20. Geishe's Shoes sells shoes at $33\frac{1}{3}$% off the regular price. Find the price of a pair of shoes normally priced at $54, after the discount is given.

20. _____

21. At the end of the season, a swim suit is on sale at 75% off. The regular price is $56. Find the amount of the discount and the sale price.

21. _____

Name: Date:
Instructor: Section:

Chapter 1 THE REAL NUMBER SYSTEM

1.1 Exponents, Order of Operations, and Inequality

Learning Objectives
1. Use exponents.
2. Use the rules for order of operations.
3. Use more than one grouping symbol.
4. Know the meanings of $\neq, <, >, \leq,$ and $\geq$.
5. Translate word statements to symbols.
6. Write statements that change the direction of inequality symbols.

Key Terms

Use the vocabulary terms listed below to complete each statement in exercises 1−3.

   exponent   base   exponential expression

1. A number written with an exponent is an _____.

2. The _____ is the number that is a repeated factor when written with an exponent.

3. An _____ is a number that indicates how many times a factor is repeated.

Objective 1   Use exponents.

Video Examples

Review this example for Objective 1:
1. Find the value of the exponential expression.

   6^2

   6^2 means $6 \cdot 6$, which equals 36.

Now Try:
1. Find the value of the exponential expression.
   7^2

   _____

Objective 1 Practice Exercises

For extra help, see Example 1 on page 28 of your text.

Find the value of each exponential expression.

1. 3^3

2. $\left(\dfrac{2}{3}\right)^4$

3. $(0.4)^2$

1. _____

2. _____

3. _____

Name: Date:
Instructor: Section:

Objective 2   Use the rules for order of operations.

Video Examples

Review this example for Objective 2:
2. Find the value of the expression.

   $6 + 7 \cdot 3$

   Multiply, then add.
   $6 + 7 \cdot 3 = 6 + 21$
   $= 27$

Now Try:
2. Find the value of the expression.

   $5 + 2 \cdot 9$

   2. _____

Objective 2 Practice Exercises

For extra help, see Example 2 on pages 29–30 of your text.

Find the value of each expression.

4. $20 \div 5 - 3 \cdot 1$ 4. _____

5. $3 \cdot 5^2 - 3 \cdot 7 - 9$ 5. _____

6. $6^2 \div 3^2 - 4 \cdot 3 - 2 \cdot 5$ 6. _____

Objective 3   Use more than one grouping symbol.

Video Examples

Review this example for Objective 3:
3. Find the value of the expression.

   $3[9 + 4(7 + 8)]$

   Start by adding inside the parentheses.
   $3[9 + 4(7 + 8)] = 3[9 + 4(15)]$ Add.
   $= 3[9 + 60]$ Multiply.
   $= 3[69]$ Add.
   $= 207$ Multiply.

Now Try:
3. Find the value of the expression.

   $5[3 + 4(8 + 2)]$

   _____

Name: Date:
Instructor: Section:

Objective 3 Practice Exercises

For extra help, see Example 3 on pages 30–31 of your text.

Find the value of each expression.

7. $\dfrac{10(5-3)-9(6-2)}{2(4-1)-2^2}$ 7. _____

8. $19-3[8(5-2)+6]$ 8. _____

9. $4[5+2(8-6)]+12$ 9. _____

Objective 4 Know the meanings of $\neq$, $<$, $>$, $\leq$, and $\geq$.

Video Examples

Review this example for Objective 4:
4. Determine whether the statement is true or false.

   $5 \cdot 6 - 12 \leq 22$
   $5 \cdot 6 - 12 \leq 22$
   $30 - 12 \leq 22$
   $18 \leq 22$

   The statement is true.

Now Try:
4. Determine whether the statement is true or false.
   $7 \cdot 4 - 15 \leq 13$

   _____

Objective 4 Practice Exercises

For extra help, see Example 4 on page 32 of your text.

Tell whether each statement is true *or* false.

10. $3 \cdot 4 \div 2^2 \neq 3$ 10. _____

11. $3.25 > 3.52$ 11. _____

12. $2[7(4)-3(5)] \leq 45$ 12. _____

Name: Date:
Instructor: Section:

Objective 5 Translate word statements to symbols.

Video Examples

Review this example for Objective 5:
5. Write each word statement in symbols.

   Thirteen is greater than or equal to nine plus four.

   $13 \geq 9 + 4$

Now Try:
5. Write each word statement in symbols.
   Nineteen is less than or equal to eleven plus 8.

   _____

Objective 5 Practice Exercises

For extra help, see Example 5 on page 32 of your text.

Write each word statement in symbols.

13. Seven equals thirteen minus six. 13. _____

14. Five times the sum of two and nine is less than one hundred six. 14. _____

15. Twenty is greater than or equal to the product of two and seven. 15. _____

Objective 6 Write statements that change the direction of inequality symbols.

Video Examples

Review this example for Objective 6:
6. Write each statement as another true statement with the inequality symbol reversed.

   a. $9 > 7$

   $7 < 9$

Now Try:
6. Write each statement as another true statement with the inequality symbol reversed.
   a. $15 > 11$

   _____

Objective 6 Practice Exercises

For extra help, see Example 6 on page 32 of your text.

Write each statement with the inequality symbol reversed.

16. $\frac{3}{4} > \frac{2}{3}$ 16. _____

17. $12 \geq 8$ 17. _____

18. $0.002 > 0.0002$ 18. _____

Name: Date:
Instructor: Section:

Chapter 1 THE REAL NUMBER SYSTEM

1.2 Variables, Expressions, and Equations

Learning Objectives
1. Evaluate algebraic expressions, given values for the variables.
2. Translate word phrases to algebraic expressions.
3. Identify solutions of equations.
4. Identify solutions of equations from a set of numbers.
5. Distinguish between *equations* and *expressions*.

Key Terms

Use the vocabulary terms listed below to complete each statement in exercises 1−7.

variable   constant   algebraic expression
equation   solution   set   elements

1. A(n) _____ is a statement that says two expressions are equal.

2. The objects that belong to a set are its _____ .

3. A _____ is a symbol, usually a letter, used to represent an unknown number.

4. A collection of numbers, variables, operation symbols, and grouping symbols is an _____.

5. A _____ is collection of objects.

6. Any value of a variable that makes an equation true is a(n) _____ of the equation.

7. A _____ is a fixed, unchanging number.

Objective 1 Evaluate algebraic expressions, given values for the variables.

Video Examples

Review these examples for Objective 1:	Now Try:
1. Find the value of each algebraic expression for $x = 4$ and then $x = 7$. $5x^2$ For $x = 4$, $\quad 5x^2 = 5 \cdot 4^2 \quad$ Let $x = 4$. $\quad\quad\, = 5 \cdot 16 \quad$ Square 4. $\quad\quad\, = 80 \quad\quad$ Multiply.	1. Find the value of each algebraic expression for $x = 6$ and then $x = 9$. $7x^2$ _____

Copyright © 2016 Pearson Education, Inc.

Name: Date:
Instructor: Section:

For $x = 7$,
$$5x^2 = 5 \cdot 7^2 \quad \text{Let } x = 7.$$
$$= 5 \cdot 49 \quad \text{Square 7.}$$
$$= 245 \quad \text{Multiply.}$$

2. Find the value of each expression for $x = 7$ and $y = 6$.

   $3x + 4y + 2$

   Replace x with 7 and y with 6.
   $$3x + 4y + 2 = 3 \cdot 7 + 4 \cdot 6 + 2$$
   $$= 21 + 24 + 2 \quad \text{Multiply.}$$
   $$= 47 \quad \text{Add.}$$

2. Find the value of each expression for $x = 8$ and $y = 4$.
   $5x + 6y + 1$

   2. _____

Objective 1 Practice Exercises

For extra help, see Examples 1–2 on page 37 of your text.

Find the value of each expression if $x = 2$ and $y = 4$.

1. $9x - 3y + 2$

   1. _____

2. $\dfrac{2x + 3y}{3x - y + 2}$

   2. _____

3. $\dfrac{3y^2 + 2x^2}{5x + y^2}$

   3. _____

Name: _____ Date: _____
Instructor: _____ Section: _____

Objective 2 Translate word phrases to algebraic expressions.

Video Examples

Review this example for Objective 2:
3. Write the word phrase as an algebraic expression, using x as the variable.

    The product of 15 and a number

    $15 \cdot x$, or $15x$

Now Try:
3. Write the word phrase as an algebraic expression, using x as the variable.
    The product of 20 and a number

    _____

Objective 2 Practice Exercises

For extra help, see Example 3 on page 38 of your text.

Write each word phrase as an algebraic expression. Use x as the variable.

4. Ten times a number, added to 21

4. _____

5. 11 fewer than eight times a number

5. _____

6. Half a number subtracted from two-thirds of the number

6. _____

Objective 3 Identify solutions of equations.

Video Examples

Review this example for Objective 3:
4. Decide whether the given number is a solution of the equation.

    $8n - 7(n-4) = 41$; 9

    $8n - 7(n-4) = 41$
    $8 \cdot 9 - 7(9-4) \stackrel{?}{=} 41$
    $8 \cdot 9 - 7 \cdot 5 \stackrel{?}{=} 41$
    $72 - 35 \stackrel{?}{=} 41$
    $37 = 41$ False – the left side does not equal the right side.

    The number 9 is not a solution of the equation.

Now Try:
4. Decide whether the given number is a solution of the equation.
    $9m - 4(m-3) = 41$; 5

    _____

Copyright © 2016 Pearson Education, Inc.

Name: Date:
Instructor: Section:

Objective 3 Practice Exercises

For extra help, see Example 4 on pages 38–39 of your text.

Decide whether the given number is a solution of the equation.

7. $5 + 3x^2 = 19;\ 2$ 7. _____

8. $\dfrac{m+2}{3m-10} = 1;\ 8$ 8. _____

9. $3y + 5(y-5) = 7;\ 4$ 9. _____

Objective 4 Identify solutions of equations from a set of numbers.

Video Examples

Review this example for Objective 4:
5. Write the word sentence as an equation. Use *x* as the variable. Then find the solution of the equation from the following set.
   {0, 2, 4, 6, 8, 10}

   The sum of a number and five is eleven.

   the sum of a
   number and five   is   eleven.
   ↓   ↓   ↓
   $x + 5$ =   11

   Because 6 + 5 = 11 is true, 6 is the only solution.

Now Try:
5. Write the word sentence as an equation. Use *x* as the variable. Then find the solution of the equation from the following set.
   {0, 2, 4, 6, 8, 10}
   The sum of a number and seven is eleven.

   _____

Objective 4 Practice Exercises

For extra help, see Example 5 on page 39 of your text.

Write each word sentence as an equation. Use x as the variable. Then find the solution of the equation from the following set. {1, 3, 5, 7, 9, 11}

10. Ten divided by a number is nine more than the number. 10. _____

11. Five more than a number is fourteen. 11. _____

12. Five times a number is 12 plus the number. 12. _____

Name: Date:
Instructor: Section:

Objective 5 Distinguish between *equations* and *expressions*.

Video Examples

Review this example for Objective 5:

6. Decide whether each is an equation or an expression.

   $5x - 4(x - 6)$

   Ask, "Is there an equality symbol?" The answer is no, so this is an expression.

Now Try:

6. Decide whether each is an equation or an expression.

   $2(x - 4) - 4x$

   _____

Objective 5 Practice Exercises

For extra help, see Example 6 on page 40 of your text.

Identify each as an **expression** *or an* **equation**.

13. $y^2 - 4y - 3$ 13. _____

14. $\dfrac{x+4}{5}$ 14. _____

15. $8x = 2y$ 15. _____

Name: Date:
Instructor: Section:

Chapter 1 THE REAL NUMBER SYSTEM

1.3 Real Numbers and the Number Line

Learning Objectives
1. Classify numbers and graph them on number lines.
2. Tell which of two real numbers is less than the other.
3. Find the additive inverse of a real number.
4. Find the absolute value of a real number.
5. Interpret meanings of real numbers from a table of data.

Key Terms

Use the vocabulary terms listed below to complete each statement in exercises 1–14.

    natural numbers   whole numbers   number line   additive inverse

    integers   negative number   positive number   signed numbers

    rational number   set-builder notation   coordinate

    irrational number   real numbers   absolute value

1. The set {0, 1, 2, 3, …} is called the set of _____.

2. The _____ of a number is the same distance from 0 on the number line as the original number, but located on the opposite side of 0.

3. The whole numbers together with their opposites and 0 are called _____.

4. The set { 1, 2, 3, …} is called the set of _____.

5. The _____ of a number is the distance between 0 and the number on the number line.

6. A _____ shows the ordering of the real numbers on a line.

7. A real number that is not a rational number is called a(n) _____.

8. The number that corresponds to a point on the number line is the _____ of that point.

9. A number located to the left of 0 on a number line is a _____.

10. A number located to the right of 0 on a number line is a _____.

11. Numbers that can be represented by points on the number line are _____.

Name: _____ Date: _____
Instructor: _____ Section: _____

12. _____ uses a variable and a description to describe a set.

13. A number that can be written as the quotient of two integers is a _____.

14. Positive numbers and negative numbers are _____.

Objective 1 Classify numbers and graph them on number lines.

Video Examples

Review these examples for Objective 1:

1. Use an integer to express the boldface italic number in the application.

   In August, 2012, the National Debt was approximately $*16* trillion.

   Use –$16 trillion because "debt" indicates a negative number.

2. Graph each number on a number line.
   $-3\frac{1}{2}, -\frac{3}{2}, 0, \frac{7}{2}, 1$

   To locate the improper fractions on the number line, write them as mixed numbers or decimals.

   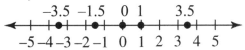

3. List the numbers in the following set that belong to each set of numbers.
   $\left\{-6, -\frac{5}{6}, 0, 0.\overline{3}, \sqrt{3}, 4\frac{1}{5}, 6, 6.7\right\}$

   a. Whole numbers

   Answer: 0 and 6

   b. Integers

   Answer: –6, 0, and 6

   c. Rational numbers

   Answer: $-6, -\frac{5}{6}, 0, 0.\overline{3}, 4\frac{1}{5}, 6, 6.7$

   d. Irrational numbers

   Answer: $\sqrt{3}$

Now Try:

1. Use an integer to express the boldface italic number in the application.
   Death Valley is *282* feet below sea level.

   _____

2. Graph each number on a number line.
   $\frac{1}{2}, 0, -3, -\frac{5}{2}$

   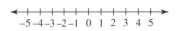

3. List the numbers in the following set that belong to each set of numbers.
   $\left\{-10, -\frac{5}{8}, 0, 0.\overline{4}, \sqrt{5}, 5\frac{1}{2}, 7, 9.9\right\}$

   a. Whole numbers

   _____

   b. Integers

   _____

   c. Rational numbers

   _____

   d. Irrational numbers

   _____

Name: Date:
Instructor: Section:

Objective 1 Practice Exercises

For extra help, see Examples 1–3 on pages 43–45 of your text.

Use a real number to express each number in the following applications.

1. Last year Nina lost 75 pounds.   1. _____

2. Between 1970 and 1982, the population of Norway increased by 279,867.   2. _____

Graph the group of rational numbers on a number line.

3. $-4.5, -2.3, 1.7, 4.2$ 3.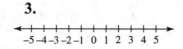

Objective 2   Tell which of two real numbers is less than the other.

Video Examples

Review this example for Objective 2:

4. Is the statement $-4 < -2$ true or false?

    Because -4 is to the left of -2 on the number line, -4 is less than -2. The statement $-4 < -2$ is true.

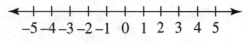

Now Try:

4. Is the statement $-10 < -8$ true or false?

    _____

Objective 2 Practice Exercises

For extra help, see Example 4 on page 46 of your text.

*Decide whether each statement is **true** or **false**.*

4. $-76 < 45$ 4. _____

5. $-5 > -5$ 5. _____

6. $-12 > -10$ 6. _____

Name: Date:
Instructor: Section:

Objective 3 Find the additive inverse of a real number.

Objective 3 Practice Exercises

For extra help, see pages 46–47 of your text.

Find the additive inverse of each number.

7. -25 7. _____

8. $\dfrac{3}{8}$ 8. _____

9. 4.5 9. _____

Objective 4 Find the absolute value of a real number.

Video Examples

Review these examples for Objective 4:
5. Simplify by finding the absolute value.

    a. $|16|$

    $|16| = 16$

    b. $|-16|$

    $|-16| = -(-16) = 16$

    c. $-|-16|$

    $-|-16| = -(16) = -16$

Now Try:
5. Simplify by finding the absolute value.

    a. $|-10|$

    _____

    b. $-|10|$

    _____

    c. $|10-7|$

    _____

Objective 4 Practice Exercises

For extra help, see Example 5 on page 48 of your text.

Simplify.

10. $-|49-39|$ 10. _____

11. $|-7.52+6.3|$ 11. _____

12. $|16-14|$ 12. _____

Name: Date:
Instructor: Section:

Objective 5   Interpret meanings of real numbers from a table of data.

Video Examples

Review this example for Objective 5:

6. In the table, which category represents a decrease for both years?

Category	Change from 2012 to 2013	Change from 2013 to 2014
Eggs	−0.3	3.9
Milk	1.6	0.7
Orange Juice	−8.8	−3.9
Electricity	0.8	3.9

Source: U.S. Bureau of Labor and Statistics

Since a decrease implies a negative number, the category Orange Juice has a negative number for both years. So the answer is Orange Juice.

Now Try:

6. In the table to the left, which category represents an increase for both years?

_____

Objective 5 Practice Exercises

For extra help, see Example 6 on page 48 of your text.

The Consumer Price Index (CPI) measures the average change in prices of goods and services purchased by urban consumers in the United States. The table shows the percent change in CPI for selected categories of goods and services from 2012 to 2013 and from 2013 to 2014. Use the table to answer each question.

Category	Change from 2012 to 2013	Change from 2013 to 2014
Gasoline	−1.2	−0.9
Eggs	−0.3	3.9
Milk	1.6	0.7
Electricity	0.8	3.9

13. Which category represents a decrease for both years?   13. _____

14. Which category in which year represents the greatest percent decrease?   14. _____

15. Which category in which year represents the least change?   15. _____

Name: Date:
Instructor: Section:

Chapter 1 THE REAL NUMBER SYSTEM

1.4 Adding and Subtracting Real Numbers

Learning Objectives
1. Add two numbers with the same sign.
2. Add numbers with different signs.
3. Use the definition of subtraction.
4. Use the rules for order of operations when adding and subtracting signed numbers.
5. Translate words and phrases involving addition and subtraction.
6. Use signed numbers to interpret data.

Key Terms

Use the vocabulary terms listed below to complete each statement in exercises 1−5.

    sum   **addends**   **minuend**

    subtrahend   **difference**

1. The number from which another number is being subtracted is called the _____.

2. The _____ is the number being subtracted.

3. The answer to a subtraction problem is called the _____.

4. The answer to an addition problem is called the _____.

5. In an addition problem, the numbers being added are the _____.

Objective 1   Add two numbers with the same sign.

Video Examples

Review these examples for Objective 1:	Now Try:
1. Use a number line to find the sum. $-3 + (-5)$. *Step 1* Start at 0 and draw an arrow 3 units to the left. *Step 2* From the left end of that arrow, draw another arrow 5 units to the left. The number below the end of this second arrow is −5, so $-3 + (-5) = -8$. 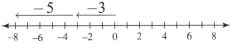	1. Use a number line to find the sum. $-4 + (-1)$ 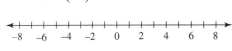 _____

Name: Date:
Instructor: Section:

2. Find the sum.

   $-3 + (-7)$

   $-3 + (-7) = -10$

2. Find the sum.

   $-8 + (-4)$

   _____

Objective 1 Practice Exercises

For extra help, see Examples 1–2 on pages 52–53 of your text.

Find each sum.

1. $-7 + (-11)$

2. $-9 + (-9)$

3. $-2\frac{3}{8} + \left(-3\frac{1}{4}\right)$

1. _____

2. _____

3. _____

Objective 2 Add numbers with different signs.

Video Examples

Review these examples for Objective 2:

3. Use the number line to find the sum $-3 + 4$.

   Step 1 Start at 0 and draw an arrow 3 units to the left.

   Step 2 From the left end of that arrow, draw a second arrow 4 units to the right.

   The number below the end of this second arrow is 1, so $-3 + 4 = 1$.

Now Try:

3. Use the number line to find the sum $7 + (-4)$.

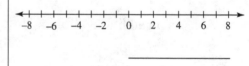

_____

4. Find the sum.

   $\frac{5}{8} + \left(-1\frac{1}{4}\right)$

   $\frac{5}{8} + \left(-1\frac{1}{4}\right) = \frac{5}{8} + \left(-\frac{5}{4}\right)$

   $= \frac{5}{8} + \left(-\frac{10}{8}\right) = +\left(\frac{5}{8} - \frac{10}{8}\right)$

   $= -\frac{5}{8}$

4. Find the sum.

   $\frac{3}{5} + \left(-1\frac{3}{10}\right)$

   _____

34 Copyright © 2016 Pearson Education, Inc.

Name: Date:
Instructor: Section:

Objective 2 Practice Exercises

For extra help, see Examples 3–4 on pages 53–54 of your text.

Use a number line to find the sum.

4. $-8+5$ 4. _____

Find each sum.

5. $\dfrac{7}{12}+\left(-\dfrac{3}{4}\right)$ 5. _____

6. $-\dfrac{4}{7}+\dfrac{3}{5}$ 6. _____

Objective 3 Use the definition of subtraction.

Video Examples

Review these examples for Objective 3:

6. Subtract.

   a. $2-8$

   $2-8=2+(-8)=-6$

   b. $-6-(-9)$

   $-6-(-9)=-6+(9)=3$

Now Try:

6. Subtract.

   a. $3-6$

   _____

   b. $-5-(-7)$

   _____

Objective 3 Practice Exercises

For extra help, see Examples 5–6 on pages 55–56 of your text.

Subtract.

7. $-14-11$ 7. _____

8. $15-(-2)$ 8. _____

9. $-\dfrac{3}{10}-\left(-\dfrac{3}{10}\right)$ 9. _____

Name: Date:
Instructor: Section:

Objective 4 Use the rules for order of operations with real numbers.

Video Examples

Review this example for Objective 4:
7. Perform each operation.

   $(9+6)-7$

   $(9+6)-7 = 15-7$
   $ = 8$

Now Try:
7. Perform each operation.

   $(7+4)-10$

   _____

Objective 4 Practice Exercises

For extra help, see Example 7 on pages 56–57 of your text.

Find each sum.

10. $-2+[4+(-18+13)]$

10. _____

11. $[(-7)+14]+[(-16)+3]$

11. _____

12. $-8.9+[6.8+(-4.7)]$

12. _____

Objective 5 Translate words and phrases that indicate addition.

Video Examples

Review these examples for Objective 5:
8. Write a numerical expression for the phrase, and simplify the expression.

   The sum of –9 and 5 and 3

   –9 + 5 + 3 simplifies to –4 + 3, which equals –1.

9. Write a numerical expression for the phrase, and simplify the expression.

   The difference between –10 and 7

   –10 –7 simplifies to $-10+(-7)$, which equals –17

Now Try:
8. Write a numerical expression for the phrase, and simplify the expression.
   The sum of –10 and 11 and 2

   _____

9. Write a numerical expression for the phrase, and simplify the expression.
   The difference between –17 and 9

   _____

Name: Date:
Instructor: Section:

10. The early morning temperature on a mountain in California was –8°F. At noon the temperature was 38°F. What was the rise in temperature?

We must subtract the lowest temperature from the highest temperature.
$$38-(-8)=38+8=46$$
The rise was 46°F.

10. The floor of Death Valley is 282 ft below sea level. A nearby mountain has an elevation of 5182 ft above sea level. Find the difference between the highest and lowest elevations.

_____

Objective 5 Practice Exercises

For extra help, see Examples 8–10 on pages 57–59 of your text.

Write a numerical expression for each phrase, and then simplify the expression.

13.   4 less than –4   **13.** _____

14.   The sum of –4 and 12, decreased by 9   **14.** _____

Solve the problem.

15.   Dr. Somers runs an experiment at –43.3°C. He then lowers the temperature by 7.9°C. What is the new temperature for the experiment?   **15.** _____

Objective 6   **Use signed numbers to interpret data.**

Video Examples

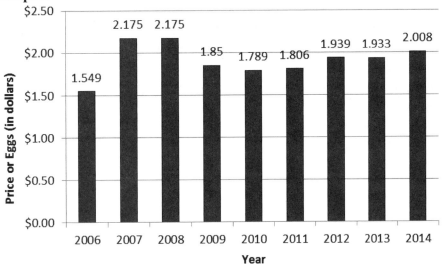

The bar graph above shows the Consumer Price Index (CPI) for a dozen of grade A large Eggs between 2006 and 2014.   *Source: U.S. Bureau of Labor and Statistics*

Review this example for Objective 6:
11. Use a signed number to represent the change in CPI from 2008 to 2009.

$$\$1.85 - \$2.175 = -\$0.325$$

Now Try:
11. Use a signed number to represent the change in CPI from 2012 to 2013.

_____

Name:
Instructor:
Date:
Section:

Objective 6 Practice Exercises

For extra help, see Example 11 on page 59 of your text.

The bar graph below shows the Consumer Price Index (CPI) for a dozen of grade A large Eggs between 2006 and 2014. Source: U.S. Bureau of Labor and Statistics

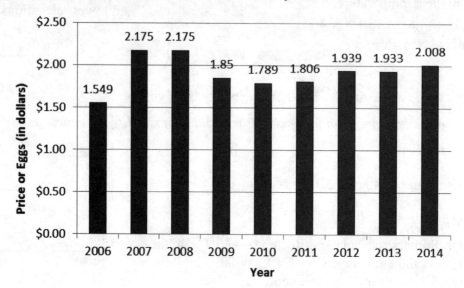

16. Use a signed number to represent the change in CPI from 2006 to 2007.

16. _____

17. Use a signed number to represent the change in CPI from 2009 to 2010.

17. _____

18. Use a signed number to represent the change in CPI from 2013 to 2014.

18. _____

Name: Date:
Instructor: Section:

Chapter 1 THE REAL NUMBER SYSTEM

1.5 Multiplying and Dividing Real Numbers

Learning Objectives
1. Find the product of a positive number and a negative number.
2. Find the product of two negative numbers.
3. Identify factors of integers.
4. Use the reciprocal of a number to apply the definition of division.
5. Use the rules for order of operations when multiplying and dividing signed numbers.
6. Evaluate expressions given values for the variables.
7. Translate words and phrases involving multiplication and division.
8. Translate simple sentences into equations.

Key Terms

Use the vocabulary terms listed below to complete each statement in exercises 1–3.

product   quotient   reciprocals

1. The answer to a division problem is called the _____.

2. Pairs of numbers whose product is 1 are called _____.

3. The answer to a multiplication problem is called the _____.

Objective 1   Find the product of a positive number and a negative number.

Video Examples

Review this example for Objective 1:
1. Find the product using the multiplication rule.

    $9(-6)$

    $9(-6) = -(9 \cdot 6) = -54$

Now Try:
1. Find the product using the multiplication rule.
    $8(-7)$

    _____

Objective 1 Practice Exercises

For extra help, see Example 1 on page 66 of your text.

Find each product.

1. $7(-4)$ 1. _____

2. $\left(\dfrac{1}{5}\right)\left(-\dfrac{2}{3}\right)$ 2. _____

3. $(-3.2)(4.1)$ 3. _____

Name: Date:
Instructor: Section:

Objective 2   Find the product of two negative numbers.

Video Examples

Review this example for Objective 2:
2. Find the product using the multiplication rule.

   a. $-7(-3)$

   $-7(-3) = 21$

Now Try:
2. Find the product using the multiplication rule.

   a. $-5(-6)$

   _____

Objective 2 Practice Exercises

For extra help, see Example 2 on page 67 of your text.

Find each product.

4. $(-4)(-10)$ 4. _____

5. $\left(-\dfrac{2}{7}\right)\left(-\dfrac{14}{5}\right)$ 5. _____

6. $(-0.4)(-3.4)$ 6. _____

Objective 3   Identify factors of integers.

Video Examples

Review this example for Objective 3:
2b. Find all integer factors of the number 52.

   $1 \times 52 = 52$
   $2 \times 26 = 52$
   $4 \times 13 = 52$
   $(-1) \times (-52) = 52$
   $(-2) \times (-26) = 52$
   $(-4) \times (-13) = 52$

   The integer factors of 52 are –52, –26, –13, –4, –2, –1, 1, 2, 4, 13, 26, and 52.

Now Try:
2b. Find all integer factors of the number 12.

   _____

Objective 3 Practice Exercises

For extra help, see page 68 of your text.

Find all integer factors of each number.

7.   8   7. _____

Name: Date:
Instructor: Section:

8. 38

8. _____

9. 42

9. _____

Objective 4   Use the reciprocal of a number to apply the definition of division.

Video Examples

Review these examples for Objective 4:
3. Find each quotient.

    a. $\dfrac{30}{6}$

    $\dfrac{30}{6} = 5$

    b. $\dfrac{15}{-3}$

    $\dfrac{15}{-3} = -5$

Now Try:
3. Find each quotient.

    a. $\dfrac{16}{8}$

    _____

    b. $\dfrac{-18}{-6}$

    _____

Objective 4 Practice Exercises

For extra help, see Example 3 on page 69 of your text.

Find each quotient.

10. $\dfrac{-120}{-20}$

10. _____

11. $\dfrac{0}{-2}$

11. _____

12. $\dfrac{10}{0}$

12. _____

Name: Date:
Instructor: Section:

Objective 5 Use the rules for order of operations when multiplying and dividing signed numbers.

Video Examples

Review these examples for Objective 5:

4. Simplify.

   a. $-6(-1-4)$

   $-6(-1-4) = -6(-5)$
   $= 30$

   b. $\dfrac{6(-4)-5(3)}{3(2-7)}$

   $\dfrac{6(-4)-5(3)}{3(2-7)} = \dfrac{-24-15}{3(-5)}$
   $= \dfrac{-39}{-15}$
   $= \dfrac{13}{5}$

Now Try:

4. Simplify.

   a. $-4(-5-2)$

   _____

   b. $\dfrac{-9(-3)+4(-8)}{-4(5-6)}$

   _____

Objective 5 Practice Exercises

For extra help, see Example 4 on pages 70–71 of your text.

Perform the indicated operations.

13. $-4[(-2)(7)-2]$

13. _____

14. $\dfrac{-7(2)-(-3)}{5+(-3)}$

14. _____

15. $\dfrac{-4[8-(-3+7)]}{-6[3-(-2)]-3(-3)}$

15. _____

Name: Date:
Instructor: Section:

Objective 6   Evaluate algebraic expressions given values for the variables.

Video Examples

Review this example for Objective 6:

5. Evaluate the expression for $x=-2$, $y=-4$, and $m=-5$.

$$(5x+6y)(-3m)$$

Substitute the given values for the variables. Then simplify.
$$(5x+6y)(-3m)$$
$$=[5(-2)+6(-4)][-3(-5)]$$
$$=[-10+(-24)][15]$$
$$=[-34]15$$
$$=-510$$

Now Try:

5. Evaluate the expression for $x=-5$, $y=-3$, and $p=-4$.

$$(6x+2y)(-3p)$$

_____

Objective 6 Practice Exercises

For extra help, see Example 5 on page 71 of your text.

Evaluate the following expressions if $x = -3$, $y = 2$, and $a = 4$.

16. $-x+[(-a+y)-2x]$ 16. _____

17. $(-4+x)(-a)-|x|$ 17. _____

18. $\dfrac{4a-x}{y^2}$ 18. _____

Objective 7   Translate words and phrases involving multiplication and division.

Video Examples

Review this example for Objective 7:

6. Write a numerical expression for the phrase, and simplify the expression.

Three fifths of the sum of –6 and –7

$\dfrac{3}{5}[-6+(-7)]$ simplifies to $\dfrac{3}{5}[-13]$,

which equals $-\dfrac{39}{5}$.

Now Try:

6. Write a numerical expression for the phrase, and simplify the expression.

Five-sixths of the sum of –8 and –4

_____

Name: Date:
Instructor: Section:

Objective 7 Practice Exercises

For extra help, see Examples 6–7 on pages 72–73 of your text.

Write a numerical expression for each phrase and simplify.

19. The product of –7 and 3, added to –7   19. _____

20. Three-tenths of the difference between 50 and –10, subtracted from 85   20. _____

21. The sum of –12 and the quotient of 49 and –7   21. _____

Objective 8 Translate simple sentences into equations.

Video Examples

Review this example for Objective 8:
8. Write the sentence in symbols, using x to represent the number.

   The quotient of 27 and a number is –3.

   $\dfrac{27}{x} = -3$

Now Try:
8. Write the sentence in symbols, using x to represent the number.

   The quotient of 36 and a number is –4

   _____

Objective 8 Practice Exercises

For extra help, see Example 8 on page 73 of your text.

Write each statement in symbols, using x as the variable.

22. Two-thirds of a number is –7.   22. _____

23. –8 times a number is 72.   23. _____

24. When a number is divided by –4, the result is 1.   24. _____

Name: Date:
Instructor: Section:

Chapter 1 THE REAL NUMBER SYSTEM

1.6 Properties of Real Numbers

Learning Objectives
1 Use the commutative properties.
2 Use the associative properties.
3 Use the identity properties.
4 Use the inverse properties.
5 Use the distributive property.

Key Terms

Use the vocabulary terms listed below to complete each statement in exercises 1–2.

identity element for addition

identity element for multiplication

1. When the _____, which is 0, is added to a number, the number is unchanged.

2. When a number is multiplied by the _____, which is 1, the number is unchanged.

Objective 1   Use the commutative properties.

Video Examples

Review these examples for Objective 1:
1. Use a commutative property to complete each statement.

   a. $-7 + 6 = 6 + $ _____

   Using the commutative property of addition,
   $-7 + 6 = 6 + (-7)$

   b. $(-3)5 = $ ____ (-3)

   Using the commutative property of multiplication,
   $(-3)5 = 5(-3)$

Now Try:
1. Use a commutative property to complete each statement.

   a. $-12 + 8 = 8 + $ _____

   _____

   b. $(-4)2 = $ ____ (-4)

   _____

Objective 1 Practice Exercises

For extra help, see Example 1 on page 79 of your text.

Complete each statement. Use a commutative property.

1. $y + 4 = $ _____ $+ y$

1. _____

Copyright © 2016 Pearson Education, Inc.

Name: Date:
Instructor: Section:

2. $5(2) = \underline{}(5)$ 2. _____

3. $-4(4+z) = \underline{}(-4)$ 3. _____

Objective 2   Use the associative properties.

Video Examples

Review these examples for Objective 2:

2. Use an associative property to complete each statement.

    a. $-5+(3+7) = (-5+\underline{})+7$

    Using the associative property of addition,
    $-5+(3+7) = (-5+3)+7$

    b. $[4 \cdot (-9)] \cdot 2 = 4 \cdot \underline{}$

    Using the associative property of multiplication,
    $[4 \cdot (-9)] \cdot 2 = 4 \cdot [(-9) \cdot 2]$

3. Decide whether each statement is an example of a commutative property, an associative property, or both.

    a. $(5+9)+11 = 5+(9+11)$

    The order of the three numbers is the same, but the change is in grouping. This is an example of the associative property.

    b. $7 \cdot (9 \cdot 11) = 7 \cdot (11 \cdot 9)$

    The only change involves the order of the number, so this is an example of the commutative property.

    c. $(12+3)+6 = 12+(6+3)$

    Both the order and the grouping are changed. This is an example of both the associative and commutative properties.

Now Try:

2. Use an associative property to complete each statement.

    a. $-8+(4+6) = (-8+\underline{})+6$

    _____

    b. $[8 \cdot (-3)] \cdot 4 = 8 \cdot \underline{}$

    _____

3. Decide whether each statement is an example of a commutative property, an associative property, or both.

    a. $(13+8)+25 = 13+(8+25)$

    _____

    b. $4 \cdot (15 \cdot 30) = 4 \cdot (30 \cdot 15)$

    _____

    c. $(21+19)+4 = 21+(4+19)$

    _____

Name: Date:
Instructor: Section:

4. Find the sum.

$21x+3+17x+29$

$21x+3+17x+29$
$=21x+(3+17x)+29$
$=(21x+17x)+(3+29)$
$=38x+32$

4. Find the sum

$15x+12+24x+8$

_____

Objective 2 Practice Exercises

For extra help, see Examples 2–4 on pages 79–80 of your text.

Complete each statement. Use an associative property.

4. $4(ab)=$ _____ $\cdot b$

4. _____

5. $[x+(-4)]+3y=x+$ _____

5. _____

6. $4r+(3s+14t)=$ _____ $+14t$

6. _____

Objective 3   Use the identity properties.

Video Examples

Review these examples for Objective 3:

5. Use an identity property to complete each statement.

  a. $-6+$ _____ $=-6$

  Use the identity property for addition.
  $-6+0=-6$

  b. _____ $\cdot \dfrac{1}{6}=\dfrac{1}{6}$

  Use the identity property for multiplication.
  $1 \cdot \dfrac{1}{6}=\dfrac{1}{6}$

Now Try:

5. Use an identity property to complete each statement.

  a. $8+$ _____ $=8$

  _____

  b. $-9 \cdot$ _____ $=-9$

  _____

Copyright © 2016 Pearson Education, Inc.   47

Name: Date:
Instructor: Section:

6.

    a. Write $\dfrac{56}{35}$ in lowest terms.

$$\dfrac{56}{35} = \dfrac{8 \cdot 7}{5 \cdot 7}$$
$$= \dfrac{8}{5} \cdot \dfrac{7}{7}$$
$$= \dfrac{8}{5} \cdot 1$$
$$= \dfrac{8}{5}$$

    b. Perform the operation: $\dfrac{5}{6} - \dfrac{7}{18}$

$$\dfrac{5}{6} - \dfrac{7}{18} = \dfrac{5}{6} \cdot 1 - \dfrac{7}{18}$$
$$= \dfrac{5}{6} \cdot \dfrac{3}{3} - \dfrac{7}{18}$$
$$= \dfrac{15}{18} - \dfrac{7}{18}$$
$$= \dfrac{8}{18}$$
$$= \dfrac{4}{9}$$

6.

    a. Write $\dfrac{49}{63}$ in lowest terms.

    _____

    b. Perform the operation:
$\dfrac{3}{7} + \dfrac{5}{21}$

    _____

Objective 3 Practice Exercises

For extra help, see Examples 5–6 on pages 80–81 of your text.

Use an identity property to complete each statement.

7. $4 + 0 = $ _____   **7.** _____

8. _____ $\cdot 1 = 12$ **8.** _____

Use an identity property to simplify the expression.

9. $\dfrac{30}{35}$ **9.** _____

Name: Date:
Instructor: Section:

Objective 4 Use the inverse properties.

Video Examples

Review these examples for Objective 4:	**Now Try:**

7. Use an inverse property to complete each statement.

a. ___ $\cdot \dfrac{6}{7} = 1$

Use the inverse property of multiplication.
$\dfrac{7}{6} \cdot \dfrac{6}{7} = 1$

b. $5 + $ ___ $= 0$

Use the inverse property of addition.
$5 + (-5) = 0$

c. $-9($ ___ $) = 1$

Use the inverse property of multiplication.
$-9\left(-\dfrac{1}{9}\right) = 1$

d. ___ $+ \dfrac{1}{4} = 0$

Use the inverse property of addition.
$-\dfrac{1}{4} + \dfrac{1}{4} = 0$

8. Simplify $-4x + 1 + 4x$.

$-4x + 1 + 4x$
$= (-4x + 1) + 4x$ Order of operations
$= [1 + (-4x)] + 4x$ Commutative property
$= 1 + [(-4x) + 4x]$ Associative property
$= 1 + 0$ Inverse property
$= 1$ Identity property

Now Try:

7. Use an inverse property to complete each statement.

a. $\dfrac{8}{5} \cdot$ ___ $= 1$

___

b. $8 + $ ___ $= 0$

___

c. $-\dfrac{1}{10}($ ___ $) = 1$

___

d. $-11 + $ ___ $= 0$

___

8. Simplify $-\dfrac{1}{4}x + 6 + \dfrac{1}{4}x$.

___

Objective 4 Practice Exercises

For extra help, see Examples 7–8 on pages 81–82 of your text.

Complete the statements so that they are examples of either an identity property or an inverse property. Identify which property is used.

10. $-4 + $ ___ $= 0$ 10. ___

11. $-9 + $ ___ $= -9$ 11. ___

Name: Date:
Instructor: Section:

12. $-\dfrac{3}{5} \cdot \underline{} = 1$

12. _____

Objective 5 Use the distributive property.
Video Examples

Review these examples for Objective 5:
9. Use the distributive property to rewrite each expression.

   a. $7(p-6)$

   $$7(p-6) = 7[p+(-6)]$$
   $$= 7p + 7(-6)$$
   $$= 7p - 42$$

   b. $-3(5x-2)$

   $$-3(5x-2) = -3[5x+(-2)]$$
   $$= -3(5x)+(-3)(-2)$$
   $$= (-3 \cdot 5)x + (-3)(-2)$$
   $$= -15x + 6$$

   c. $4 \cdot 8 + 4 \cdot 7$

   $$4 \cdot 8 + 4 \cdot 7 = 4(8+7)$$

   d. $5 \cdot 3 + 5x + 5m$

   $$5 \cdot 3 + 5x + 5m = 5(3+x+m)$$

10. Rewrite each expression.

    a. $-(5x+7)$

    $$-(5x+7) = -1 \cdot (5x+7)$$
    $$= -1 \cdot 5x + (-1) \cdot 7$$
    $$= -5x - 7$$

    b. $-(-p-5r+9x)$

    $$-(-p-5r+9x)$$
    $$= -1 \cdot (-1p - 5r + 9x)$$
    $$= -1 \cdot (-1p) - 1 \cdot (-5r) - 1 \cdot (9x)$$
    $$= p + 5r - 9x$$

Now Try:
9. Use the distributive property to rewrite each expression.

   a. $17(x-6)$

   _____

   b. $-4(2x-5)$

   _____

   c. $3 \cdot 11 + 3 \cdot 7$

   _____

   d. $12y + 12 \cdot 6 + 12x$

   _____

10. Rewrite each expression.

    a. $-(3x+4)$

    _____

    b. $-(-4x-5y+z)$

    _____

50 Copyright © 2016 Pearson Education, Inc.

Name: Date:
Instructor: Section:

c. $6a + 6b + 6$ **c.** $3x + 3y + 3$

$6a + 6b + 6 = 6a + 6b + 6 \cdot 1$
$= 6(a + b + 1)$

_____

Objective 5 Practice Exercises

For extra help, see Examples 9–10 on pages 83–84 of your text.

Use the distributive property to rewrite each expression. Simplify if possible.

13. $n(2a - 4b + 6c)$ 13. _____

14. $-2(5y - 9z)$ 14. _____

15. $-(-2k + 7)$ 15. _____

Name: Date:
Instructor: Section:

Chapter 1 THE REAL NUMBER SYSTEM

1.7   Simplifying Expressions

Learning Objectives
1   Simplify expressions.
2   Identify terms and numerical coefficients.
3   Identify like terms.
4   Combine like terms.
5   Simplify expressions from word phrases.

Key Terms

Use the vocabulary terms listed below to complete each statement in exercises 1–3.

   term   numerical coefficient   like terms

1. In the term $4x^2$, "4" is the _____.

2. A number, a variable, or a product or quotient of a number and one or more variables raised to powers is called a _____.

3. Terms with exactly the same variables, including the same exponents, are called _____.

Objective 1   Simplify expressions.

Video Examples

Review these examples for Objective 1:
1. Simplify each expression.

   a. $8(4m - 6n)$

   Use the distributive property.
   $8(4m - 6n) = 8(4m) + 8(-6n)$
   $\qquad\qquad\quad = 32m - 48n$

   b. $9 - (4y - 6)$

   $9 - (4y - 6) = 9 - 1(4y - 6)$
   $\qquad\qquad\quad\;\; = 9 - 4y + 6$
   $\qquad\qquad\quad\;\; = 15 - 4y$

Now Try:
1. Simplify each expression.

   a. $7(5x - 3y)$

   _____

   b. $8 - (7x - 3)$

   _____

52   Copyright © 2016 Pearson Education, Inc.

Name: Date:
Instructor: Section:

Objective 1 Practice Exercises

For extra help, see Example 1 on page 88 of your text.

Simplify each expression.

1. $4(2x+5)+7$

    1. _____

2. $-4+s-(12-21)$

    2. _____

3. $-2(-5x+2)+7$

    3. _____

Objective 2   Identify terms and numerical coefficients.

Objective 2 Practice Exercises

For extra help, see pages 88–89 of your text.

Give the numerical coefficient of each term.

4. $-2y^2$

    4. _____

5. $\dfrac{7x}{9}$

    5. _____

6. $5.6r^5$

    6. _____

Objective 3   Identify like terms.

Objective 3 Practice Exercises

For extra help, see page 89 of your text.

Identify each group of terms as **like** *or* **unlike**.

7. $4x^2, -7x^2$

    7. _____

8. $-8m, -8m^2$

    8. _____

9. $7xy, -6xy^2$

    9. _____

Name: Date:
Instructor: Section:

Objective 4   Combine like terms.

Video Examples

Review these examples for Objective 4:

2. Combine like terms in each expression.

    a. $8r + 5r + 4r$

    $8r + 5r + 4r = (8 + 5 + 4)r$
                           $= 17r$

    b. $9x + x$

    $9x + x = 9x + 1x$
               $= (9 + 1)x$
               $= 10x$

    c. $15x^2 - 8x^2$

    $15x^2 - 8x^2 = (15 - 8)x^2$
                          $= 7x^2$

3. Simplify each expression.

    a. $8k - 5 - 4(7 - 3k)$

    $8k - 5 - 4(7 - 3k) = 8k - 5 - 4(7) - 4(-3k)$
                                 $= 8k - 5 - 28 + 12k$
                                 $= 20k - 33$

    b. $-\frac{3}{5}(x - 10) - \frac{1}{10}x$

    $-\frac{3}{5}(x - 10) - \frac{1}{10}x = -\frac{3}{5}x - \frac{3}{5}(-10) - \frac{1}{10}x$
                                  $= -\frac{3}{5}x + 6 - \frac{1}{10}x$
                                  $= -\frac{6}{10}x + 6 - \frac{1}{10}x$
                                  $= -\frac{7}{10}x + 6$

Now Try:

2. Combine like terms in each expression.

    a. $14r + 7r + 2r$

    _____

    b. $18x + x$

    _____

    c. $17x^2 - 9x^2$

    _____

3. Simplify each expression.

    a. $7k - 9 - 5(3 - 6k)$

    _____

    b. $-\frac{3}{4}(x - 8) - \frac{1}{2}x$

    _____

Objective 4 Practice Exercises

For extra help, see Examples 2–3 on pages 89–91 of your text.

Simplify.

10. $12y - 7y^2 + 4y - 3y^2$ 10. _____

Name: Date:
Instructor: Section:

11. $-4(x+4)+2(3x+1)$ 11. _____

12. $2.5(3y+1)-4.5(2y-3)$ 12. _____

Objective 5 Simplify expressions from word phrases.

Video Examples

Review this example for Objective 5:
4. Translate the phrase into a mathematical expression and simplify.
   The sum of 8, three times a number, nine times a number, and seven times a number.

   Use *x* for the number.
   $8+3x+9x+7x$ simplifies to $8+19x$.

Now Try:
4. Translate the phrase into a mathematical expression and simplify.
   The sum of 11, ten times a number, eight times a number, and four times a number

   _____

Objective 5 Practice Exercises

For extra help, see Example 4 on page 91 of your text.

Write each phrase as a mathematical expression and simplify by combining like terms. Use x as the variable.

13. The sum of six times a number and 12, added to four times the number.   13. _____

14. The sum of seven times a number and 2, subtracted from three times the number.   14. _____

15. Four times the difference between twice a number and six times the number, added to six times the sum of the number and 9.   15. _____

Name: Date:
Instructor: Section:

Chapter 2 LINEAR EQUATIONS AND INEQUALITIES IN ONE VARIABLE

2.1 The Addition Property of Equality

Learning Objectives
1. Identify linear equations.
2. Use the addition property of equality.
3. Simplify, and then use the addition property of equality.

Key Terms

Use the vocabulary terms listed below to complete each statement in exercises 1–3.

    linear equation   solution set   equivalent equations

1. Equations that have exactly the same solutions sets are called _____.

2. An equation that can be written in the form $Ax + B = C$, where A, B, and C are real numbers and $A \neq 0$, is called a _____.

3. The set of all numbers that satisfy an equation is called its _____.

Objective 1   Identify linear equations.

Objective 1 Practice Exercises

For extra help, see page 104 of your text.

Tell whether each of the following is a linear equation.

1. $3x^2 + 4x + 3 = 0$ 1. _____

2. $\dfrac{5}{x} - \dfrac{3}{2} = 0$ 2. _____

3. $4x - 2 = 12x + 9$ 3. _____

Name: Date:
Instructor: Section:

Objective 2 Use the addition property of equality.

Video Examples

Review these examples for Objective 2:
1. Solve $x - 15 = 8$.

$$x - 15 = 8$$
$$x - 15 + 15 = 8 + 15$$
$$x = 23$$

Check $x - 15 = 8$
$23 - 15 \stackrel{?}{=} 8$
$8 = 8$ True

The solution is 23, and the solution set is {23}.

3. Solve $-5 = x + 17$.

$$-5 = x + 17$$
$$-5 - 17 = x + 17 - 17$$
$$-22 = x$$

Check $-5 = x + 17$
$-5 \stackrel{?}{=} -22 + 17$
$-5 = -5$ True

The solution set is {−22}.

5. Solve $\frac{5}{6}k + 9 = \frac{11}{6}k$.

$$\frac{5}{6}k + 9 = \frac{11}{6}k$$
$$\frac{5}{6}k + 9 - \frac{5}{6}k = \frac{11}{6}k - \frac{5}{6}k$$
$$9 = 1k$$
$$9 = k$$

Check by substituting 9 in the original equation. The solution set is {9}.

6. Solve $9 - 5p = -6p + 3$.

$$9 - 5p = -6p + 3$$
$$9 - 5p + 6p = -6p + 3 + 6p$$
$$9 + p - 9 = 3 - 9$$
$$p = -6$$

Check by substituting −6 in the original equation. The solution set is {−6}.

Now Try:
1. Solve $x - 12 = 9$.

_____

3. Solve $-10 = x + 9$.

_____

5. Solve $\frac{4}{7}k + 13 = \frac{11}{7}k$.

_____

6. Solve $10 - 8p = -9p + 7$.

_____

Name: Date:
Instructor: Section:

Objective 2 Practice Exercises

For extra help, see Examples 1–6 on pages 105–107 of your text.

Solve each equation by using the addition property of equality. Check each solution.

4. $y - 4 = 16$

4. _____

5. $\frac{9}{8}p - \frac{1}{2} = \frac{1}{8}p$

5. _____

6. $9.5y - 2.4 = 10.5y$

6. _____

Objective 3 Simplify, and then use the addition property of equality.

Video Examples

Review these examples for Objective 3:

7. Solve $5t - 16 + t + 4 = 9 + 5t + 6$.

$$5t - 16 + t + 4 = 9 + 5t + 6$$
$$6t - 12 = 15 + 5t$$
$$6t - 12 - 5t = 15 + 5t - 5t$$
$$t - 12 = 15$$
$$t - 12 + 12 = 15 + 12$$
$$t = 27$$

Check by substituting 27 in the original equation. The solution set is {27}.

8. Solve $4(3 + 6x) - (5 + 23x) = 19$.

$$4(3 + 6x) - (5 + 23x) = 19$$
$$4(3) + 4(6x) - 1(5) - 1(23x) = 19$$
$$12 + 24x - 5 - 23x = 19$$
$$x + 7 = 19$$
$$x + 7 - 7 = 19 - 7$$
$$x = 12$$

Check by substituting 12 in the original equation. The solution set is {12}.

Now Try:

7. Solve
$8t - 9 + t + 7 = 12 + 8t + 15$.

_____

8. Solve
$5(7 + 8x) - (29 + 39x) = 14$.

_____

Name: Date:
Instructor: Section:

Objective 3 Practice Exercises

For extra help, see Examples 7–8 on page 108 of your text.

Solve each equation. First simplify each side of the equation as much as possible. Check each solution.

7. $3(t+3)-(2t+7)=9$ 7. _____

8. $-4(5g-7)+3(8g-3)=15-4+3g$ 8. _____

9. $3.6p+4.8+4.0p=8.6p-3.1+0.7$ 9. _____

Name: Date:
Instructor: Section:

Chapter 2 LINEAR EQUATIONS AND INEQUALITIES IN ONE VARIABLE

2.2 The Multiplication Property of Equality

Learning Objectives
1 Use the multiplication property of equality.
2 Simplify, and then use the multiplication property of equality.

Key Terms

Use the vocabulary terms listed below to complete each statement in exercises 1–2.

multiplication property of equality   **addition property of equality**

1. The _____ states that multiplying both sides of an equation by the same nonzero number will not change the solution.

2. When the same quantity is added to both sides of an equation, the _____ is being applied.

Objective 1 Use the multiplication property of equality.

Video Examples

Review these examples for Objective 1:
1. Solve $6x = 78$.

    $6x = 78$

    $\dfrac{6x}{6} = \dfrac{78}{6}$

    $x = 13$

    Check $6x = 78$

    $6(13) \stackrel{?}{=} 78$

    $78 = 78$ True

    The solution set is $\{13\}$.

3. Solve $4.3x = 10.32$.

    $4.3x = 10.32$

    $\dfrac{4.3x}{4.3} = \dfrac{10.32}{4.3}$

    $x = 2.4$

    Check by substituting 2.4 in the original equation. The solution set is $\{2.4\}$.

Now Try:
1. Solve $4x = 56$.

    _____

3. Solve $3.6x = 20.52$.

    _____

Name: Date:
Instructor: Section:

4. Solve $\dfrac{x}{7} = 5$.

$$\dfrac{x}{7} = 5$$
$$\dfrac{1}{7}x = 5$$
$$7 \cdot \dfrac{1}{7}x = 7 \cdot 5$$
$$x = 35$$

Check by substituting 35 in the original equation. The solution set is {35}.

5. Solve $\dfrac{5}{6}x = 15$.

$$\dfrac{5}{6}x = 15$$
$$\dfrac{6}{5} \cdot \dfrac{5}{6}x = \dfrac{6}{5} \cdot 15$$
$$1 \cdot x = \dfrac{6}{5} \cdot \dfrac{15}{1}$$
$$x = 18$$

Check by substituting 18 in the original equation. The solution set is {18}.

6. Solve $-x = 5$.

$$-x = 5$$
$$-1 \cdot x = 5$$
$$-1(-1 \cdot x) = -1(5)$$
$$[-1(-1)] \cdot x = -5$$
$$1 \cdot x = -5$$
$$x = -5$$

Check by substituting −5 in the original equation. The solution set is {−5}.

4. Solve $\dfrac{x}{8} = 3$.

_____

5. Solve $\dfrac{7}{9}h = 28$.

_____

6. Solve $-x = 3$.

_____

Objective 1 Practice Exercises

For extra help, see Examples 1–6 on pages 113–115 of your text.

Solve each equation and check your solution.

1. $-3w = 51$

1. _____

Name: Date:
Instructor: Section:

2. $\dfrac{3p}{7} = -6$ 2. _____

3. $-2.7v = -17.28$ 3. _____

Objective 2 Simplify, and then use the multiplication property of equality.

Video Examples

Review this example for Objective 2:
7. Solve $9m + 4m = 39$.

$$9m + 4m = 39$$
$$13m = 39$$
$$\dfrac{13m}{13} = \dfrac{39}{13}$$
$$m = 3$$

Check by substituting 3 in the original equation.
The solution set is {3}.

Now Try:
7. Solve $12m + 8m = 80$.

_____

Objective 2 Practice Exercises

For extra help, see Example 7 on page 115 of your text.

Solve each equation and check your solution.

4. $-7b + 12b = 125$ 4. _____

5. $3w - 7w = 20$ 5. _____

6. $-11h - 6h + 14h = -21$ 6. _____

Name: Date:
Instructor: Section:

Chapter 2 LINEAR EQUATIONS AND INEQUALITIES IN ONE VARIABLE

2.3 More on Solving Linear Equations

Learning Objectives
1. Learn and use the four steps for solving a linear equation.
2. Solve equations that have no solution or infinitely many solutions.
3. Solve equations with fractions or decimals as coefficients.
4. Write expressions for two related unknown quantities.

Key Terms

Use the vocabulary terms listed below to complete each statement in exercises 1–3.

conditional equation **identity** **contradiction**

1. An equation with no solution is called a(n) _____.

2. A(n) _____ is an equation that is true for some values of the variable and false for other values.

3. An equation that is true for all values of the variable is called a(n) _____.

Objective 1 Learn and use the four steps for solving a linear equation.

Video Examples

Review these examples for Objective 1:
1. Solve $-5x + 8 = 23$.

   Step 1 There are no parentheses, fractions, or decimals in this equation, so this step is not necessary.

   $$-5x + 8 = 23$$

   Step 2 $-5x + 8 - 8 = 23 - 8$
   $$-5x = 15$$

   Step 3 $\dfrac{-5x}{-5} = \dfrac{15}{-5}$
   $$x = -3$$

   Step 4 Check by substituting -3 for x in the original equation.
   $$-5x + 8 = 23$$
   $$-5(-3) + 8 \stackrel{?}{=} 23$$
   $$15 + 8 \stackrel{?}{=} 23$$
   $$23 = 23 \quad \text{True}$$

   The solution, -3, checks, so the solution set is $\{-3\}$.

Now Try:
1. Solve $-8x + 11 = 59$.

   _____

Name: Date:
Instructor: Section:

2. Solve $4x+3=6x-11$.

   Step 1 There are no parentheses, fractions, or decimals in this equation, so begin with Step 2.
   $$4x+3=6x-11$$
   Step 2 $\quad 4x+3-4x=6x-11-4x$
   $$3=2x-11$$
   $$3+11=2x-11+11$$
   $$14=2x$$
   Step 3 $\quad\dfrac{14}{2}=\dfrac{2x}{2}$
   $$7=x$$
   Step 4 Check by substituting 7 for x in the original equation.
   $$4x+3=6x-11$$
   $$4(7)+3\stackrel{?}{=}6(7)-11$$
   $$28+3\stackrel{?}{=}42-11$$
   $$31=31 \quad \text{True}$$
   The solution, 7, checks, so the solution set is $\{7\}$.

2. Solve $5x+4=8x-20$.

   _____

4. Solve $9a-(4+3a)=2a+5$.

   $$9a-(4+3a)=2a+5$$
   Step 1 $\quad 9a-4-3a=2a+5$
   $$6a-4=2a+5$$
   Step 2 $\quad 6a-4-2a=2a+5-2a$
   $$4a-4=5$$
   $$4a-4+4=5+4$$
   $$4a=9$$
   Step 3 $\quad\dfrac{4a}{4}=\dfrac{9}{4}$
   $$a=\dfrac{9}{4}$$
   Step 4 Check that the solution set is $\left\{\dfrac{9}{4}\right\}$.

4. Solve $10a-(11+3a)=5a+4$.

   _____

Objective 1 Practice Exercises

For extra help, see Examples 1–5 on pages 117–120 of your text.

Solve each equation and check your solution.

1. $7t+6=11t-4$

1. _____

Name: Date:
Instructor: Section:

2. $3a - 6a + 4(a-4) = -2(a+2)$ 2. _____

3. $3(t+5) = 6 - 2(t-4)$ 3. _____

Objective 2 Solve equations that have no solution or infinitely many solutions.

Video Examples

Review these examples for Objective 2:

6. Solve $6x - 18 = 6(x-3)$.

$$6x - 18 = 6(x-3)$$
$$6x - 18 = 6x - 18$$
$$6x - 18 - 6x = 6x - 18 - 6x$$
$$-18 = -18$$
$$-18 + 18 = -18 + 18$$
$$0 = 0$$

The solution set is {all real numbers}.

7. Solve $3x + 4(x-5) = 7x + 5$.

$$3x + 4(x-5) = 7x + 5$$
$$3x + 4x - 20 = 7x + 5$$
$$7x - 20 = 7x + 5$$
$$7x - 20 - 7x = 7x + 5 - 7x$$
$$-20 = 5 \quad \text{False}$$

There is no solution. The solution set is $\varnothing$.

Now Try:

6. Solve $3x + 4(x-5) = 7x - 20$.

   _____

7. Solve $-5x + 17 = x - 6(x+3)$.

   _____

Objective 2 Practice Exercises

For extra help, see Examples 6–7 on page 121 of your text.

Solve each equation and check your solution.

4. $3(6x - 7) = 2(9x - 6)$ 4. _____

5. $6y - 3(y+2) = 3(y-2)$ 5. _____

Name:
Instructor:
Date:
Section:

6. $3(r-2)-r+4=2r+6$ 6. _____

Objective 3 Solve equations with fractions or decimals as coefficients.

Video Examples

Review these examples for Objective 3:

9. Solve $\frac{1}{4}(x+3)-\frac{2}{5}(x+1)=2$.

   To clear fractions, multiply by 20, the LCD.
   $$\frac{1}{4}(x+3)-\frac{2}{5}(x+1)=2$$

   Step 1 $\quad 20\left[\frac{1}{4}(x+3)-\frac{2}{5}(x+1)\right]=20(2)$

   $\quad\quad 20\left[\frac{1}{4}(x+3)\right]+20\left[-\frac{2}{5}(x+1)\right]=20(2)$

   $\quad\quad\quad 5(x+3)-8(x+1)=40$

   $\quad\quad\quad 5x+15-8x-8=40$

   $\quad\quad\quad -3x+7=40$

   Step 2 $\quad\quad -3x+7-7=40-7$

   $\quad\quad\quad -3x=33$

   Step 3 $\quad\quad \frac{-3x}{-3}=\frac{33}{-3}$

   $\quad\quad\quad x=-11$

   Step 4 Check to confirm that $\{-11\}$ is the solution set.

Now Try:

9. Solve
$\frac{1}{7}(x+5)-\frac{1}{2}(x+4)=-2$.

_____

10. Solve $0.2x+0.04(10-x)=0.06(4)$.

   To clear decimals, multiply by 100.
   $$0.2x+0.04(10-x)=0.06(4)$$

   Step 1 $\quad 100[0.2x+0.04(10-x)]=100[0.06(4)]$

   $\quad\quad 100(0.2x)+100[0.04(10-x)]=100[0.06(4)]$

   $\quad\quad\quad 20x+4(10)+4(-x)=24$

   $\quad\quad\quad 20x+40-4x=24$

   $\quad\quad\quad 16x+40=24$

   Step 2 $\quad\quad 16x+40-40=24-40$

   $\quad\quad\quad 16x=-16$

   Step 3 $\quad\quad \frac{16x}{16}=\frac{-16}{16}$

   $\quad\quad\quad x=-1$

   Step 4 Check to confirm that $\{-1\}$ is the solution set.

10. Solve
$0.5x+0.04(5-8x)=0.07(8)$.

_____

Name: Date:
Instructor: Section:

Objective 3 Practice Exercises

For extra help, see Examples 8–10 on pages 122–124 of your text.

Solve each equation and check your solution.

7. $\dfrac{3}{8}x - \dfrac{1}{3}x = \dfrac{1}{12}$

7. _____

8. $\dfrac{1}{3}(2m - 1) - \dfrac{3}{4}m = \dfrac{5}{6}$

8. _____

9. $0.45a - 0.35(20 - a) = 0.02(50)$

9. _____

Objective 4 Write expressions for two related unknown quantities.

Video Examples

Review this example for Objective 4:

11. Two numbers have a sum of 51. If one of the numbers is represented by x, find an expression for the other number.

    If one number is x, then the other number is obtained by subtracting x from 51.
    $51 - x$.
    To check, we find the sum of the two numbers.
    $x + (51 - x) = 51$

Now Try:

11. Two numbers have a sum of 67. If one of the numbers is represented by t, find an expression for the other number.

_____

Objective 4 Practice Exercises

For extra help, see Example 11 on page 124 of your text.

Write an expression for the two related unknown quantities.

10. Two numbers have a sum of 36. One is m. Find the other number.

10. _____

11. The product of two numbers is 17. One number is p. What is the other number?

11. _____

12. Admission to the circus costs x dollars for an adult and y dollars for a child. Find the total cost of 6 adults and 4 children.

12. _____

Name: Date:
Instructor: Section:

Chapter 2 LINEAR EQUATIONS AND INEQUALITIES IN ONE VARIABLE

2.4 Applications of Linear Equations

Learning Objectives
1. Learn the six steps for solving applied problems.
2. Solve problems involving unknown numbers.
3. Solve problems involving sums of quantities.
4. Solve problems involving consecutive integers.
5. Solve problems involving complementary and supplementary angles.

Key Terms

Use the vocabulary terms listed below to complete each statement in exercises 1–5.

complementary angles   **right angle**   **supplementary angles**

straight angle   **consecutive integers**

1. Two angles whose measures sum to 180º are _____.

2. Two angles whose measures sum to 90º are _____.

3. An angle whose measure is exactly 90º is a _____.

4. An angle whose measure is exactly 180º is a _____.

5. Two integers that differ by 1 are _____.

Objective 1   Learn the six steps for solving applied problems.

Objective 1 Practice Exercises

For extra help, see page 128 of your text.

1. Write the six problem-solving steps.   1. _____

   _____

   _____

   _____

   _____

   _____

Name: Date:
Instructor: Section:

Objective 2 Solve problems involving unknown numbers.

Video Examples

Review this example for Objective 2:

2. The product of 5, and a number decreased by 8, is 150. What is the number?

   Step 1 Read the problem carefully. We are asked to find a number.

   Step 2 Assign a variable to represent the unknown quantity.
   Let x = the number.

   Step 3 Write an equation.

   The product of 5, and a number decreased by 8, is 150.
   $$5 \cdot (x - 8) = 150$$

   Step 4 Solve the equation.
   $$5(x-8) = 150$$
   $$5x - 40 = 150$$
   $$5x - 40 + 40 = 150 + 40$$
   $$5x = 190$$
   $$\frac{5x}{5} = \frac{190}{5}$$
   $$x = 38$$

   Step 5 State the answer. The number is 38.

   Step 6 Check. The number 38 decreased by 8 is 30. The product of 5 and 30 is 150. The answer, 38, is correct.

Now Try:

2. The product of 8, and a number decreased by 11, is 40. What is the number?

   _____

Objective 2 Practice Exercises

For extra help, see Examples 1–2 on pages 128–129 of your text.

Write an equation for each of the following and then solve the problem. Use x as the variable.

2. If 4 is added to 3 times a number, the result is 7. Find the number.

   2. _____

   _____

Name: Date:
Instructor: Section:

3. If -2 is multiplied by the difference between 4 and a number, the result is 24. Find the number.

3. _____

    _____

4. If four times a number is added to 7, the result is five less than six times the number. Find the number.

4. _____

    _____

Objective 3   Solve problems involving sums of quantities.

Video Examples

Review these examples for Objective 3:

3. George and Al were opposing candidates in the school board election. George received 21 more votes than Al, with 439 votes cast. How many votes did Al receive?

Step 1 Read the problem carefully. We are given total votes and asked to find the number of votes Al received.

Step 2 Assign a variable.
Let x = the number of votes Al received.
Then $x + 21$ = the number of votes George received.

Step 3 Write an equation.

The total is votes for Al plus votes for George
$$439 = x + (x+21)$$

Step 4 Solve the equation.
$$439 = x + (x+21)$$
$$439 = 2x + 21$$
$$439 - 21 = 2x + 21 - 21$$
$$418 = 2x$$
$$\frac{418}{2} = \frac{2x}{2}$$
$$209 = x \quad \text{or} \quad x = 209$$

Step 5 State the answer. Al received 209 votes.

Now Try:

3. On a psychology test, the highest grade was 38 points more than the lowest grade. The sum of the two grades was 142. Find the lowest grade.

    _____

Name: Date:
Instructor: Section:

Step 6 Check. George won 209 + 21 = 230 votes. The total number of votes is 209 + 230 = 439. The answer checks.

6. Penny is making punch for a party. The recipe requires twice as much orange juice as cranberry juice and 8 times as much ginger ale as cranberry juice. If she plans to make 176 ounces of punch, how much of each ingredient should she use?

    Step 1 Read the problem. The three amounts of ingredients must be found.

    Step 2 Assign a variable.
    Let x = the number of ounces of cranberry juice.
    Then $2x$ = the number of ounces of orange juice, and $8x$ = the number of ounces of ginger ale.

    Step 3 Write an equation.
    Cranberry plus orange plus ginger ale is total
    $$x + 2x + 8x = 176$$

    Step 4 Solve the equation.
    $$x + 2x + 8x = 176$$
    $$11x = 176$$
    $$\frac{11x}{11} = \frac{176}{176}$$
    $$x = 16$$

    Step 5 State the answer. There are 16 ounces of cranberry juice, 2(16) = 32 ounces of orange juice, and 8(16) = 128 ounces of ginger ale.

    Step 6 Check. The sum is 176. All conditions of the problem are satisfied.

6. Linda wishes to build a rectangular dog pen using 52 feet of fence and the back of her house, which is 36 feet long to enclose the pen. How wide will the dog pen be if the pen is 36 feet long?

    6. _____

Objective 3 Practice Exercises

For extra help, see Examples 3–6 on pages 130–133 of your text.

Write an equation for each of the following and then solve the problem. Use x as the variable.

5. Mount McKinley in Alaska is 5910 feet higher than Mount Rainier in Washington. Together, their heights total 34,730 feet. How high is each mountain?

    5. _____

    Mt. Rainier _____

    Mt. McKinley _____

Name: Date:
Instructor: Section:

6. Charles bought five general admission tickets and four student tickets for a movie. He paid $35.25. If each student ticket cost $3.50, how much did each general admission ticket cost?

6. _____

_____

7. Pablo, Faustino, and Mark swim at a public pool each day for exercise. One day Pablo swam five more than three times as many laps as Mark, and Faustino swam four times as many laps as Mark. If the men swam 29 laps altogether, how many laps did each one swim?

7. _____

Mark _____

Pablo _____

Faustino _____

Objective 4 Solve problems involving consecutive integers.

Video Examples

Review this example for Objective 4:

8. Find two consecutive odd integers such that if three times the smaller is added to twice the larger, the sum is 69.

Step 1 Read the problem. We must find two consecutive odd integers.

Step 2 Assign a variable.
Let x = the lesser consecutive odd integer.
Then $x + 2$ = the greater consecutive odd integer.

Step 3 Write an equation.
Three times is added twice the
the smaller to larger is 69.
$$3x + 2(x+2) = 69$$

Step 4 Solve the equation.
$$3x + 2x + 4 = 69$$
$$5x + 4 = 69$$
$$5x = 65$$
$$x = 13$$

Step 5 State the answer. The lesser integer is 13. The greater is $13 + 2 = 15$.

Now Try:

8. The sum of four consecutive even integers is 4. Find the integers.

_____

Name: Date:
Instructor: Section:

Step 6 Check. Three times the smaller is 39, added to twice the larger, 30, is a sum of 69. The answers check.

Objective 4 Practice Exercises

For extra help, see Examples 7–8 on pages 133–135 of your text.

Solve each problem.

8. Find two consecutive even integers such that the smaller, added to twice the larger, is 292.

8. _____

9. Find two consecutive integers such that the larger, added to three times the smaller, is 109.

9. _____

10. Find three consecutive odd integers whose sum is 363.

10. _____

Objective 5 Solve problems involving complementary and supplementary angles.

Video Examples

Review this example for Objective 5:

10. Find the measure of an angle if its supplement measures 4° less than three times its complement.

Step 1 Read the problem. We must find the measure of an angle.

Step 2 Assign a variable.
Let x = the degree measure of the angle
Then $90 - x$ = the degree measure of its complement,
and $180 - x$ = the degree measure of its supplement.

Step 3 Write an equation.

The supplement	is	Three times the complement	minus	4
↓	↓	↓	↓	↓
$180 - x$	=	$3(90 - x)$	−	4

Now Try:

10. Find the measure of an angle such that the sum of the measures of its complement and its supplement is 138°.

_____

Name: Date:
Instructor: Section:

Step 4 Solve the equation.
$$180 - x = 270 - 3x - 4$$
$$180 - x = 266 - 3x$$
$$180 - x + 3x = 266 - 3x + 3x$$
$$180 + 2x = 266$$
$$180 + 2x - 180 = 266 - 180$$
$$2x = 86$$
$$x = 43$$

Step 5 State the answer. The angle is 43°.

Step 6 Check. If the angle measures 43°, then its complement measures 90° − 43° = 47°, and the supplement measures 180° − 43° = 137°. Also, 137 is equal to 4 less than 3 times 47° (that is, 137° = 3(47) − 4). The answer is correct.

Objective 5 Practice Exercises

For extra help, see Examples 9–10 on pages 135–136 of your text.

Solve each problem.

11. Find the measure of an angle if the measure of the angle is 8° less than three times the measure of its supplement.

11. _____

12. Find the measure of an angle whose supplement measures 20° more than twice its complement.

12. _____

13. Find the measure of an angle whose complement is 9° more than twice its measure.

13. _____

Copyright © 2016 Pearson Education, Inc.

Name: Date:
Instructor: Section:

Chapter 2 LINEAR EQUATIONS AND INEQUALITIES IN ONE VARIABLE

2.5 Formulas and Additional Applications from Geometry

Learning Objectives
1. Solve a formula for one variable, given the values of the other variables.
2. Use a formula to solve an applied problem.
3. Solve problems involving vertical angles and straight angles.
4. Solve a formula for a specified variable.

Key Terms

Use the vocabulary terms listed below to complete each statement in exercises 1–4.

    formula   area   perimeter   vertical angles

1. The nonadjacent angles formed by two intersecting lines are called _____.

2. An equation in which variables are used to describe a relationship is called a(n) _____.

3. The distance around a figure is called its _____.

4. A measure of the surface covered by a figure is called its _____.

Objective 1 Solve a formula for one variable, given the values of the other variables.

Video Examples

Review this example for Objective 1:
1. Find the value of the remaining variable in each formula.

   $A = LW$; $A = 54, L = 8$

   Substitute the given values for A and L into the formula.
   $$A = LW$$
   $$54 = 8W$$
   $$\frac{54}{8} = \frac{8W}{8}$$
   $$6.75 = W$$
   The width is 6.75. Since $8(6.75) = 54$, the answer checks.

Now Try:
1. Find the value of the remaining variable in each formula.

   $A = LW$; $A = 88, L = 16$

   _____

Name: Date:
Instructor: Section:

Objective 1 Practice Exercises

For extra help, see Example 1 on page 142 of your text.

In the following exercises, a formula is given, along with the values of all but one of the variables in the formula. Find the value of the variable that is not given.

1. $S = \dfrac{a}{1-r}$; $S = 60$, $r = 0.4$
    1. _____

2. $I = prt$; $I = 288$, $r = 0.04$, $t = 3$
    2. _____

3. $A = \tfrac{1}{2}(b+B)h$; $b = 6$, $B = 16$, $A = 132$
    3. _____

Objective 2 Use a formula to solve an applied problem.

Video Examples

Review these examples for Objective 2:

2. Find the dimensions of a rectangle. The length is 4 m less than three times the width. The perimeter is 96 m.

    Step 1 Read the problem. We must find the dimensions of the rectangle.

    Step 2 Assign a variable.
    Let W = the width of the rectangle, in meters.
    Then $L = 3W - 4$ is the length, in meters.

    Step 3 Write an equation. Use the formula for the perimeter of a rectangle. Substitute $3W - 4$ for the length.
    $P = 2L + 2W$
    $96 = 2(3W - 4) + 2W$

    Step 4 Solve.
    $96 = 6W - 8 + 2W$
    $96 = 8W - 8$
    $96 + 8 = 8W - 8 + 8$
    $104 = 8W$
    $\dfrac{104}{8} = \dfrac{8W}{8}$
    $13 = W$

Now Try:

2. Ruth has 42 feet of binding for a rectangular rug that she is weaving. If the rug is 9 feet wide, how long can she make the rug if she wishes to use all the binding on the perimeter of the rug?

    _____

Name: Date:
Instructor: Section:

Step 5 State the answer. The width is 13 m. The length is $3(13) - 4 = 35$ m.

Step 6 Check. The perimeter is $2(13) + 2(35) = 96$ m. The answer checks.

3. The longest side of a triangle is 4 feet longer than the shortest side. The medium side is 2 feet longer than the shortest side. If the perimeter is 36 feet, what are the lengths of the three sides?

   Step 1 Read the problem. We must find the lengths of the sides.

   Step 2 Assign a variable.
   Let s = the length of the shortest side, in feet.
   Then $s + 2$ = the length of the medium side, in feet,
   and $s + 4$ = the length of the longest side, in feet.

   Step 3 Write an equation. Use the formula for the perimeter of a triangle.
   $$P = a + b + c$$
   $$36 = s + (s+2) + (s+4)$$

   Step 4 Solve.
   $$36 = 3s + 6$$
   $$30 = 3s$$
   $$10 = s$$

   Step 5 State the answer. Since s represents the length of the shortest side, its measure is 10 ft.
   $s + 2 = 10 + 2 = 12$ ft is the length of the medium side.
   $s + 4 = 10 + 4 = 14$ ft is the length of the longest side.

   Step 6 Check. The perimeter is $10 + 12 + 14 = 36$ ft, as required.

3. The longest side of a triangle is twice as long as the shortest side. The medium side is 5 feet longer than the shortest side. If the perimeter is 65 feet, what are the lengths of the three sides?

   _____

Objective 2 Practice Exercises

For extra help, see Examples 2–4 on pages 143–144 of your text.

Use a formula to write an equation for each of the following applications; then solve the application. (Use 3.14 as an approximation for π.)

4. Find the height of a triangular banner whose area is 48 square inches and base is 12 inches.

4. _____

Name: Date:
Instructor: Section:

5. Linda invests $5000 at 6% simple interest and earns $450. How long did Linda invest her money?

5. _____

6. The circumference of a circular garden is 628 feet. Find the area of the garden. (Hint: First find the radius of the garden.)

6. _____

Objective 3   Solve problems involving vertical angles and straight angles.

Video Examples

Review this example for Objective 3:

5.

Find the measure of the marked angles in the figure below.

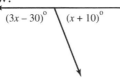

The measures of the marked angles must add to 180° because together they form a straight angle. The angles are supplements of each other.

$$(3x-30)+(x+10)=180$$
$$4x-20=180$$
$$4x=200$$
$$x=50$$

Replace x with 50 in the measure of each marked angle.

$$3x-30=3(50)-30=150-30=120$$
$$x+10=50+10=60$$

The two angles measure 120° and 60°.

Now Try:

5.

Find the measure of the marked angles in the figure below.

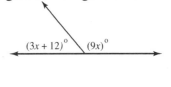

_____

Name: Date:
Instructor: Section:

Objective 3 Practice Exercises

For extra help, see Example 5 on page 145 of your text.

Find the measure of each marked angle.

7.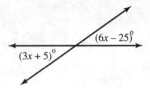

7. _____

8.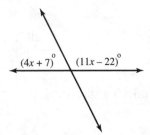

8. _____

9.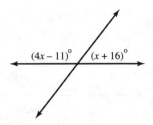

9. _____

Objective 4   **Solve a formula for a specified variable.**

Video Examples

Review these examples for Objective 4:

6. Solve $A = \frac{1}{2}bh$ for h.

$$A = \frac{1}{2}bh$$
$$2A = bh$$
$$\frac{2A}{b} = \frac{bh}{b}$$
$$\frac{2A}{b} = h \quad \text{or} \quad h = \frac{2A}{b}$$

Now Try:
6. Solve $d = rt$ for r.

_____

Name: Date:
Instructor: Section:

7. Solve $A = p + prt$ for r.

$$A = p + prt$$
$$A - p = p + prt - p$$
$$A - p = prt$$
$$\frac{A-p}{pt} = \frac{prt}{pt}$$
$$\frac{A-p}{pt} = r \quad \text{or} \quad r = \frac{A-p}{pt}$$

8. Solve $V = k + gt$ for t.

$$V = k + gt$$
$$V - k = k + gt - k$$
$$V - k = gt$$
$$\frac{V-k}{g} = \frac{gt}{g}$$
$$\frac{V-k}{g} = t$$
$$t = \frac{V-k}{g}$$

7. Solve $P = a + b + c$ for a.

   _____

8. Solve $A = \frac{1}{2}h(b+B)$ for h.

   _____

Objective 4 Practice Exercises

For extra help, see Examples 6–9 on pages 146–147 of your text.

Solve each formula for the specified variable.

10. $V = LWH$ for H

10. _____

11. $S = (n-2)180$ for n

11. _____

12. $V = \frac{1}{3}\pi r^2 h$ for h

12. _____

Name: Date:
Instructor: Section:

Chapter 2 LINEAR EQUATIONS AND INEQUALITIES IN ONE VARIABLE

2.6 Ratio, Proportion, and Percent

Learning Objectives
1. Write ratios.
2. Solve proportions.
3. Solve applied problems using proportions.
4. Find percents and percentages.

Key Terms

Use the vocabulary terms listed below to complete each statement in exercises 1–4.

    ratio   proportion   cross products   terms

1. A _____ is a statement that two ratios are equal.

2. A _____ is a comparison of two quantities using a quotient.

3. In the proportion, $\frac{a}{b} = \frac{c}{d}$, a, b, c, and d are called the _____.

4. To see whether a proportion is true, determine if the _____ are equal.

Objective 1 Write ratios.

Video Examples

Review these examples for Objective 1:
1. Write a ratio for each word phrase.

    a. 7 hr to 9 hr

    $\frac{7 \text{ hr}}{9 \text{ hr}} = \frac{7}{9}$

    b. 15 hr to 4 days

    First convert 4 days to hours.
        4 days $= 4 \cdot 24 = 96$ hr

    Now write the ratio using the common unit of measure, hours.

    $\frac{15 \text{ hr}}{4 \text{ days}} = \frac{15 \text{ hr}}{96 \text{ hr}} = \frac{15}{96}$, or $\frac{5}{32}$

Now Try:
1. Write a ratio for each word phrase.

    a. 11 hr to 17 hr

    _____

    b. 32 hr to 5 days

    _____

Name: Date:
Instructor: Section:

2. The local grocery store charges the following prices for a bottle of olive oil.
     16-ounce bottle: $6.99
     25.5-ounce bottle: $9.99
     32-ounce bottle: $12.99
     44-ounce bottle: $14.99
Which size is the best buy? That is, which size has the lowest unit price?

To find the best buy, write ratios comparing the price for each size bottle to the number of units (ounces) per bottle.

Size	Unit price (dollars per ounce)
16 oz	$\dfrac{\$6.99}{16} = \$0.437$
25.5 oz	$\dfrac{\$9.99}{25.5} = \$0.392$
32 oz	$\dfrac{\$12.99}{32} = \$0.406$
44 oz	$\dfrac{\$14.99}{44} = \$0.341$

Because the 44-oz size has the lowest unit price, $0.341, it is the best buy.

2. The local grocery store charges the following prices for a jar of applesauce.
     16-ounce jar: $1.19
     24-ounce jar: $1.29
     48-ounce jar: $2.69
     64-ounce jar: $3.49
Which size is the best buy? That is, which size has the lowest unit price?

_____

Objective 1 Practice Exercises

For extra help, see Examples 1–2 on pages 152–153 of your text.

Write a ratio for each word phrase. Write fractions in lowest terms.

1. 8 men to 3 men   1. _____

2. 9 dollars to 48 quarters   2. _____

A supermarket was surveyed and the following prices were charged for items in various sizes. Find the best buy (based on price per unit) for each of the following items.

3. Trash bags   3. _____
     10-count box: $2.89
     20-count box: $5.29
     45-count box: $6.69
     85-count box: $13.99

Name: Date:
Instructor: Section:

Objective 2 Solve proportions.

Video Examples

Review these examples for Objective 2:

3. Decide whether the proportion is *true* or *false*.

$$\frac{2}{5} = \frac{12}{30}$$

Check to see whether the cross product are equal.

$$5 \cdot 12 = 60$$
$$\frac{2}{5} = \frac{12}{30}$$
$$2 \cdot 30 = 60$$

The cross products are equal, so the proportion is true.

4. Solve the proportion $\frac{6}{11} = \frac{x}{88}$.

Solve for *x*.

$$\frac{6}{11} = \frac{x}{88}$$
$$6 \cdot 88 = 11 \cdot x$$
$$528 = 11x$$
$$48 = x$$

Check by substituting 48 for *x* in the proportion. The solution set is {48}.

5. Solve the equation $\frac{n-1}{3} = \frac{2n+1}{4}$.

$$\frac{n-1}{3} = \frac{2n+1}{4}$$
$$3(2n+1) = 4(n-1)$$
$$6n+3 = 4n-4$$
$$2n+3 = -4$$
$$2n = -7$$
$$n = -\frac{7}{2}$$

A check confirms that the solution is $-\frac{7}{2}$, so the solution set is $\left\{-\frac{7}{2}\right\}$.

Now Try:

3. Decide whether the proportion is *true* or *false*.

$$\frac{5}{6} = \frac{20}{24}$$

_____

4. Solve the proportion $\frac{x}{15} = \frac{42}{90}$.

_____

5. Solve the equation $\frac{2x+1}{2} = \frac{7x+3}{9}$.

_____

Name: Date:
Instructor: Section:

Objective 2 Practice Exercises

For extra help, see Examples 3–5 on pages 154–155 of your text.

Solve each equation.

4. $\dfrac{z}{20} = \dfrac{25}{125}$ 4. _____

5. $\dfrac{m}{5} = \dfrac{m-2}{2}$ 5. _____

6. $\dfrac{z+1}{4} = \dfrac{z+7}{2}$ 6. _____

Objective 3   Solve applied problems using proportions.

Video Examples

Review this example for Objective 3:

6. If four pounds of fertilizer will cover 50 square feet of garden, how many pounds would be needed for 125 square feet?

     To solve this problem, set up a proportion, with pounds in the numerator and square feet in the denominator.

$$\frac{4}{50} = \frac{x}{125}$$
$$4(125) = 50x$$
$$500 = 50x$$
$$10 = x$$

10 lb of fertilizer are needed.

Now Try:

6. Margie earns $168.48 in 26 hours. How much does she earn in 40 hours?

     _____

Objective 3 Practice Exercises

For extra help, see Example 6 on page 156 of your text.

Solve each problem.

7. On a road map, 6 inches represents 50 miles. How many inches would represent 125 miles?   7. _____

Name: Date:
Instructor: Section:

8. If 12 rolls of tape cost $4.60, how much will 15 rolls cost?

8. _____

9. A garden service charges $30 to install 50 square feet of sod. Find the charge to install 225 square feet.

9. _____

Objective 4   Find percents and percentages.

Video Examples

Review these examples for Objective 4:
7. Solve each problem.

   a. What is 18% of 700?

   Let n = the number. The word of indicates multiplication.

   What is 18% of 700
   $n = 0.18 \cdot 700$
   $n = 126$

   Thus, 126 is 18% of 700.

   b. 54% of what number is 162?

   54% of what number is 162
   $0.54 \cdot n = 162$
   $n = \dfrac{162}{0.54}$
   $n = 300$

   54% of 300 is 162.

   c. 75 is what percent of 500?

   75 is what percent of 500
   $75 = p \cdot 500$
   $\dfrac{75}{500} = p$
   $0.15 = p$

   Thus, 75 is 15% of 500.

Now Try:
7. Solve each problem.

   a. What is 35% of 400?

   _____

   b. 42% of what number is 399?

   _____

   c. 102 is what percent of 120?

   _____

Name: Date:
Instructor: Section:

8. An advertisement for a BluRay player gives a sale price of $175.50. The regular price is $225. Find the percent discount on this BluRay player.

   The savings amounted to $225 − $175.50 = $49.50. We can now restate the problem: What percent of 225 is 49.50?

   What percent of 225 is 49.50
   $$p \cdot 225 = 49.50$$
   $$p = \frac{49.50}{225}$$
   $$p = 0.22 \text{ or } 22\%$$

   The discount is 22%.

8. The number of students enrolled in a calculus course is 145. If 40% of these students are female, how many are male?

   _____

Objective 4 Practice Exercises

For extra help, see Examples 7–8 on pages 156–157 of your text.

Answer each question about percent.

10. What is 2.5% of 3500?

10. _____

11. What percent of 5200 is 104?

11. _____

Solve the problem.

12. Paul recently bought a duplex for $144,000. He expects to earn $6120 per year on this investment. What percent of the purchase price will he earn?

12. _____

Name: Date:
Instructor: Section:

Chapter 2 LINEAR EQUATIONS AND INEQUALITIES IN ONE VARIABLE

2.7   Further Applications of Linear Equations

Learning Objectives
1   Use percent in solving problems involving rates.
2   Solve problems involving mixtures
3   Solve problems involving simple interest.
4   Solve problems involving denominations of money.
5   Solve problems involving distance, rate, and time.

Objective 1   Use percent solving problems involving rates.

Video Examples

Review this example for Objective 1:

1. The purchase price of a new car is $12,500. In order to finance the car a purchaser is required to make a minimum down payment of 20% of the purchase price. What is the minimum down payment required?

    $12,500 \cdot 0.20 = 2500$
    The down payment is $2500

Now Try:

1. A certificate of deposit pays 2.6% simple interest in one year on a principal of $4500. What interest is being paid on this deposit?

    _____

Objective 1 Practice Exercises

For extra help, see Example 1 on page 162 of your text.

Solve each problem.

1. Twelve percent of a college student body has a grade point average of 3.0 or better. If there are 1250 students enrolled in the college, how many have a grade point average of less than 3.0?

    1. _____

2. In a class of freshmen and sophomores only, there are 85 students. If 60% of the class is sophomores, how many students are sophomores?

    2. _____

Name: Date:
Instructor: Section:

3. At a large university, 40% of the student body is from out of state. If the total enrollment at the university is 25,000 students, how many are from in-state?

3. _____

Objective 2 Solve problems involving mixtures.

Video Examples

Review this example for Objective 2:

2. How many gallons of a 20% alcohol solution must be mixed with 15 gallons of a 12% alcohol solution to obtain a 14% alcohol solution?

Step 1 Read the problem. Find the amount of the 20% alcohol solution.

Step 2 Assign a variable. Let x = the amount of the 20% alcohol solution.

Liters of Solution	Rate (as a decimal)	Liters of Pure Acid
x	0.20	$0.2x$
15	0.12	0.12(15)
$x+15$	0.14	$0.14(x+15)$

Step 3 Write an equation.
$0.20x + 0.12(15) = 0.14(x+15)$

Step 4 Solve.
$0.20x + 1.8 = 0.14x + 2.1$
$0.06x = 0.3$
$x = 5$

Step 5 State the answer. 5 gallons of the 20% alcohol solution must be added.

Step 6 Check.
$0.20(5) + 0.12(15) \stackrel{?}{=} 0.14(5+15)$
$2.8 = 2.8$
The answer checks.

Now Try:

2. How many ounces of a 35% solution of acid must be mixed with a 60% solution to get 20 ounces of a 50% solution?

_____

Name: Date:
Instructor: Section:

Objective 2 Practice Exercises

For extra help, see Examples 2–3 on pages 163–164 of your text.

Solve each problem.

4. How many pounds of peanuts worth $3 per pound must be mixed with mixed nuts worth $5.50 per pound to make 40 pounds of a mixture worth $5 per pound?

4. _____

5. How many pounds of candy worth $7 per pound must be mixed with candy worth $4.50 per pound to make 100 pounds of candy worth $6 per pound?

5. _____

Objective 3 Solve problems involving simple interest.

Video Examples

Review this example for Objective 3:

4. Larry invested some money at 8% simple interest and $700 less than this amount at 7%. His total annual income from the interest was $584. How much was invested at each rate?

   Step 1 Read the problem. Find the two amounts.

   Step 2 Assign a variable.
   Let x = the amount at 8%.
   Let $x - 700$ = the amount at 7%.

Amount invested (in dollars)	Rate (as a decimal)	Interest for One Year (in dollars)
x	0.08	$0.08x$
$x - 700$	0.07	$0.07(x - 700)$

   Step 3 Write an equation.
   $0.08x + 0.07(x - 700) = 584$

Now Try:

4. Tracy invested some money at 5% and $300 more than twice this amount at 7%. Her total annual income from the two investments is $325. How much is invested at each rate?

   _____

Name: Date:
Instructor: Section:

Step 4 Solve.
$$0.08x + 0.07x - 49 = 584$$
$$0.15x - 49 = 584$$
$$0.15x = 633$$
$$x = 4220$$

Step 5 State the answer. Larry must invest $4220 at 8%, and 4220 − 700 = $3520 at 7%.

Step 6 Check. The sum of the two amounts should be $584.
$$0.08(4220) + 0.07(3520) = 584$$

Objective 3 Practice Exercises

For extra help, see Example 4 on page 165 of your text.

Solve each problem.

6. Louisa invested $16,000 in bonds paying 7% simple interest. How much additional money should she invest at 4% simple interest so that the average return on the two investments is 6%?

6. _____

7. Desiree invested some money at 9% and $100 less than three times that amount at 7%. Her total annual interest was $83. How much did she invest at each rate?

7. 7%: _____

    9%: _____

8. Jacob has $48,000 invested in stocks paying 6%. How much additional money should he invest in certificates of deposit paying 2.5% so that the average return on the two investments is 4%?

8. _____

Name:
Instructor:
Date:
Section:

Objective 4 Solve problems involving denominations of money.

Video Examples

Review this example for Objective 4:

5. A collection of dimes and nickels has a total value of $2.70. The number of nickels is 2 more than twice the number of dimes. How many of each type of coin are in the collection?

   Step 1 Read the problem. Find the number of dimes and the number of nickels.

   Step 2 Assign a variable.
   Let d = the number of dimes.
   Then $2d + 2$ = the number of nickels.

Number of coins	Denomination (in dollars)	Total Value (in dollars)
d	0.10	$0.10d$
$2d+2$	0.05	$0.05(2d+2)$

   Step 3 Write an equation.
   $0.10d + 0.05(2d+2) = 2.70$

   Step 4 Solve.
   $$10d + 5(2d+2) = 270$$
   $$10d + 10d + 10 = 270$$
   $$20d + 10 = 270$$
   $$20d = 260$$
   $$d = 13$$

   Step 5 State the answer. There are 13 dimes and $2(13) + 2 = 28$ nickels.

   Step 6 Check.
   $0.10(13) + 0.05(28) = 1.30 + 1.40 = 2.70$
   The answer is correct.

Now Try:

5. Erica's piggy bank has quarters and nickels in it. The total number of coins in the piggy bank is 50. Their total value is $8.90. How many of each type are in the piggy bank?

   _____

   _____

Name: Date:
Instructor: Section:

Objective 4 Practice Exercises

For extra help, see Example 5 on page 166 of your text.

Solve each problem.

9. Anthony sells two different size jars of peanut butter. The large size sells for $2.60 and the small size sells for $1.80. He has 80 jars worth $164. How many of each size jar does he have?

9.
large _____
small _____

10. Stan has 14 bills in his wallet worth $95 altogether. If the wallet contains only $5 and $10 bills, how many bills of each denomination does he have?

10.
$5 bills _____
$10 bills _____

11. Twice as many general admission tickets to a basketball game were sold as reserved seat tickets. General admission tickets cost $10 and reserved seat tickets cost $15. If the total value of both kinds of tickets was $26,250, how many tickets of each kind were sold?

11.
general admission_____
reserved seats _____

Name: Date:
Instructor: Section:

Objective 5 Use percent involving distance, rate, and time.

Video Examples

Review these examples for Objective 5:

6. A driver averaged 58 mph and took 10 hours to drive from Little Rock to Indianapolis. What is the distance between Little Rock and Indianapolis?

   We must find the distance, given the rate and time using $rt = d$.
   $58 \cdot 10 = 580$ miles
   The distance is 580 miles.

7. A car and a truck leave Oklahoma City at the same time and travel west on the same route. The car travels at a constant rate of 62 mph. The truck travels at a constant rate of 68 mph. In how many hours will the distance between them be 30 miles?

   Step 1 Read the problem.

   Step 2 Assign a variable. We are looking for time.
   Let t = the number of hours until the distance between them is 30 miles.

	Rate	Time	Distance
Truck	68	t	$68t$
Car	62	t	$62t$

   Step 3 Write an equation.
   $68t - 62t = 30$

   Step 4 Solve.
   $$6t = 30$$
   $$t = 5$$

   Step 5 State the answer. It will take 5 hours for the truck and car to be 30 miles apart.

   Step 6 Check. After 5 hours, the truck travels $68 \cdot 5 = 340$ miles and the car travels $62 \cdot 5 = 310$ miles. The difference is $340 - 310 = 30$, as required.

Now Try:

6. A driver averaged 54 mph and took 5 hours to drive from Los Angeles to Las Vegas. What is the distance between Los Angeles and Las Vegas?

   _____

7. A car and a truck leave Dallas at the same time and travel north on the same route. The car travels at a constant rate of 67 mph. The truck travels at a constant rate of 72 mph. In how many hours will the distance between them be 25 miles?

   _____

Name: Date:
Instructor: Section:

Objective 5 Practice Exercises

For extra help, see Examples 6–8 on pages 167–169 of your text.

Solve each problem.

12. A driver averages 52 mph and took 10 hours to drive from Charlotte, North Carolina to Orlando, Florida. What is the distance between Charlotte and Orlando?

12. _____

13. A driver averages 50 mph and took 9 hours to drive from Denver, Colorado to Albuquerque, New Mexico. What is the distance between Denver and Albuquerque?

13. _____

14. A car and a truck leave Billings, Montana at the same time and travel east on the same route. The truck travels at a constant rate of 78 mph. The car travels at a constant rate of 67 mph. In how many hours will the distance between them be 33 miles?

14. _____

Chapter 2 LINEAR EQUATIONS AND INEQUALITIES IN ONE VARIABLE

2.8 Solving Linear Inequalities

Learning Objectives
1. Graph intervals on a number line.
2. Use the addition property of inequality.
3. Use the multiplication property of inequality.
4. Solve linear inequalities using both properties of inequality.
5. Solve applied problems using inequalities.
6. Solve linear inequalities with three parts.

Key Terms

Use the vocabulary terms listed below to complete each statement in exercises 1–5.

inequalities   interval   interval notation

linear inequality   three-part inequality

1. An inequality that says that one number is between two other numbers is a(n)_____.

2. A portion of a number line is called a(n) _____.

3. A(n) _____ can be written in the form $Ax + B < C$, $Ax + B \leq C$, $Ax + B > C$, or $Ax + B \geq C$, where A, B, and C are real numbers with $A \neq 0$.

4. Algebraic expressions related by $<$, $\leq$, $>$, or $\geq$ are called _____.

5. The _____ for $a \leq x < b$ is $[a, b)$.

Objective 1 Graph intervals on a number line.

Video Examples

Review this example for Objective 1:
1. Write the inequality in interval notation, and graph the interval.

    $x > -3$

    The statement $x > -3$ says that x can represent any value greater than -3, but cannot equal -3, written $(-3, \infty)$. We graph this interval by placing a parenthesis at -3 and drawing an arrow to the right. The parenthesis indicates that -3 is not part of the graph.

    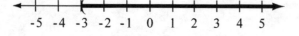

Now Try:
1. Write the inequality in interval notation, and graph the interval.

    $x > -1$

    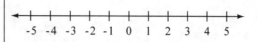

Name: Date:
Instructor: Section:

Objective 1 Practice Exercises

For extra help, see Example 1 on page 175 of your text.

Write each inequality in interval notation and graph the interval.

1. $3 < a$

2. $y \geq -2$

3. $x < -4$

Objective 2   Use the addition property of inequality.

Video Examples

Review this example for Objective 2:

2. Solve $8 + 4x \geq 3x + 3$ and graph the solution set.

$$8 + 4x \geq 3x + 3$$
$$8 + 4x - 3x \geq 3x + 3 - 3x$$
$$8 + x \geq 3$$
$$8 + x - 8 \geq 3 - 8$$
$$x \geq -5$$

The solution set, $[-5, \infty)$ is graphed below.

Now Try:

2. Solve $5 + 9x \geq 8x + 2$ and graph the solution set.

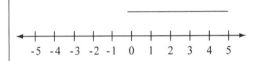

Objective 2 Practice Exercises

For extra help, see Example 2 on page 176 of your text.

Solve each inequality. Write the solution set in interval notation and then graph it.

4. $5a + 3 \leq 6a$

Copyright © 2016 Pearson Education, Inc.

Name:
Instructor:

Date:
Section:

5. $6 + 3x < 4x + 4$

5.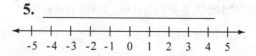

6. $3 + 5p \leq 4p + 3$

6.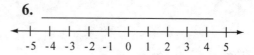

Objective 3 Use the multiplication property of inequality.

Video Examples

Review these examples for Objective 3:

3. Solve each inequality, and graph the solution set.

   a. $6x < -24$

   We divide each side by 6.
   $$6x < -24$$
   $$\frac{6x}{6} < \frac{-24}{6}$$
   $$x < -4$$

   The graph of the solution set $(-\infty, -4)$, is shown below.

   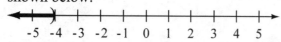

   b. $-6x \geq 30$

   Here each side of the inequality must be divided by –6, a negative number, which does require changing the direction of the inequality symbol.
   $$-6x \geq 30$$
   $$\frac{-6x}{-6} \leq \frac{30}{-6}$$
   $$x \leq -5$$

   The solution set, $(-\infty, -5]$, is graphed below.

   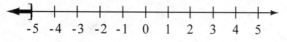

Now Try:

3. Solve each inequality, and graph the solution set.

   a. $8x \leq -40$

   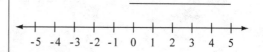

   b. $-9t > 36$

   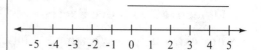

Name: Date:
Instructor: Section:

Objective 3 Practice Exercises

For extra help, see Example 3 on pages 178–179 of your text.

Solve each inequality. Write the solution set in interval notation and then graph it.

7. $-2s < 4$

7. _____

8. $4k \geq -16$

8. _____

9. $-9m \geq -36$

9. _____

Objective 4 Solve linear inequalities using both properties of inequality.

Video Examples

Review this example for Objective 4:

4. Solve $4x + 3 - 7 > -2x + 8 + 3x$. Graph the solution set.

   Step 1 Combine like terms and simplify.
   $$4x + 3 - 7 > -2x + 8 + 3x$$
   $$4x - 4 > x + 8$$

   Step 2 Use the addition property of inequality.
   $$4x - 4 - x > x + 8 - x$$
   $$3x - 4 > 8$$
   $$3x - 4 + 4 > 8 + 4$$
   $$3x > 12$$

   Step 3 Use the multiplication property of inequality.
   $$\frac{3x}{3} > \frac{12}{3}$$
   $$x > 4$$

   The solution set is $(4, \infty)$. The graph is shown below.

   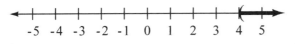

Now Try:

4. Solve $8x - 5 + 4 \geq 6x - 3x + 9$. Graph the solution set.

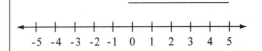

Name: Date:
Instructor: Section:

Objective 4 Practice Exercises

For extra help, see Example 4–6 on pages 179–180 of your text.

Solve each inequality. Write the solution set in interval notation and then graph it.

10. $4(y-3)+2>3(y-2)$

10. _____
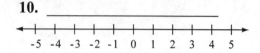

11. $-3(m+2)+3 \leq -4(m-2)-6$

11. _____
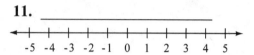

12. $7(2-x) \leq -2(x-3)-x$

12. _____
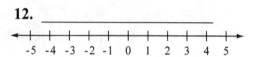

Objective 5 Solve applied problems using inequalities.

Video Examples

Review this example for Objective 5:

7. Ruth tutors mathematics in the evenings in an office for which she pays $600 per month rent. If rent is her only expense and she charges each student $40 per month, how many students must she teach to make a profit of at least $1600 per month?

   Step 1 Read the problem again.

   Step 2 Assign a variable.
   Let x = the number of students.

   Step 3 Write an inequality.
   $40x - 600 \geq 1600$

   Step 4 Solve.
   $40x - 600 + 600 \geq 1600 + 600$
   $40x \geq 2200$
   $\dfrac{40x}{40} \geq \dfrac{2200}{40}$
   $x \geq 55$

Now Try:

7. Two sides of a triangle are equal in length, with the third side 8 feet longer than one of the equal sides. The perimeter of the triangle cannot be more than 38 feet. Find the largest possible value for the length of the equal sides.

   _____

Name: Date:
Instructor: Section:

Step 5 State the answer. Ruth must have 55 or more students to have at least $1600 profit.

Step 6 Check. $40(55) - 600 = 1600$ Also, any number greater than 55 makes the profit greater than $1600.

Objective 5 Practice Exercises

For extra help, see Example 7 on page 181 of your text.

Solve each problem.

13. Lauren has grades of 98 and 86 on her first two chemistry quizzes. What must she score on her third quiz to have an average of at least 91 on the three quizzes?

13. _____

14. Nina has a budget of $230 for gifts for this year. So far she has bought gifts costing $47.52, $38.98, and $26.98. If she has three more gifts to buy, find the average amount she can spend on each gift and still stay within her budget.

14. _____

15. If twice the sum of a number and 7 is subtracted from three times the number, the result is more than –9. Find all such numbers.

15. _____

Name: Date:
Instructor: Section:

Objective 6   Solve linear inequalities with three parts.

Video Examples

Review this example for Objective 6:

9. Solve the inequality, and graph the solution set.

$$3 \le 4x - 5 < 7$$

$$3 \le 4x - 5 < 7$$
$$3 + 5 \le 4x - 5 + 5 < 7 + 5$$
$$8 \le 4x < 12$$
$$\frac{8}{4} \le \frac{4x}{4} < \frac{12}{4}$$
$$2 \le x < 3$$

The solution set is $[2, 3)$. The graph is shown below.

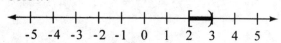

Now Try:

9. Solve the inequality, and graph the solution set.

$$8 \le 6x - 4 < 20$$

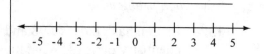

Objective 6 Practice Exercises

For extra help, see Examples 8–9 on pages 182–183 of your text.

Solve each inequality. Write the solution set in interval notation and then graph it.

16. $7 < 2x + 3 \le 13$ 16.

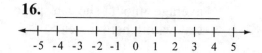

17. $-17 \le 3x - 2 < -11$ 17.

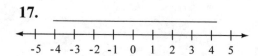

18. $1 < 3z + 4 < 19$ 18.

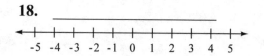

Name: Date:
Instructor: Section:

Chapter 3 LINEAR EQUATIONS AND INEQUALITIES IN TWO VARIABLES

3.1 Linear Equations and Rectangular Coordinates

Learning Objectives
1. Interpret graphs.
2. Write a solution as an ordered pair.
3. Decide whether a given ordered pair is a solution of a given equation.
4. Complete ordered pairs for a given equation.
5. Complete a table of values.
6. Plot ordered pairs.

Key Terms

Use the vocabulary terms listed below to complete each statement in exercises 1–13.

    line graph   **linear equation in two variables**

    ordered pair   **table of values**   **x-axis**

    y-axis   **rectangular (Cartesian) coordinate system**

    origin   **quadrants**   **plane**   **coordinates**

    plot   **scatter diagram**

1. A _____ uses dots connected by lines to show trends.

2. An equation that can be written in the form $Ax + By = C$, where A, B, and C are real numbers and $A, B \neq 0$, is called a _____.

3. _____ are the numbers in the ordered pair that specify the location of a point on a rectangular coordinate system.

4. In a coordinate system, the horizontal axis is called the _____.

5. In a coordinate system, the vertical axis is called the _____.

6. A pair of numbers written between parentheses in which order is important is called a(n) _____.

7. Together, the x-axis and the y-axis form a _____.

8. A coordinate system divides the plane into four regions called _____.

9. The axis lines in a coordinate system intersect at the _____.

10. To _____ an ordered pair is to find the corresponding point on a coordinate system.

Copyright © 2016 Pearson Education, Inc.

Name: Date:
Instructor: Section:

11. A graph of ordered pairs is called a _____.

12. A table showing selected ordered pairs of numbers that satisfy an equation is called a _____.

13. A flat surface determined by two intersecting lines is a _____.

Objective 1   Interpret graphs.

Video Examples

The line graph shows the number of degrees awarded by a university for the years 2000–2005.

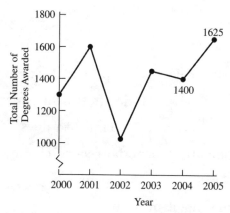

Review these examples for Objective 1:
1.
  a. Between which years did the number of degrees awarded decrease?

  The line between 2001 and 2002 and between 2003 and 2004 falls, so the number of degrees awarded decreased between 2001-2002 and 2003-2004.

  b. Estimate the total number of degrees awarded in 2002 and in 2005. About how many more degrees were awarded in 2005?

  Move up from 2002 on the horizontal scale to the point plotted for 2002. This point is about 1000. So about 1000 degrees were awarded in 2002.
  Similarly, locate the point plotted for 2005. Moving across to the vertical scale, the graph indicates that the number of degrees awarded in 2005 was 1625.
  Between 2002 and 2005, the increase was
  $1625 - 1000 = 625$.

Now Try:
1.
  a. Between which years did the number of degrees awarded increase?

  _____

  b. Estimate the total number of degrees awarded in 2000 and in 2001. About how many more degrees were awarded in 2001?

  _____

Name: Date:
Instructor: Section:

Objective 1 Practice Exercises

For extra help, see Example 1 on page 200 of your text.

The line graph shows the number of degrees awarded by a university for the years 2000–2005. Use this graph to answer exercises 1–2.

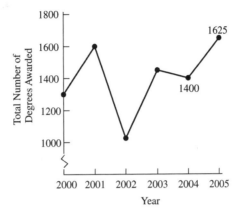

1. Between what pairs of consecutive years did the number of degrees decrease?

1. _____

2. If 20% of the degrees awarded in 2005 were MBA degrees, how many MBAs were awarded in 2005?

2. _____

Objective 2 Write a solution as an ordered pair.

Objective 2 Practice Exercises

For extra help, see page 201 of your text.

Write each solution as an ordered pair.

3. $x = 4$ and $y = 7$

3. _____

4. $y = \frac{1}{3}$ and $x = 0$

4. _____

5. $x = 0.2$ and $y = 0.3$

5. _____

Name: Date:
Instructor: Section:

Objective 3 Decide whether a given ordered pair is a solution of a given equation.

Video Examples

Review these examples for Objective 3:

2. Decide whether each ordered pair is a solution of the equation $4x+5y=40$.

   a. $(5, 4)$

   Substitute 5 for x and 4 for y in the given equation.
   $$4x+5y=40$$
   $$4(5)+5(4) \stackrel{?}{=} 40$$
   $$20+20 \stackrel{?}{=} 40$$
   $$40 = 40 \quad \text{True}$$
   This result is true, so $(5, 4)$ is a solution of $4x+5y=40$.

   b. $(-3, 6)$

   Substitute -3 for x and 6 for y in the given equation.
   $$4x+5y=40$$
   $$4(-3)+5(6) \stackrel{?}{=} 40$$
   $$-12+30 \stackrel{?}{=} 40$$
   $$18 = 40 \quad \text{False}$$
   This result is false, so $(-3, 6)$ is not a solution of $4x+5y=40$.

Now Try:

2. Decide whether each ordered pair is a solution of the equation $3x-4y=12$.

   a. $(8, 3)$

   _____

   b. $(5,-4)$

   _____

Objective 3 Practice Exercises

For extra help, see Example 2 on page 202 of your text.

Decide whether the given ordered pair is a solution of the given equation.

6. $4x-3y=10;\ (1,2)$ 6. _____

7. $2x-3y=1;\ \left(0,\frac{1}{3}\right)$ 7. _____

8. $x=-7;\ (-7,9)$ 8. _____

Name: Date:
Instructor: Section:

Objective 4 Complete ordered pairs for a given equation.

Video Examples

Review this example for Objective 4:
3. Complete the ordered pair for the equation $y = 5x + 8$.

   $(3, \underline{})$

   Replace x with 3.
   $y = 5x + 8$
   $y = 5(3) + 8$
   $y = 15 + 8$
   $y = 23$
   The ordered pair is (3, 23).

Now Try:
3. Complete the ordered pair for the equation $y = 4x - 7$.

   $(5, \underline{})$

   _____

Objective 4 Practice Exercises

For extra help, see Example 3 on pages 202–203 of your text.

For each of the given equations, complete the ordered pairs beneath it.

9. $y = 2x - 5$
   (a) (2,)
   (b) (0,)
   (c) (, 3)
   (d) (, −7)
   (e) (, 9)

9.
   (a) _____
   (b) _____
   (c) _____
   (d) _____
   (e) _____

10. $y = 3 + 2x$
    (a) (−4,)
    (b) (2,)
    (c) (, 0)
    (d) (−2,)
    (e) (, −7)

10.
    (a) _____
    (b) _____
    (c) _____
    (d) _____
    (e) _____

Copyright © 2016 Pearson Education, Inc.

Name: Date:
Instructor: Section:

Objective 5 Complete a table of values.

Video Examples

Review this example for Objective 5:
4. Complete the table of values for the equation. Then write the results as ordered pairs.

$2x - 3y = 6$

x	y
9	
6	
	−2
	8

From the table, we can write the ordered pairs: (9, ___), (6, ___), (___, −2), (___, 8).
From the first row of the table, let $x = 9$ in the equation. From the second row of the table, let $x = 6$.

If $x = 9$,   If $x = 6$,
$2x - 3y = 6$ $2x - 3y = 6$
$2(9) - 3y = 6$ $2(6) - 3y = 6$
$18 - 3y = 6$ $12 - 3y = 6$
$-3y = -12$ $-3y = -6$
$y = 4$ $y = 2$

The first two ordered pairs are (9, 4) and (6, 2).

From the third and fourth rows of the table, let $y = -2$ and $y = 8$, respectively.

If $y = -2$,   If $y = 8$,
$2x - 3y = 6$ $2x - 3y = 6$
$2x - 3(-2) = 6$ $2x - 3(8) = 6$
$2x + 6 = 6$ $2x - 24 = 6$
$2x = 0$ $2x = 30$
$x = 0$ $x = 15$

The last two ordered pairs are (0, −2) and (15, 8). The completed table and corresponding ordered pairs follow.

x	y	Ordered pairs
9	4	→ (9, 4)
6	2	→ (6, 2)
0	−2	→ (0, −2)
15	8	→ (15, 8)

Now Try:
4. Complete the table of values for the equation. Then write the results as ordered pairs.

$4x - y = 8$

x	y
1	
5	
	0
	4

_____

108 Copyright © 2016 Pearson Education, Inc.

Name: Date:
Instructor: Section:

Objective 5 Practice Exercises

For extra help, see Example 4 on pages 203–204 of your text.

Complete each table of values. Write the results as ordered pairs.

11. $2x + 5 = 7$

x	y
	−3
	0
	5

11. _____

12. $y − 4 = 0$

x	y
−4	
0	
6	

12. _____

13. $4x + 3y = 12$

x	y
0	
	0
	−1

13. _____

Objective 6 Plot ordered pairs.

Video Examples

Review these examples for Objective 6:

5. Plot the given points in a coordinate system.
   (5, 4)   (−2,−1)   (−3, 5)   (4,−2)
   (1,−2.5)   (3, 0)   (0, 4)

   Step 1 Move right or left the number of units that correspond to the *x*-coordinate in the ordered pair—right if the *x*-coordinate is positive and left if it is negative.

   Step 2 Then turn and move up or down the number of units that corresponds to the *y*-coordinate—up if the *y*-coordinate is

Now Try:

5. Plot the given points in a coordinate system.
   (2, 4)   (−5, 1)   (−4,−2)
   (3,−5)   (2,−1.5)   (6, 0)
   (0,−6)

Name: Date:
Instructor: Section:

positive or down if it is negative.

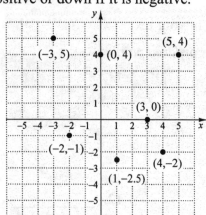
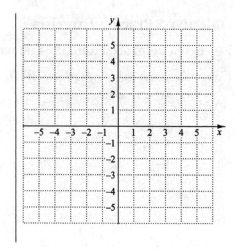

Objective 6 Practice Exercises

For extra help, see Examples 5–6 on pages 205–206 of your text.

Plot the each ordered pair on a coordinate system.

14. $(0, -2)$ 14.

15. $(-3, 4)$ 15.

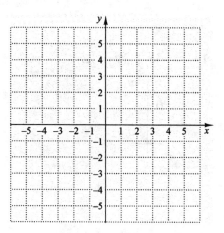

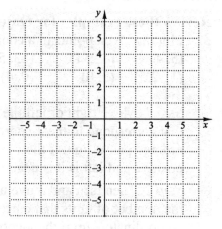

Name: Date:
Instructor: Section:

16. $(2,-5)$

16.

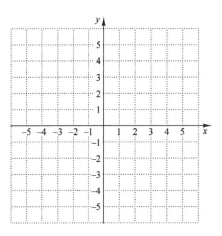

Name: Date:
Instructor: Section:

Chapter 3 LINEAR EQUATIONS AND INEQUALITIES IN TWO VARIABLES

3.2 Graphing Linear Equations in Two Variables

Learning Objectives
1. Graph linear equations by plotting ordered pairs.
2. Find intercepts.
3. Graph linear equations of the form $Ax + By = 0$.
4. Graph linear equations of the form $y = b$ or $x = a$.
5. Use a linear equation to model data.

Key Terms

Use the vocabulary terms listed below to complete each statement in exercises 1–4.

    graph   graphing   y-intercept   x-intercept

1. If a graph intersects the y-axis at k, then the _____ is (0, k).

2. If a graph intersects the x-axis at k, then the _____ is (k, 0).

3. The process of plotting the ordered pairs that satisfy a linear equation and drawing a line through them is called _____.

4. The set of all points that correspond to the ordered pairs that satisfy the equation is called the _____ of the equation.

Objective 1 Graph linear equations by plotting ordered pairs.

Video Examples

Review this example for Objective 1:

2. Graph $2x + 3y = 6$.

First let $x = 0$ and then let $y = 0$ to determine two ordered pairs.

$$
\begin{array}{l|l}
2(0)+3y=6 & 2x+3(0)=6 \\
0+3y=6 & 2x+0=6 \\
3y=6 & 2x=6 \\
y=2 & x=3
\end{array}
$$

The ordered pairs are (0, 2) and (3, 0). Find a third ordered pair by choosing a number other than 0 for x or y. We choose $y = 4$.

$$2x + 3(4) = 6$$
$$2x + 12 = 6$$
$$2x = -6$$
$$x = -3$$

Now Try:

2. Graph $x + y = 3$.

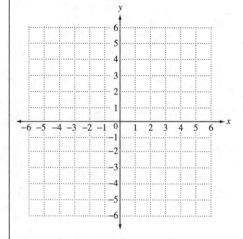

112 Copyright © 2016 Pearson Education, Inc.

Name: Date:
Instructor: Section:

This gives the ordered pair (−3, 4). We plot the three ordered pairs (0, 2), (3, 0), and (−3, 4) and draw a line through them.

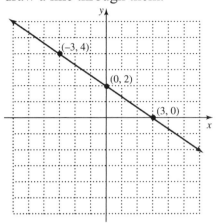

Objective 1 Practice Exercises

For extra help, see Examples 1–2 on pages 213–214 of your text.

Complete the ordered pairs for each equation. Then graph the equation by plotting the points and drawing a line through them.

1. $y = 3x - 2$

    (0,   )

    (, 0)

    (2,   )

1.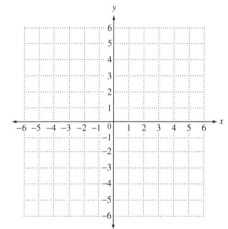

2. $x - y = 4$

    (0,   )

    (, 0)

    (−2,   )

2.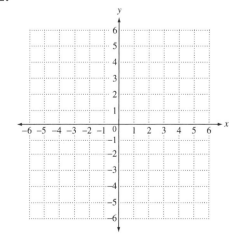

Copyright © 2016 Pearson Education, Inc.

Name: Date:
Instructor: Section:

3. $x = 2y + 1$

  (0,)
  (, 0)
  (, −2)

3.
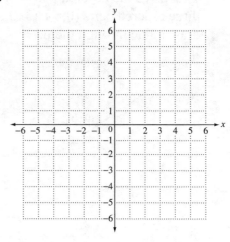

Objective 2 Find intercepts.

Video Examples

Review this example for Objective 2:

3. Graph $3x + y = 6$ using intercepts.

To find the y-intercept, let $x = 0$.
To find the x-intercept, let $y = 0$.

$3(0) + y = 6 \quad | \quad 3x + 0 = 6$
$0 + y = 6 \quad | \quad 3x = 6$
$y = 6 \quad | \quad x = 2$

The intercepts are (0, 6) and (2, 0). To find a third point, as a check, we let $x = 1$.

$3(1) + y = 6$
$3 + y = 6$
$y = 3$

This gives the ordered pair (1, 3).

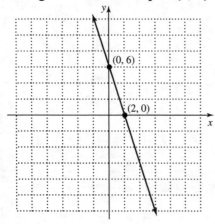

Now Try:

3. Graph $5x - 2y = -10$ using intercepts.

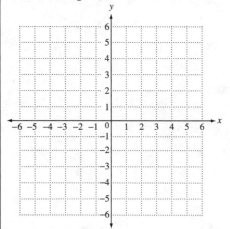

114 Copyright © 2016 Pearson Education, Inc.

Name: Date:
Instructor: Section:

4. Graph $y = \frac{3}{2}x - 4$.

   To find the y-intercept, let $x = 0$.
   To find the x-intercept, let $y = 0$.

   $y = \frac{3}{2}(0) - 4 \quad \bigg| \quad 0 = \frac{3}{2}x - 4$

   $y = 0 - 4 \quad \bigg| \quad 4 = \frac{3}{2}x$

   $y = -4 \quad \bigg| \quad \frac{8}{3} = x$

   The intercepts are $(0, -4)$ and $\left(\frac{8}{3}, 0\right)$. To find a third point, as a check, we let $x = 2$.

   $y = \frac{3}{2}(2) - 4$

   $y = 3 - 4$

   $y = -1$

   This gives the ordered pair $(2, -1)$.

   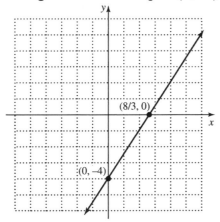

4. Graph $y = 4x - 4$.

   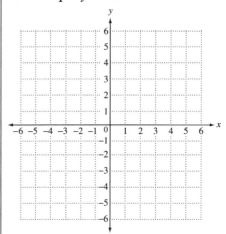

Copyright © 2016 Pearson Education, Inc.

Name:
Instructor:
Date:
Section:

Objective 2 Practice Exercises

For extra help, see Examples 3–4 on pages 214–216 of your text.

Find the intercepts for each equation. Then graph the equation.

4. $y = \dfrac{2}{3}x - 2$

4.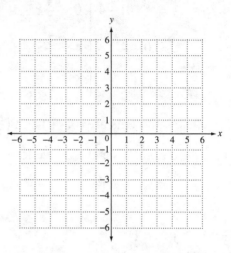

5. $4x - 7y = -8$

5.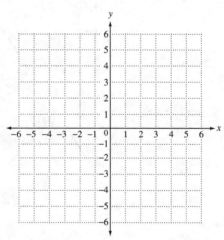

Name: Date:
Instructor: Section:

Objective 3 Graph linear equations of the form $Ax + By = 0$.

Video Examples

Review this example for Objective 3:

5. Graph $x + 5y = 0$.

   To find the y-intercept, let $x = 0$.
   To find the x-intercept, let $y = 0$.

   $$\begin{array}{l|l} 0 + 5y = 0 & x + 5(0) = 0 \\ 5y = 0 & x + 0 = 0 \\ y = 0 & x = 0 \end{array}$$

   The x- and y-intercepts are the same point $(0, 0)$. We must select two other values for x or y to find two other points. We choose $y = 1$ and $y = -1$.

   $$\begin{array}{l|l} x + 5(1) = 0 & x + 5(-1) = 0 \\ x + 5 = 0 & x - 5 = 0 \\ x = -5 & x = 5 \end{array}$$

   We use $(-5, 1)$, $(0, 0)$, and $(5, -1)$ to draw the graph.

   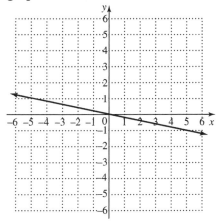

Now Try:

5. Graph $3x - y = 0$.

   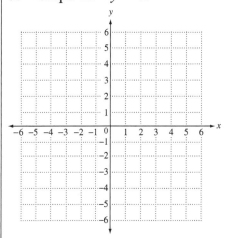

Copyright © 2016 Pearson Education, Inc.

Name: Date:
Instructor: Section:

Objective 3 Practice Exercises

For extra help, see Example 5 on page 216 of your text.

Graph each equation.

6. $-3x - 2y = 0$

6.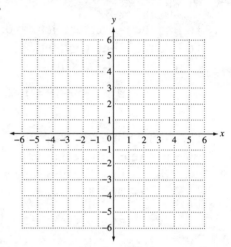

7. $x + y = 0$

7.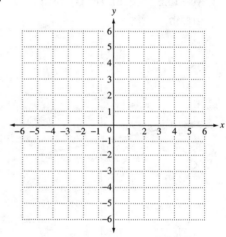

8. $y = 2x$

8.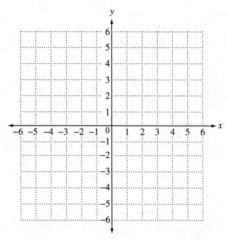

Name: Date:
Instructor: Section:

Objective 4   Graph linear equations of the form $y = b$ or $x = a$.

Video Examples

Review these examples for Objective 4:

6. Graph $y = -2$.

   For any value of x, y is always -2. Three ordered pairs that satisfy the equation are $(-4, -2)$, $(0, -2)$ and $(2, -2)$. Drawing a line through these points gives the horizontal line. The y-intercept is $(0, -2)$. There is no x-intercept.

   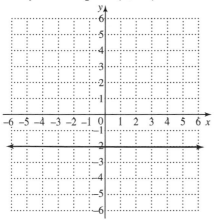

Now Try:

6. Graph $y = 4$.

   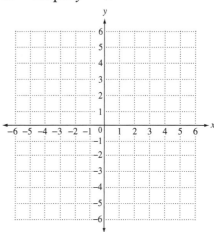

7. Graph $x + 4 = 0$.

   First we subtract 4 from each side of the equation to get the equivalent equation $x = -4$. All ordered-pair solutions of this equation have x-coordinate -4.
   Three ordered pairs that satisfy the equation are $(-4, -1)$, $(-4, 0)$, and $(-4, 3)$. The graph is a vertical line. The x-intercept is $(-4, 0)$. There is no y-intercept.

   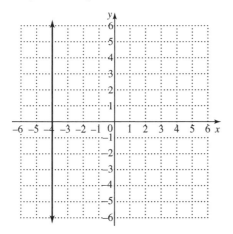

7. Graph $x = 0$.

   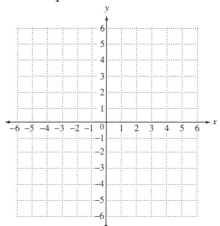

Copyright © 2016 Pearson Education, Inc.

Name: Date:
Instructor: Section:

Objective 4 Practice Exercises

For extra help, see Examples 6–7 on page 217 of your text.

Graph each equation.

9. $x - 1 = 0$

9.
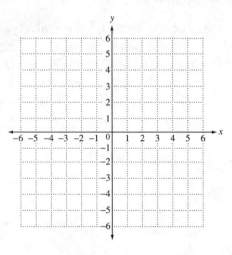

10. $y + 3 = 0$

10.
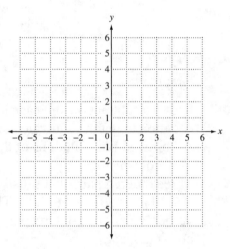

Name: Date:
Instructor: Section:

Objective 5 Use a linear equation to model data.

Video Examples

Review these examples for Objective 5:

8. Every year sea turtles return to a certain group of islands to lay eggs. The number of turtle eggs that hatch can be approximated by the equation $y = -70x + 3260$, where y is the number of eggs that hatch and $x = 0$ representing 1990.

   a. Use this equation to find the number of eggs that hatched in 1995, 2000, and 2005, and 2015.

   Substitute the appropriate value for each year x to find the number of eggs hatched in that year.

   For 1995:
   $y = -70(5) + 3260$ $1995 - 1990 = 5$
   $y = 2910$ eggs   Replace x with 5.

   For 2000:
   $y = -70(10) + 3260$ $2000 - 1990 = 10$
   $y = 2560$ eggs   Replace x with 10.

   For 2005:
   $y = -70(15) + 3260$ $2005 - 1990 = 15$
   $y = 2210$ eggs   Replace x with 15.

   For 2015:
   $y = -70(25) + 3260$ $2015 - 1990 = 25$
   $y = 1510$ eggs   Replace x with 25.

   b. Write the information from part (a) as four ordered pairs, and use them to graph the given linear equation.

   Since x represents the year and y represents the number of eggs, the ordered pairs are (5, 2910), (10, 2560), (15, 2210), and (25, 1510).

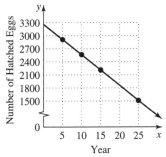

Now Try:

8. Suppose that the demand and price for a certain model of calculator are related by the equation $y = 45 - \frac{3}{5}x$, where y is the price (in dollars) and x is the demand (in thousands of calculators).

   a. Assuming that this model is valid for a demand up to 50,000 calculators, use this equation to find the price of calculators at each level of demand.

   0 calculators _____

   5000 calculators _____

   20,000 calculators _____

   45,000 calculators _____

   b. Write the information from part (a) as four ordered pairs, and use them to graph the given linear equation.

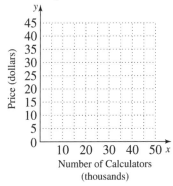

Copyright © 2016 Pearson Education, Inc. 121

Name: Date:
Instructor: Section:

c. Use this graph and the equation to estimate the number of eggs that will hatch in 2010.

For 2010, $x = 20$. On the graph, find 20 on the horizontal axis, move up to the graphed line and then across to the vertical axis. It appears that in 2010, there were about 1900 eggs.

To use the equation, substitute 20 for x.

$y = -70(20) + 3260$

$y = 1860$ eggs

This result for 2020 is close to our estimate of 1900 eggs from the graph.

c. Use this graph and the equation to estimate the price of 30,000 calculators.

_____

Objective 5 Practice Exercises

For extra help, see Example 8 on pages 218–219 of your text.

Solve each problem. Then graph the equation.

11. The profit y in millions of dollars earned by a small computer company can be approximated by the linear equation $y = 0.63x + 4.9$, where $x = 0$ corresponds to 2004, $x = 1$ corresponds to 2005, and so on. Use this equation to approximate the profit in each year from 2004 through 2007.

11. 2004 _____

    2005 _____

    2006 _____

    2007 _____

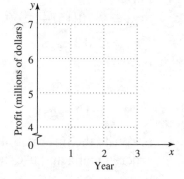

Name: Date:
Instructor: Section:

12. The number of band instruments sold by Elmer's Music Shop can be approximated by the equation $y = 325 + 42x$, where y is the number of instruments sold and x is the time in years, with $x = 0$ representing 2003. Use this equation to approximate the number of instruments sold in each year from 2003 through 2006.

12. 2003 _____

    2004 _____

    2005 _____

    2006 _____

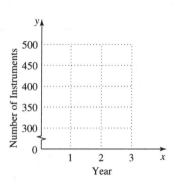

13. According to *The Old Farmer's Almanac*, the temperature in degrees Celsius can be determined by the equation $y = \frac{1}{3}x + 4$, where x is the number of cricket chirps in 25 seconds and y is the temperature in degrees Celsius. Use this equation to find the temperature when there are 48 chirps, 54 chirps, 60 chirps, and 66 chirps.

13. 48 _____

    54 _____

    60 _____

    66 _____

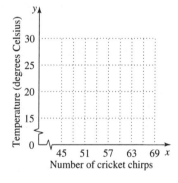

Name: Date:
Instructor: Section:

Chapter 3 LINEAR EQUATIONS AND INEQUALITIES IN TWO VARIABLES

3.3   The Slope of a Line

Learning Objectives
1. Find the slope of a line given two points.
2. Find the slope from the equation of a line.
3. Use slope to determine whether two lines are parallel, perpendicular, or neither.

Key Terms

Use the vocabulary terms listed below to complete each statement in exercises 1–5.

   rise   run   slope   parallel lines

   perpendicular lines

1. Two lines that intersect in a 90° angle are called _____.

2. The _____ of a line is the ratio of the change in y compared to the change in x when moving along the line from one point to another.

3. The vertical change between two different points on a line is called the _____.

4. Two lines in a plane that never intersect are called _____.

5. The horizontal change between two different points on a line is called the _____.

Objective 1   Find the slope of a line given two points.

Video Examples

Review these examples for Objective 1:
1. Find the slope of the line.

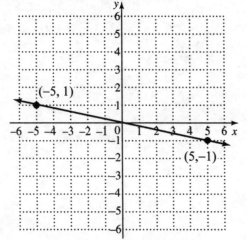

Now Try:
1. Find the slope of the line.

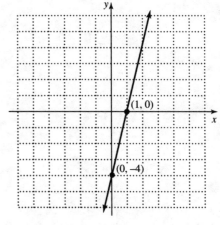

124   Copyright © 2016 Pearson Education, Inc.

Name: Date:
Instructor: Section:

We use the two points shown on the line. The vertical change is the difference in the *y*-values, or $-1 - 1 = -2$, and the horizontal change is the difference in the *x*-values or $5 - (-5) = 10$. Thus, the line has

$$\text{slope} = \frac{-2}{10}, \text{ or } -\frac{1}{5}.$$

2. Find the slope of the line.

The line passing through $(-5, 4)$ and $(2, -6)$

Apply the slope formula.

$(x_1, y_1) = (-5, 4)$ and $(x_2, y_2) = (2, -6)$

$$\text{slope } m = \frac{y_2 - y_1}{x_2 - x_1} = \frac{-6 - 4}{2 - (-5)}$$

$$= \frac{-10}{7}, \text{ or } -\frac{10}{7}$$

3. Find the slope of the line passing through $(-9, 3)$ and $(4, 3)$.

$(x_1, y_1) = (-9, 3)$ and $(x_2, y_2) = (4, 3)$

$$m = \frac{y_2 - y_1}{x_2 - x_1} = \frac{3 - 3}{4 - (-9)} = \frac{0}{13} = 0$$

4. Find the slope of the line passing through $(-5, 3)$ and $(-5, 8)$.

$(x_1, y_1) = (-5, 3)$ and $(x_2, y_2) = (-5, 8)$

$$m = \frac{y_2 - y_1}{x_2 - x_1} = \frac{8 - 3}{-5 - (-5)} = \frac{5}{0} \text{ undefined slope}$$

2. Find the slope of the line.

The line passing through $(-6, 7)$ and $(3, -9)$

_____

3. Find the slope of the line passing through $(8, -5)$ and $(-7, -5)$.

_____

4. Find the slope of the line passing through $(9, 11)$ and $(9, -7)$.

_____

Objective 1 Practice Exercises

For extra help, see Examples 1–4 on pages 225–228 of your text.

Find the slope of the line through the given points.

    1.   (4, 3) and (3, 5)   **1.** _____

    2.   (−4, 6) and (−4, −1)   **2.** _____

Name: Date:
Instructor: Section:

3. (−3, 3) and (6, 3) 3. _____

Objective 2 Find the slope from the equation of a line.

Video Examples

Review this example for Objective 2:
5. Find the slope of the line.
$$4x - 3y = 7$$

Step 1 Solve the equation for y.
$$4x - 3y = 7$$
$$-3y = -4x + 7$$
$$y = \frac{4}{3}x - \frac{7}{3}$$

Step 2 The slope is given by the coefficient of x, so the slope is $\frac{4}{3}$.

Now Try:
5. Find the slope of the line.
$$7x - 4y = 8$$

_____

Objective 2 Practice Exercises

For extra help, see Example 5 on pages 229–230 of your text.

Find the slope of each line.

4. $7y - 4x = 11$ 4. _____

5. $3y = 2x - 1$ 5. _____

6. $y = -\frac{2}{5}x - 4$ 6. _____

Name: Date:
Instructor: Section:

Objective 3   Use slope to determine whether two lines are parallel, perpendicular, or neither.

Video Examples

Review these examples for Objective 3:

6. Decide whether each pair of lines is parallel, perpendicular, or neither.

   a. $5x - y = 3$
         $15x - 3y = 12$

   Solve each equation for y.
   $y = 5x - 3$
   $y = 5x - 4$
   Both lines have slope 5, so the lines are parallel.

   b. $x + 3y = 8$ and $-3x + y = 5$

   Find the slope of each line by first solving each equation for y.

   $3y = -x + 8$ | $y = 3x + 5$
   $y = -\frac{1}{3}x + \frac{8}{3}$
   The slope is $-\frac{1}{3}$. | The slope is 3.

   Because the slopes are not equal, the lines are not parallel.

   Check the product of the slopes: $-\frac{1}{3}(3) = -1$.

   The two lines are perpendicular because the product of their slopes is -1.

Now Try:

6. Decide whether each pair of lines is parallel, perpendicular, or neither.

   a. $2x - 4y = 7$
       $3x - 6y = 8$

   _____

   b. $9x - y = 7$ and $x + 9y = 11$

   _____

Objective 3 Practice Exercises

For extra help, see Example 6 on pages 231–232 of your text.

*In each pair of equations, give the slope of each line, and then determine whether the two lines are **parallel**, **perpendicular**, or **neither**.*

7. $-x + y = -7$ 7. _____
        $x - y = -3$

Name: Date:
Instructor: Section:

8. $4x + 2y = 8$
    $x + 4y = -3$

8. _____

9. $9x + 3y = 2$
    $x - 3y = 5$

9. _____

Name: Date:
Instructor: Section:

Chapter 3 LINEAR EQUATIONS AND INEQUALITIES IN TWO VARIABLES

3.4 Slope-Intercept Form of a Linear Equation

Learning Objectives
1. Use the slope-intercept form of the equation of a line.
2. Graph a line by using its slope and a point on the line.
3. Write an equation of a line by using its slope and any point on the line.
4. Graph and write equations of horizontal and vertical lines.

Key Terms

Use the vocabulary terms listed below to complete each statement in exercises 1–3.

slope-intercept form   **point-slope form**   **standard form**

1. A linear equation in the form $y - y_1 = m(x - x_1)$ is written in
   _____.

2. A linear equation in the form $Ax + By = C$ is written in
   _____.

3. A linear equation in the form $y = mx + b$ is written in
   _____.

Objective 1 Use the slope-intercept form of the equation of a line.

Video Examples

Review these examples for Objective 1:	Now Try:
1. Identify the slope and y-intercept of the line with each equation.	1. Identify the slope and y-intercept of the line with each equation.
a. $y = -8x + 7$	**a.** $y = -12x + 6$
The slope is –8, and the y-intercept is (0, 7).	_____
b. $y = \dfrac{x}{9} - \dfrac{5}{4}$	**b.** $y = -\dfrac{x}{7} - \dfrac{7}{5}$
The equation can be written as $y = \dfrac{1}{9}x + \left(-\dfrac{5}{4}\right)$.	_____
The slope is $\dfrac{1}{9}$, and the y-intercept is $\left(0, -\dfrac{5}{4}\right)$.	

Name: Date:
Instructor: Section:

Objective 1 Practice Exercises

For extra help, see Example 1 on page 238 of your text.

Identify the slope and y-intercept of the line with each equation.

1. $y = \dfrac{3}{2}x - \dfrac{2}{3}$

1. _____

   _____

2. $y = -4x$

2. _____

   _____

Objective 2   Graph a line by using its slope and a point on the line.

Video Examples

Review this example for Objective 2:
2. Graph the equation by using the slope and y-intercept.

$2x - 3y = 6$

Step 1 Solve for y to write the equation in slope-intercept form.
$$2x - 3y = 6$$
$$-3y = -2x + 6$$
$$y = \dfrac{2}{3}x - 2$$

Step 2 The y-intercept is (0, –2). Graph this point.

Step 3 The slope is $\dfrac{2}{3}$. By definition,

$$\text{slope } m = \dfrac{\text{change in } y \text{ (rise)}}{\text{change in } x \text{ (run)}} = \dfrac{2}{3}$$

From the y-intercept, count up 2 units and to the right 3 units to obtain the point (3, 0).

Step 4 Draw the line through the points (0, –2) and (3, 0) to obtain the graph.

Now Try:
2. Graph the equation by using the slope and y-intercept.

$y = \dfrac{2}{3}x$

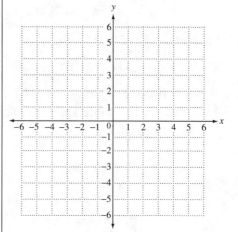

Name: Date:
Instructor: Section:

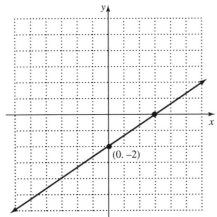

Objective 2 Practice Exercises

For extra help, see Examples 2–3 on pages 239–240 of your text.

Graph each equation by using the slope and y-intercept.

3. $4x - y = 4$

3.
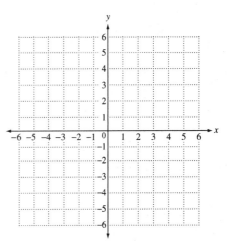

4. $y = -3x + 6$

4.
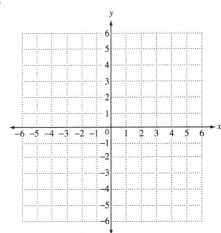

Name: Date:
Instructor: Section:

Graph the line passing through the given point and having the given slope.

5. $(4, -2)$; $m = -1$ 5.

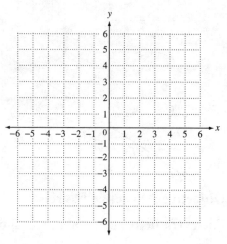

Objective 3 Write an equation of a line using its slope and any point on the line.

Video Examples

Review these examples for Objective 3:

4. Write an equation in slope-intercept form of the line passing through the given point and having the given slope.

  a. $(0, 2)$, $m = -3$

  Because the point $(0, 2)$ is the y-intercept, $b = 2$. Substitute $b = 2$ and $m = -3$ directly in the slope-intercept form.
  $y = mx + b$
  $y = -3x + 2$

  b. $(2, 9)$, $m = 5$

  Since the line passes through the point $(2, 9)$, we can substitute $x = 2$, $y = 9$, and slope $m = 5$ into $y = mx + b$ and solve for b.
  $y = mx + b$
  $9 = 5(2) + b$
  $-1 = b$
  Now substitute the values of m and b into slope-intercept form.
  $y = mx + b$
  $y = 5x - 1$

Now Try:

4. Write an equation in slope-intercept form of the line passing through the given point and having the given slope.

  a. $(0, -5)$, $m = \dfrac{3}{4}$

  _____

  b. $(-1, 4)$, $m = 6$

  _____

Name: Date:
Instructor: Section:

Objective 3 Practice Exercises

For extra help, see Example 4 on pages 240–241 of your text.

Write an equation in slope-intercept form of the line passing through the given point and having the given slope.

6. $(0,-4)$, $m = \dfrac{2}{3}$

6. _____

7. $(3, 6)$, $m = -2$

7. _____

8. $(-2, 0)$, $m = 1$

8. _____

Objective 4   Graph and write equations of horizontal and vertical lines.

Video Examples

Review these examples for Objective 4:

5. Graph each line passing through the given point and having the given slope.

   a. $(-5,-2)$, $m = 0$

   Horizontal lines have slope 0. Plot the point $(-5,-2)$ and draw a horizontal line through it.

   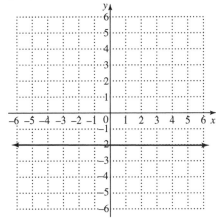

Now Try:

5. Graph each line passing through the given point and having the given slope.

   a. $(-2,-2)$, $m = 0$

   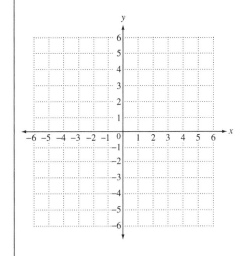

Name: Date:
Instructor: Section:

b. (−4, 3), undefined slope

Vertical lines have undefined slope. Plot the point (−4, 3) and draw a vertical line through it.

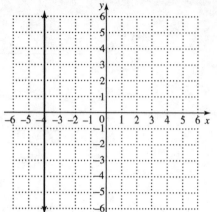

b. (−3, −1), undefined slope

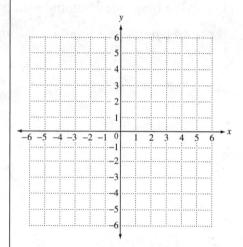

6. Write an equation of the line passing through the point (2, −2) that satisfies the given condition.

   a. The line has slope 0.

   Since the slope is 0, this is a horizontal line. $y = -2$.

   b. The line has undefined slope.

   This is a vertical line, since the slope is undefined. $x = 2$

6. Write an equation of the line passing through the point (−5, 5) that satisfies the given condition.

   a. The line has slope 0.

   _____

   b. The line has undefined slope.

   _____

Objective 4 Practice Exercises

For extra help, see Examples 5–6 on pages 241–242 of your text.

Graph the line passing through the given point and having the given slope.

9. (−1, 4); $m = 0$

9.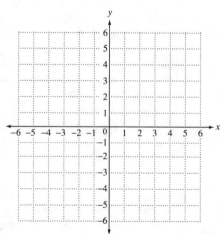

134 Copyright © 2016 Pearson Education, Inc.

Name: Date:
Instructor: Section:

Write an equation of the line passing through (–3, 3) and having the given slope.

10. Slope 0

10. _____

11. Undefined slope

11. _____

Name: Date:
Instructor: Section:

Chapter 3 LINEAR EQUATIONS AND INEQUALITIES IN TWO VARIABLES

3.5 Point-Slope Form of a Linear Equation and Modeling

Learning Objectives
1. Use point-slope form to write an equation of a line.
2. Write an equation of a line using two points on the line.
3. Write an equation of a line that fits a data set.

Key Terms

Use the vocabulary terms listed below to complete each statement in exercises 1–3.

slope-intercept form   point-slope form   standard form

1. A linear equation in the form $Ax + By = C$ is written in _____.

2. A linear equation in the form $y = mx + b$ is written in _____.

3. A linear equation in the form $y - y_1 = m(x - x_1)$ is written in _____.

Objective 1 Use point-slope form to write an equation of a line.

Video Examples

Review this example for Objective 1:
1. Write an equation of each line. Give the final answer in slope-intercept form.

    The line passing through (5, –3) with slope $\frac{5}{4}$.

    $$y - y_1 = m(x - x_1)$$
    $$y - (-3) = \frac{5}{4}(x - 5)$$
    $$y + 3 = \frac{5}{4}x - \frac{25}{4}$$
    $$y = \frac{5}{4}x - \frac{37}{4}$$

Now Try:
1. Write an equation of each line. Give the final answer in slope-intercept form.
    The line passing through (6, 11) with slope $-\frac{2}{3}$.

    _____

Name: Date:
Instructor: Section:

Objective 1 Practice Exercises

For extra help, see Example 1 on page 246 of your text.

Write an equation for the line passing through the given point and having the given slope. Write the equations in slope-intercept form, if possible.

1. $(-3, 4);\ m = -\dfrac{3}{5}$ 1. _____

2. $(-4, -3);\ m = -2$ 2. _____

3. $(2, 2);\ m = -\dfrac{3}{2}$ 3. _____

Objective 2   Write an equation of a line by using two points on the line.

Video Examples

Review this example for Objective 2:

2. Write the equation of the line passing through the points (6, 8) and (−3, 5). Give the final answer in slope-intercept form and then in standard form.

    First, find the slope of the line.

    $(x_1, y_1) = (6, 8)$ and $(x_2, y_2) = (-3, 5)$

    slope $m = \dfrac{y_2 - y_1}{x_2 - x_1} = \dfrac{5 - 8}{-3 - 6} = \dfrac{-3}{-9} = \dfrac{1}{3}$

    Now use (x_1, y_1), here (6, 8) and point-slope form.

Now Try:

2. Write the equation of the line passing through the points (7, 15) and (15, 9). Give the final answer in slope-intercept form and then in standard form.

    _____

Name: Date:
Instructor: Section:

$$y - y_1 = m(x - x_1)$$
$$y - 8 = \frac{1}{3}(x - 6)$$
$$y - 8 = \frac{1}{3}x - 2$$
$$y = \frac{1}{3}x + 6 \quad \text{Slope-intercept form}$$
$$3y = x + 18$$
$$-x + 3y = 18$$
$$x - 3y = -18 \quad \text{Standard form}$$

Objective 2 Practice Exercises

For extra help, see Example 2 on page 247 of your text.

Write an equation for the line passing through each pair of points. Write the equations in standard form.

4. $(-2, 1)$ and $(3, 11)$ 4. _____

5. $(2, 3)$ and $(-2, -3)$ 5. _____

6. $(3, -4)$ and $(2, 7)$.   6. _____

Name: Date:
Instructor: Section:

Objective 3 Write an equation of a line that fits a data set.

Video Examples

Review this example for Objective 3:

3. The table shows the number of internet users in the world from 1998 to 2005, where year 0 represents 1998.

Year	Number of Internet Users (millions)
0	147
2	361
4	587
6	817
8	1093

Plot the data and find an equation that approximates it.

Letting y represent the number of internet users in year x, we plot the data.

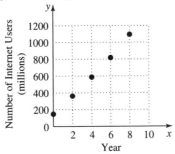

The points appear to lie approximately in a straight line. To find an equation of the line, we choose the ordered pairs (0, 147) and (8, 1093) from the table and find the slope of the line through these points.

$(x_1, y_1) = (0, 147)$ and $(x_2, y_2) = (8, 1093)$

$$\text{slope } m = \frac{y_2 - y_1}{x_2 - x_1} = \frac{1093 - 147}{8 - 0} = \frac{946}{8}$$
$$= 118.25$$

Use the slope, 118.25, and the point (0, 147) in slope-intercept form.

$y = mx + b$
$147 = 118.25(0) + b$
$147 = b$

Thus, $m = 118.25$ and $b = 147$, so the equation of the line is $y = 118.25x + 147$.

Now Try:

3. The table shows the average annual telephone expenditures for residential and pay telephones from 2001 to 2006, where year 0 represents 2001.

Year	Annual Telephone Expenditures
0	$686
2	$620
3	$592
4	$570
5	$542

Plot the data and find an equation that approximates it.

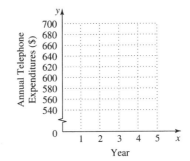

_____

Name:
Instructor:

Date:
Section:

Objective 3 Practice Exercises

For extra help, see Example 3 on pages 248–249 of your text.

Plot the data and find an equation that approximates it.

7. The table shows the U.S. municipal solid waste recycling percents since 1985, where year 0 represents 1985.

Year	Recycling Percent
0	10.1
5	16.2
10	26.0
15	29.1
20	32.5

7.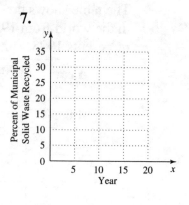

8. The table shows the approximate consumer expenditures for food in the U.S. in billions of dollars for selected years, where year 0 represents 1985.

Year	Food Expenditures (billions of dollars)
0	233
5	298
10	343
15	417
20	515

8.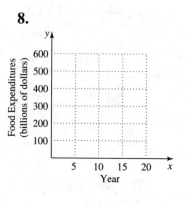

Name: Date:
Instructor: Section:

Chapter 4 EXPONENTS AND POLYNOMIALS

4.1 The Product Rule and Power Rules for Exponents

Learning Objectives
1. Use exponents.
2. Use the product rule for exponents.
3. Use the rule $(a^m)^n = a^{mn}$.
4. Use the rule $(ab)^m = a^m b^m$.
5. Use the rule $\left(\dfrac{a}{b}\right)^m = \dfrac{a^m}{b^m}$.
6. Use combinations of the rules for exponents.
7. Use the rules for exponents in a geometry application.

Key Terms

Use the vocabulary terms listed below to complete each statement in exercises 1–3.

exponential expression **base** **power**

1. 2^5 is read "2 to the fifth _____".

2. A number written with an exponent is called a(n) _____.

3. The _____ is the number being multiplied repeatedly.

Objective 1 Use exponents.

Video Examples

Review these examples for Objective 1:

1. Write $5 \cdot 5 \cdot 5$ in exponential form.

    Since 5 occurs as a factor three times, the base is 5 and the exponent is 3.
    $5 \cdot 5 \cdot 5 = 5^3$

2. Name the base and exponent of each expression. Then evaluate.

    a. 3^4

    Base: 3
    Exponent: 4
    Value: $3^4 = 3 \cdot 3 \cdot 3 \cdot 3 = 81$

Now Try:

1. Write $4 \cdot 4 \cdot 4 \cdot 4 \cdot 4$ in exponential form.

    _____

2. Name the base and exponent of each expression. Then evaluate.

    a. 2^6

    _____

Name: Date:
Instructor: Section:

b. $(-3)^4$

Base: –3
Exponent: 4
Value: $(-3)^4 = (-3)(-3)(-3)(-3) = 81$

b. $(-2)^6$

b. _____

Objective 1 Practice Exercises

For extra help, see Examples 1–2 on page 262 of your text.

Write the expression in exponential form and evaluate, if possible.

1. $\left(\frac{1}{3}\right)\left(\frac{1}{3}\right)\left(\frac{1}{3}\right)\left(\frac{1}{3}\right)\left(\frac{1}{3}\right)$

1. _____

Evaluate each exponential expression. Name the base and the exponent.

2. $(-4)^4$

2. _____
base _____
exponent _____

3. -3^8

3. _____
base _____
exponent _____

Objective 2 Use the product rule for exponents.

Video Examples

Review these examples for Objective 2:

3. Use the product rule for exponents to simplify, if possible.

   a. $8^4 \cdot 8^5$

   $8^4 \cdot 8^5 = 8^{4+5}$
   $= 8^9$

   b. $m^7 m^8 m^9$

   $m^7 m^8 m^9 = m^{7+8+9}$
   $= m^{24}$

Now Try:

3. Use the product rule for exponents to simplify, if possible.

   a. $9^6 \cdot 9^7$

   _____

   b. $m^{11} m^9 m^7$

   _____

Name: Date:
Instructor: Section:

c. $(5x^4)(6x^9)$

$$5x^4 \cdot 6x^9 = (5 \cdot 6) \cdot (x^4 \cdot x^9)$$
$$= 30x^{4+9}$$
$$= 30x^{13}$$

d. $5^2 + 5^3$

$$5^2 + 5^3 = 25 + 125$$
$$= 150$$

c. $(6x^5)(3x^6)$

_____

d. $3^4 + 3^3$

_____

Objective 2 Practice Exercises

For extra help, see Example 3 on page 363 of your text.

Use the product rule to simplify each expression, if possible. Write each answer in exponential form.

4. $7^4 \cdot 7^3$

4. _____

5. $(-2c^7)(-4c^8)$

5. _____

6. $(3k^7)(-8k^2)(-2k^9)$

6. _____

Objective 3 Use the rule $(a^m)^n = a^{mn}$.

Video Examples

Review these examples for Objective 3:
4. Use power rule (a) for exponents to simplify.

 a. $(5^6)^3$

 $$(5^6)^3 = 5^{6 \cdot 3}$$
 $$= 5^{18}$$

 b. $(x^3)^4$

 $$(x^3)^4 = x^{3 \cdot 4}$$
 $$= x^{12}$$

Now Try:
4. Use power rule (a) for exponents to simplify.

 a. $(7^2)^4$

 _____

 b. $(x^5)^6$

 _____

Name: Date:
Instructor: Section:

Objective 3 Practice Exercises

For extra help, see Example 4 on page 264 of your text.

Simplify each expression. Write all answers in exponential form.

7. $\left(7^3\right)^4$

7. _____

8. $-\left(v^4\right)^9$

8. _____

9. $\left[(-3)^3\right]^7$

9. _____

Objective 4 Use the rule $(ab)^m = a^m b^m$.

Video Examples

Review this example for Objective 4:
5. Use power rule (b) for exponents to simplify.

$$(5xy)^3$$
$$(5xy)^3 = 5^3 x^3 y^3$$
$$= 125 x^3 y^3$$

Now Try:
5. Use power rule (b) for exponents to simplify.
$(4ab)^3$

_____

Objective 4 Practice Exercises

For extra help, see Example 5 on page 265 of your text.

Simplify each expression.

10. $\left(5r^3 t^2\right)^4$

10. _____

11. $\left(-0.2 a^4 b\right)^3$

11. _____

12. $\left(-2 w^3 z^7\right)^4$

12. _____

Name: Date:
Instructor: Section:

Objective 5 Use the rule $\left(\dfrac{a}{b}\right)^m = \dfrac{a^m}{b^m}$.

Video Examples

Review this example for Objective 5:
6. Use power rule (c) for exponents to simplify.

$$\left(\dfrac{1}{8}\right)^3$$

$$\left(\dfrac{1}{8}\right)^3 = \dfrac{1^3}{8^3} = \dfrac{1}{512}$$

Now Try:
6. Use power rule (c) for exponents to simplify.

$$\left(\dfrac{1}{4}\right)^5$$

_____

Objective 5 Practice Exercises

For extra help, see Example 6 on page 265 of your text.

Simplify each expression.

13. $\left(-\dfrac{2x}{5}\right)^3$

13. _____

14. $\left(\dfrac{xy}{z^2}\right)^4$

14. _____

15. $\left(\dfrac{-2a}{b^2}\right)^7$

15. _____

Objective 6 Use combinations of the rules for exponents.

Video Examples

Review these examples for Objective 6:
7. Simplify each expression.

a. $\left(\dfrac{3}{4}\right)^3 \cdot 3^2$

$$\left(\dfrac{3}{4}\right)^3 \cdot 3^2 = \dfrac{3^3}{4^3} \cdot \dfrac{3^2}{1}$$
$$= \dfrac{3^3 \cdot 3^2}{4^3 \cdot 1}$$
$$= \dfrac{3^{3+2}}{4^3}$$
$$= \dfrac{3^5}{4^3}, \text{ or } \dfrac{243}{64}$$

Now Try:
7. Simplify each expression.

a. $\left(\dfrac{5}{2}\right)^3 \cdot 5^2$

_____

b. $(-x^5y)^4(-x^6y^5)^3$

$(-x^5y)^4(-x^6y^5)^3$

$=(-1x^5y)^4(-1x^6y^5)^3$

$=(-1)^4(x^5)^4(y^4)\cdot(-1)^3(x^6)^3(y^5)^3$

$=(-1)^4(x^{20})(y^4)\cdot(-1)^3(x^{18})(y^{15})$

$=(-1)^7 x^{20+18} y^{4+15}$

$=-1x^{38}y^{19}$

$=-x^{38}y^{19}$

b. $(-x^5y)^3(-x^6y^5)^2$

_____

Objective 6 Practice Exercises

For extra help, see Example 7 on page 266–267 of your text.

Simplify. Write all answers in exponential form.

16. $(-x^3)^2(-x^5)^4$

16. _____

17. $(2ab^2c)^5(ab)^4$

17. _____

18. $(5x^2y^3)^7(5xy^4)^4$

18. _____

Name: Date:
Instructor: Section:

Objective 7   Use the rules for exponents in a geometry application.

Video Examples

Review this example for Objective 7:   **Now Try:**

8. Find the area of the figure.   8. Find the area of the figure.

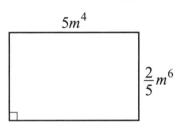   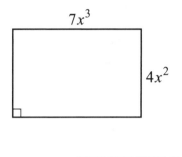

Use the formula for the area of a rectangle.
$A = LW$

$A = (5m^4)\left(\dfrac{2}{5}m^6\right)$

$A = 5 \cdot \dfrac{2}{5} \cdot m^{4+6}$

$A = 2m^{10}$

Objective 7 Practice Exercises

For extra help, see Example 8 on page 267 of your text.

Find a polynomial that represents the area of each figure.

19. 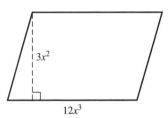   **19.**

20. 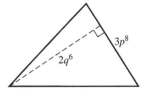   **20.** _____

21. 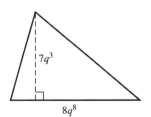   **21.** _____

Name: Date:
Instructor: Section:

Chapter 4 EXPONENTS AND POLYNOMIALS

4.2 Integer Exponents and the Quotient Rule

Learning Objectives
1. Use 0 as an exponent.
2. Use negative numbers as exponents.
3. Use the quotient rule for exponents.
4. Use combinations of the rules for exponents.

Key Terms

Use the vocabulary terms listed below to complete each statement in exercises 1–3.

    exponent   base   product rule for exponents

    power rule for exponents

1. The statement "If m and n are any integers, then $(a^m)^n = a^{mn}$" is an example of the _____.

2. In the expression a^m, a is the _____ and m is the _____.

3. The statement "If m and n are any integers, then $a^m \cdot a^n = a^{m+n}$" is an example of the _____.

Objective 1   Use 0 as an exponent.

Video Examples

Review these examples for Objective 1:	Now Try:

1. Evaluate.
 	
    a. $75^0 = 1$
 	
    b. $-75^0 = -(1)$ or -1
 	
    c. $(-9x)^0 = 1 \quad (x \neq 0)$
 	
    d. $3^0 + 12^0 = 1 + 1 = 2$
	1. Evaluate.
	
	    a. 88^0
	    _____
	
	    b. -88^0
	    _____
	
	    c. $(-88a)^0 \quad (a \neq 0)$
	    _____
	
	    d. $5^0 - 16^0$
	    _____

Name: Date:
Instructor: Section:

Objective 1 Practice Exercises

For extra help, see Example 1 on page 271 of your text.

Evaluate each expression.

1. -12^0

2. $-15^0 - (-15)^0$

3. $\dfrac{0^8}{8^0}$

1. _____

2. _____

3. _____

Objective 2 Use negative numbers as exponents.

Video Examples

Review these examples for Objective 2:
2. Simplify by writing with positive exponents. Assume that all variables represent nonzero real numbers.

a. 4^{-3}

$4^{-3} = \dfrac{1}{4^3}$, or $\dfrac{1}{64}$

b. $\left(\dfrac{1}{3}\right)^{-3}$

$\left(\dfrac{1}{3}\right)^{-3} = 3^3$, or 27

c. $\left(\dfrac{2}{3}\right)^{-5}$

$\left(\dfrac{2}{3}\right)^{-5} = \left(\dfrac{3}{2}\right)^5$

$= \dfrac{3^5}{2^5}$

$= \dfrac{243}{32}$

Now Try:
2. Simplify by writing with positive exponents. Assume that all variables represent nonzero real numbers.

a. 3^{-3}

_____

b. $\left(\dfrac{1}{5}\right)^{-2}$

_____

c. $\left(\dfrac{3}{2}\right)^{-3}$

_____

Name: Date:
Instructor: Section:

d. $5^{-1} - 3^{-1}$

$$5^{-1} - 3^{-1} = \frac{1}{5} - \frac{1}{3}$$

$$= \frac{3}{15} - \frac{5}{15}$$

$$= -\frac{2}{15}$$

e. q^{-3} $(q \neq 0)$

$$q^{-3} = \frac{1}{q^3}$$

3. Simplify. Assume that all variables represent nonzero real numbers.

  a. $\dfrac{3^{-4}}{7^{-2}} = \dfrac{7^2}{3^4}$, or $\dfrac{49}{81}$

  b. $a^{-6}b^4 = \dfrac{b^4}{a^6}$

  c. $\dfrac{x^{-3}y}{4z^{-4}} = \dfrac{yz^4}{4x^3}$

d. $4^{-1} - 8^{-1}$

  _____

e. p^{-5} $(p \neq 0)$

  _____

3. Simplify. Assume that all variables represent nonzero real numbers.

  a. $\dfrac{6^{-2}}{5^{-3}}$

  _____

  b. $x^{-7}y^2$

  _____

  c. $\dfrac{p^{-3}q}{4r^{-5}}$

  _____

Objective 2 Practice Exercises

For extra help, see Examples 2–3 on pages 272–273 of your text.

Evaluate or simplify each expression, and write it using only positive exponents. Assume that all variables represent nonzero real numbers.

4. $-2k^{-4}$ 4. _____

5. $(m^2 n)^{-9}$ 5. _____

6. $\dfrac{2x^{-4}}{3y^{-7}}$ 6. _____

Name: Date:
Instructor: Section:

Objective 3   Use the quotient rule for exponents.

Video Examples

Review these examples for Objective 3:
4. Simplify. Assume that all variables represent nonzero real numbers.

   a. $\dfrac{4^9}{4^6} = 4^{9-6} = 4^3 = 64$

   b. $\dfrac{p^6}{p^{-4}} = p^{6-(-4)} = p^{10}$

   c. $\dfrac{(x+7)^{-3}}{(x+7)^{-5}} \quad x \neq -7$

   $\dfrac{(x+7)^{-3}}{(x+7)^{-5}} = (x+7)^{-3-(-5)}$
   $= (x+7)^{-3+5}$
   $= (x+7)^2$

   d. $\dfrac{8x^{-4}y^3}{5^{-1}x^3y^{-4}}$

   $\dfrac{8x^{-4}y^3}{5^{-1}x^3y^{-4}} = \dfrac{8 \cdot 5 y^3 y^4}{x^3 x^4}$

   $= \dfrac{40 y^7}{x^7}$

Now Try:
4. Simplify. Assume that all variables represent nonzero real numbers.

   a. $\dfrac{3^{18}}{3^{16}}$

   _____

   b. $\dfrac{z^6}{z^{-4}}$

   _____

   c. $\dfrac{(a-b)^{-8}}{(a-b)^{-10}} \quad a \neq b$

   _____

   d. $\dfrac{9a^{-5}b^3}{4^{-1}a^3b^{-4}}$

   _____

Objective 3 Practice Exercises

For extra help, see Example 4 on pages 274–275 of your text.

Use the quotient rule to simplify each expression, and write it using only positive exponents. Assume that all variables represent nonzero real numbers.

7. $\dfrac{4k^7 m^{10}}{8k^3 m^5}$ 7. _____

Name: Date:
Instructor: Section:

8. $\dfrac{a^4 b^3}{a^{-2} b^{-3}}$

8. _____

9. $\dfrac{3^{-1} m^{-4} p^6}{3^4 m^{-1} p^{-2}}$

9. _____

Objective 4 Use combinations of the rules for exponents.

Video Examples

Review these examples for Objective 4:
5. Simplify each expression. Assume that all variables represent nonzero real numbers.

a. $\dfrac{(5^4)^2}{5^6}$

$\dfrac{(5^4)^2}{5^6} = \dfrac{5^8}{5^6}$

$= 5^{8-6}$

$= 5^2$

$= 25$

b. $(3a)^4 (3a)^2$

$(3a)^4 (3a)^2 = (3a)^6$

$= 3^6 a^6$

$= 729 a^6$

c. $\left(\dfrac{3x^4}{4}\right)^{-5}$

$\left(\dfrac{3x^4}{4}\right)^{-5} = \left(\dfrac{4}{3x^4}\right)^5$

$= \dfrac{4^5}{3^5 x^{20}}$

$= \dfrac{1024}{243 x^{20}}$

Now Try:
5. Simplify each expression. Assume that all variables represent nonzero real numbers.

a. $\dfrac{(6^3)^2}{6^5}$

_____

b. $(5b)^3 (5b)^2$

_____

c. $\left(\dfrac{2p^4}{3}\right)^{-5}$

_____

152 Copyright © 2016 Pearson Education, Inc.

Name: Date:
Instructor: Section:

d. $\dfrac{\left(k^2 m^{-3} n\right)^{-5}}{\left(3km^2 n^{-4}\right)^{-6}}$

$\dfrac{\left(k^2 m^{-3} n\right)^{-5}}{\left(3km^2 n^{-4}\right)^{-6}} = \dfrac{(k^2)^{-5}(m^{-3})^{-5} n^{-5}}{3^{-6} k^{-6} (m^2)^{-6} (n^{-4})^{-6}}$

$= \dfrac{k^{-10} m^{15} n^{-5}}{3^{-6} k^{-6} m^{-12} n^{24}}$

$= \dfrac{3^6 m^{15+12}}{k^{-6+10} n^{24+5}}$

$= \dfrac{729 m^{27}}{k^4 n^{29}}$

d. $\dfrac{\left(7xy^{-2} z^3\right)^{-3}}{\left(x^{-4} yz^{-2}\right)^4}$

_____

Objective 4 Practice Exercises

For extra help, see Example 5 on page 276 of your text.

Simplify each expression, and write it using only positive exponents. Assume that all variables represent nonzero real numbers.

10. $(9xy)^7 (9xy)^{-8}$

10. _____

11. $\dfrac{\left(a^{-1} b^{-2}\right)^{-4} \left(ab^2\right)^6}{\left(a^3 b\right)^{-2}}$

11. _____

12. $\left(\dfrac{k^3 t^4}{k^2 t^{-1}}\right)^{-4}$

12. _____

Copyright © 2016 Pearson Education, Inc.

Name: Date:
Instructor: Section:

Chapter 4 EXPONENTS AND POLYNOMIALS

4.3 Scientific Notation

Learning Objectives
1. Express numbers in scientific notation.
2. Convert numbers in scientific notation to standard notation.
3. Use scientific notation in calculations.

Key Terms

Use the vocabulary terms listed below to complete each statement in exercises 1–3.

scientific notation   quotient rule   power rule

1. A number written as $a \times 10^n$, where $1 \leq |a| < 10$ and n is an integer, is written in _____.

2. The statement "If m and n are any integers and $b \neq 0$, then $\left(\dfrac{a}{b}\right)^m = \dfrac{a^m}{b^m}$" is an example of the _____.

3. The statement "If m and n are any integers and $b \neq 0$, then $\dfrac{a^m}{a^n} = a^{m-n}$" is an example of the _____.

Objective 1 Express numbers in scientific notation.

Video Examples

Review these examples for Objective 1:
1. Write each number in scientific notation.

   a. 84,300,000,000

   Move the decimal point 10 places to the left.
   $84{,}300{,}000{,}000 = 8.43 \times 10^{10}$

   b. 0.00573

   The first nonzero digit is 5. Count the places. Move the decimal point 3 places to the right.
   $0.00573 = 5.73 \times 10^{-3}$

Now Try:
1. Write each number in scientific notation.

   a. 47,710,000,000

   _____

   b. 0.0463

   _____

Name: Date:
Instructor: Section:

Objective 1 Practice Exercises

For extra help, see Example 1 on page 281 of your text.

Write each number in scientific notation.

1. 23,651

1. _____

2. −429,600,000,000

2. _____

3. −0.0002208

3. _____

Objective 2 Convert numbers in scientific notation to standard notation.

Video Examples

Review these examples for Objective 2:
2. Write each number without exponents.

   a. 3.57×10^6

Move the decimal point 6 places to the right, and attach four zeros.
$3.57 \times 10^6 = 3,570,000$

   b. 8.98×10^{-3}

Move the decimal point 3 places to the left.
$8.98 \times 10^{-3} = 0.00898$

Now Try:
2. Write each number without exponents.

   a. 2.796×10^7

_____

   b. 1.64×10^{-4}

_____

Objective 2 Practice Exercises

For extra help, see Example 2 on page 282 of your text.

Write each number in standard notation.

4. -2.45×10^6

4. _____

5. 6.4×10^{-3}

5. _____

6. -4.02×10^4

6. _____

Copyright © 2016 Pearson Education, Inc.

Name: Date:
Instructor: Section:

Objective 3   Use scientific notation in calculations.

Video Examples

Review these examples for Objective 3:

3. Perform each calculation. Write answers in scientific notation and also without exponents.

   a. $(8 \times 10^4)(7 \times 10^3)$

   $(8 \times 10^4)(7 \times 10^3) = (8 \times 7)(10^4 \times 10^3)$
   $= 56 \times 10^7$
   $= (5.6 \times 10^1) \times 10^7$
   $= 5.6 \times 10^8$
   $= 560,000,000$

   b. $\dfrac{6 \times 10^{-4}}{3 \times 10^2}$

   $\dfrac{6 \times 10^{-4}}{3 \times 10^2} = \dfrac{6}{3} \times \dfrac{10^{-4}}{10^2}$
   $= 2 \times 10^{-6}$
   $= 0.000002$

4. The Sahara desert covers approximately 3.5×10^6 square miles. Its sand is, on average, 12 feet deep. Find the volume, in cubic feet, of sand in the Sahara. (Hint: $1 \text{ mi}^2 = 5280^2 \text{ ft}^2$) Round your answer to two decimal places.

   $(3.5 \times 10^6)(5280^2)(12)$
   $= 97574400(12) \times 10^6$
   $= 1170892800 \times 10^6$
   $\approx 1.17 \times 10^{15} \text{ ft}^3$

   The volume is 1.17×10^{15} cubic feet.

Now Try:

3. Perform each calculation. Write answers in scientific notation and also without exponents.

   a. $(9 \times 10^5)(3 \times 10^2)$

   _____

   _____

   b. $\dfrac{39 \times 10^{-3}}{13 \times 10^5}$

   _____

   _____

4. The Sahara desert covers approximately 3.5×10^6 square miles. Its sand is, on average, 12 feet deep. The volume of a single grain of sand is approximately 1.3×10^{-9} cubic feet. About how many grains of sand are in the Sahara?

   _____

Name: Date:
Instructor: Section:

Objective 3 Practice Exercises

For extra help, see Examples 3–5 on pages 282–283 of your text.

Perform the indicated operations, and write the answers in scientific notation.

7. $(2.3 \times 10^4) \times (1.1 \times 10^{-2})$ 7. _____

8. $\dfrac{9.39 \times 10^1}{3 \times 10^3}$ 8. _____

Work the problem. Give answer in scientific notation.

9. There are about 6×10^{23} atoms in a mole of atoms. How many atoms are there in 8.1×10^{-5} mole?   9. _____

Name: Date:
Instructor: Section:

Chapter 4 EXPONENTS AND POLYNOMIALS

4.4 Adding, Subtracting, and Graphing Polynomials

Learning Objectives
1 Identify terms and coefficients.
2 Combine like terms.
3 Know the vocabulary for polynomials.
4 Evaluate polynomials.
5 Add and subtract polynomials.
6 Graph equations defined by polynomials of degree 2.

Key Terms

Use the vocabulary terms listed below to complete each statement in exercises 1−9.

    term   **like terms**   **polynomial**

    descending powers   **degree of a term**

    degree of a polynomial   **monomial**

    binomial   **trinomial**

    parabola   **vertex**   **axis**   **line of symmetry**

1. The _____ is the sum of the exponents on the variables in that term.

2. A polynomial in *x* is written in _____ if the exponents on *x* in its terms are decreasing order.

3. A _____ is a number, a variable, or a product or quotient of a number and one or more variables raised to powers.

4. A polynomial with exactly three terms is called a _____.

5. A _____ is a term, or the sum of a finite number of terms with whole number exponents.

6. A polynomial with exactly one term is called a _____.

7. The _____ is the greatest degree of any term of the polynomial.

8. A _____ is a polynomial with exactly two terms.

9. Terms with exactly the same variables (including the same exponents) are called _____.

Name: Date:
Instructor: Section:

10. If a graph is folded on its _____, the two sides coincide.

11. The _____ of a parabola that opens upward or downward is the lowest or highest point on the graph.

12. The _____ of a parabola that opens upward or downward is a vertical line through the vertex.

13. The graph of the quadratic equation $y = ax^2 + bx + c$ is called a _____.

Objective 1 Identify terms and coefficients.

Video Examples

Review this example for Objective 1:
1. For each expression, determine the number of terms and name the coefficients of the terms.

   $6 - 3x^4 - x^2$

   Rewrite the expression as $6x^0 - 3x^4 - 1x^2$.
   There are three terms: $6, -3x^4$, and $-x^2$.
   The coefficients are 6, –3, and –1.

Now Try:
1. For each expression, determine the number of terms and name the coefficients of the terms.

   $x^2 + 7 - 2x$

   _____

Objective 1 Practice Exercises

For extra help, see Example 1 on page 288 of your text.

For each expression, determine the number of terms and name the coefficients of the terms.

1. $3x^2 - 2 + x$ 1. _____

2. $5 + 6z^3$ 2. _____

3. $8y - y^3 - 1$ 3. _____

Name: Date:
Instructor: Section:

Objective 2 Combine like terms.

Video Examples

Review this example for Objective 2:
2. Simplify the expression by combining like terms.

$$19m^3 + 6m + 5m^3$$

$$19m^3 + 6m + 5m^3 = (19+5)m^3 + 6m$$
$$= 24m^3 + 6m$$

Now Try:
2. Simplify the expression by combining like terms.
$$22m^2 + 15m^3 + 7m^2$$

_____

Objective 2 Practice Exercises

For extra help, see Example 2 on pages 288–289 of your text.

In each polynomial, combine like terms whenever possible. Write the result with descending powers.

4. $7z^3 - 4z^3 + 5z^3 - 11z^3$

4. _____

5. $-1.3z^7 + 0.4z^7 + 2.6z^8$

5. _____

6. $6c^3 - 9c^2 - 2c^2 + 14 + 3c^2 - 6c - 8 + 2c^3$

6. _____

Objective 3 Know the vocabulary for polynomials.

Video Examples

Review these examples for Objective 3:
3. Simplify each polynomial, if possible, and write in descending powers of the variable. Then give the degree and tell whether the polynomial is a monomial, a binomial, a trinomial, or none of these.

Now Try:
3. Simplify each polynomial, if possible, and write in descending powers of the variable. Then give the degree and tell whether the polynomial is a monomial, a binomial, a trinomial, or none of these.

a. $5x + 6x^3 - 7x - 2x^2 + 4x$

$5x + 6x^3 - 7x - 2x^2 + 4x = 6x^3 - 2x^2 + 2x$
The degree is 3. The simplified polynomial is a trinomial.

a. $2x - 7x^2 - 6x + 3x^3 + 8x$

_____

Name: Date:
Instructor: Section:

b. $9y^3 - 7y^5 + 3y^3 + 2y^5$

$9y^3 - 7y^5 + 3y^3 + 2y^5 = -5y^5 + 12y^3$
The degree is 5. The simplified polynomial is a binomial.

b. $w^5 - 4w^2 + 3w^5 + w^2$

_____

Objective 3 Practice Exercises

For extra help, see Example 3 on page 290 of your text.

For each polynomial, first simplify, if possible, and write the resulting polynomial in descending powers of the variable. Then give the degree of this polynomial, and tell whether it is a monomial, a binomial, a trinomial, or none of these.

7. $3n^8 - n^2 - 2n^8$

7. _____

degree: _____

type: _____

8. $-d^2 + 3.2d^3 - 5.7d^8 - 1.1d^5$

8. _____

degree: _____

type: _____

9. $-6c^4 - 6c^2 + 9c^4 - 4c^2 + 5c^5$

9. _____

degree: _____

type: _____

Objective 4 Evaluate polynomials.

Video Examples

Review this example for Objective 4:

4. Find the value of $4x^3 + 6x^2 - 5x - 5$ for

$x = 2$

$4x^3 + 6x^2 - 5x - 5$
$= 4(2)^3 + 6(2)^2 - 5(2) - 5$
$= 4(8) + 6(4) - 5(2) - 5$
$= 32 + 24 - 10 - 5$
$= 41$

Now Try:

4. Find the value of
$5x^4 + 3x^2 - 9x - 7$ for
$x = 4$

_____

Name: Date:
Instructor: Section:

Objective 4 Practice Exercises

For extra help, see Example 4 on page 290 of your text.

Find the value of each polynomial (a) *when x = –2 and* (b) *when x = 3.*

10. $3x^3 + 4x - 19$

10. a._____

b._____

11. $-4x^3 + 10x^2 - 1$

11. a._____

b._____

12. $x^4 - 3x^2 - 8x + 9$

12. a._____

b._____

Objective 5   Add and subtract polynomials.

Video Examples

Review these examples for Objective 5:
6. Find each sum.

  a. Add $5x^4 - 7x^3 + 9$ and $-3x^4 + 8x^3 - 7$

  $(5x^4 - 7x^3 + 9) + (-3x^4 + 8x^3 - 7)$

  $= 5x^4 - 3x^4 - 7x^3 + 8x^3 + 9 - 7$

  $= 2x^4 + x^3 + 2$

  b. $(5x^4 - 7x^2 + 6x) + (-3x^3 + 4x^2 - 7)$

  $(5x^4 - 7x^2 + 6x) + (-3x^3 + 4x^2 - 7)$

  $= 5x^4 - 3x^3 - 7x^2 + 4x^2 + 6x - 7$

  $= 5x^4 - 3x^3 - 3x^2 + 6x - 7$

Now Try:
6. Find each sum.

  a. Add $15x^3 - 5x + 3$ and $-11x^3 + 6x + 9$

  _____

  b. $(8x^2 - 6x + 4) + (7x^3 - 8x - 5)$

  _____

Name: Date:
Instructor: Section:

7. Perform the subtraction.

   Subtract $8x^3 - 5x^2 + 8$ from $9x^3 + 6x^2 - 7$.

   $(9x^3 + 6x^2 - 7) - (8x^3 - 5x^2 + 8)$
   $= (9x^3 + 6x^2 - 7) + (-8x^3 + 5x^2 - 8)$
   $= x^3 + 11x^2 - 15$

9. Perform the indicated operations to simplify the expression
   $(5 - 2x + 9x^2) - (7 - 5x + 8x^2) + (6 + 3x - 5x^2)$

   Rewrite, changing the subtraction to adding the opposite.
   $(5 - 2x + 9x^2) - (7 - 5x + 8x^2) + (6 + 3x - 5x^2)$
   $= (5 - 2x + 9x^2) + (-7 + 5x - 8x^2) + (6 + 3x - 5x^2)$
   $= (-2 + 3x + x^2) + (6 + 3x - 5x^2)$
   $= 4 + 6x - 4x^2$

10. Add or subtract as indicated.

    $(3x^2y + 5xy + y^2) - (4x^2y + xy - 3y^2)$

    Change the signs of the terms in the parentheses and add like terms vertically.
    $3x^2y + 5xy + y^2$
    $\underline{-4x^2y - xy + 3y^2}$
    $-x^2y + 4xy + 4y^2$

7. Perform the subtraction.

   Subtract $18x^3 + 4x - 6$ from $7x^3 - 3x - 5$.

   _____

9. Perform the indicated operations to simplify the expression
   $(10 - 7x + 6x^2) - (5 - 11x + 3x^2) + (2 + 4x - 7x^2)$

   _____

10. Add or subtract as indicated.

    $(7x^2y + 3xy + 4y^2)$
    $-(6x^2y - xy + 4y^2)$

    _____

Objective 5 Practice Exercises

For extra help, see Examples 5–10 on pages 291–293 of your text.

Add or subtract as indicated.

13. $(3r^3 + 5r^2 - 6) + (2r^2 - 5r + 4)$

13. _____

Name: Date:
Instructor: Section:

14. $(-8w^3 + 11w^2 - 12) - (-10w^2 + 3)$ 14. _____

15. $(2x^2y + 2xy - 4xy^2) + (6xy + 9xy^2) - (9x^2y + 5xy)$ 15. _____

Objective 6 Graph equations defined by polynomials of degree 2.

Video Examples

Review this example for Objective 6:
11. Graph the equation.

$y = x^2 - 3$

Find several ordered pairs. Let $x = 0$ to find the y-intercept.

$y = x^2 - 3 = 0^2 - 3 = -3$

This gives the ordered pair $(0, -3)$. Select several values for x and find the corresponding values for y. Plot the ordered pairs and join them with a smooth curve.

x	y
2	1
1	-2
0	-3
-1	-2
-2	1

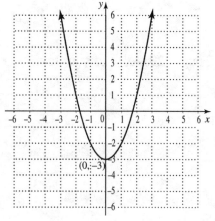

Now Try:
11. Graph the equation.

$y = 9 - x^2$

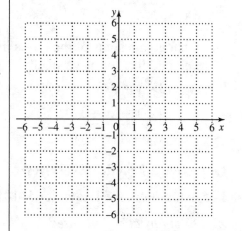

164 Copyright © 2016 Pearson Education, Inc.

Name: Date:
Instructor: Section:

Objective 6 Practice Exercises

For extra help, see Example 11 on pages 293–294 of your text.

Graph each equation.

16. $y = -x^2 - 1$

16.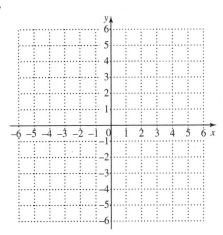

vertex: _____

17. $y = x^2 + 2$

17.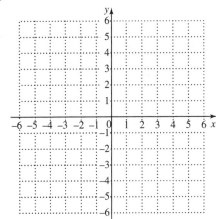

vertex: _____

Name:	Date:
Instructor:	Section:

Chapter 4 EXPONENTS AND POLYNOMIALS

4.5 Multiplying Polynomials

Learning Objectives
1. Multiply a monomial and a polynomial.
2. Multiply two polynomials.
3. Multiply binomials by the FOIL method.

Key Terms

Use the vocabulary terms listed below to complete each statement in exercises 1–3.

FOIL   **outer product**   **inner product**

1. The _____ of $(2y-5)(y+8)$ is $-5y$.

2. _____ is a shortcut method for finding the product of two binomials.

3. The _____ of $(2y-5)(y+8)$ is $16y$.

Objective 1 Multiply a monomial and a polynomial.

Video Examples

Review this example for Objective 1:
1. Find the product.

    $5x^2(7x+3)$

    Use the distributive property.
    $5x^2(7x+3) = 5x^2(7x) + 5x^2(3)$
    $= 35x^3 + 15x^2$

Now Try:
1. Find the product.

    $8x^3(4x+8)$

    _____

Objective 1 Practice Exercises

For extra help, see Example 1 on page 298 of your text.

Find each product.

1. $7z(5z^3+2)$ 1. _____

2. $2m(3+7m^2+3m^3)$ 2. _____

3. $-3y^2(2y^3+3y^2-4y+11)$ 3. _____

Name: Date:
Instructor: Section:

Objective 2   Multiply two polynomials.

Video Examples

Review these examples for Objective 2:

2. Multiply $(x^2+6)(5x^3-4x^2+3x)$.

   Multiply each term of the second polynomial by each term of the first.
   $(x^2+6)(5x^3-4x^2+3x)$
   $= x^2(5x^3)+x^2(-4x^2)+x^2(3x)$
   $\quad +6(5x^3)+6(-4x^2)+6(3x)$
   $= 5x^5-4x^4+3x^3+30x^3-24x^2+18x$
   $= 5x^5-4x^4+33x^3-24x^2+18x$

3. Multiply $(2x^3+7x^2+5x-1)(4x+6)$ vertically.

   Write the polynomials vertically.
   $$2x^3+7x^2+5x-1$$
   $$\underline{\qquad\qquad\qquad 4x+6}$$
   Begin by multiplying each term in the top row by 6.
   $$2x^3+7x^2+5x-1$$
   $$\underline{\qquad\qquad\qquad 4x+6}$$
   $$12x^3+42x^2+30x-6$$
   Now multiply each term in the top row by $4x$. Then add like terms.
   $$2x^3+7x^2+5x-1$$
   $$\underline{\qquad\qquad\qquad 4x+6}$$
   $$12x^3+42x^2+30x-6$$
   $$\underline{8x^4+28x^3+20x^2-4x\qquad}$$
   $$8x^4+40x^3+62x^2+26x-6$$

   The product is $8x^4+40x^3+62x^2+26x-6$.

Now Try:

2. Multiply
   $(x^3+9)(4x^4-2x^2+x)$

   _____

3. Multiply
   $(4x^3-3x^2+6x+5)(7x-3)$
   vertically.

   _____

Name: Date:
Instructor: Section:

4. Find the product of $-16m^3 + 12m^2 + 4$ and $\frac{1}{4}m^2 + \frac{3}{4}$.

   Multiply each term of the second polynomial by each term of the first.

   $(-16m^3 + 12m^2 + 4)\left(\frac{1}{4}m^2 + \frac{3}{4}\right)$

   $= -16m^3\left(\frac{1}{4}m^2\right) - 16m^3\left(\frac{3}{4}\right) + 12m^2\left(\frac{1}{4}m^2\right)$
   $\quad + 12m^2\left(\frac{3}{4}\right) + 4\left(\frac{1}{4}m^2\right) + 4\left(\frac{3}{4}\right)$

   $= -12m^3 + 9m^2 + 3 - 5m^5 + 3m^4 + m^2$

   $= -4m^5 + 3m^4 - 12m^3 + 10m^2 + 3$

   The product is $-4m^5 + 3m^4 - 12m^3 + 10m^2 + 3$.

4. Find the product of $12x^3 - 36x^2 + 6$ and $\frac{1}{6}x^2 + \frac{5}{6}$.

   _____

Objective 2 Practice Exercises

For extra help, see Examples 2–4 on page 299 of your text.

Find each product.

4. $(x+3)(x^2 - 3x + 9)$

4. _____

5. $(2m^2 + 1)(3m^3 + 2m^2 - 4m)$

5. _____

6. $(3x^2 + x)(2x^2 + 3x - 4)$

6. _____

Objective 3 Multiply binomials by the FOIL method.

Video Examples

Review these examples for Objective 3:
5. Use the FOIL method to find the product $(x+7)(x-5)$.

   Step 1 F Multiply the first terms: $x(x) = x^2$.

Now Try:
5. Use the FOIL method to find the product $(x+9)(x-6)$.

   _____

Name: Date:
Instructor: Section:

Step 2 O Find the outer product: $x(-5) = -5x$.

Step 3 I Find the inner product: $7(x) = 7x$.
Add the outer and inner products mentally:
$$-5x + 7x = 2x$$

Step 4 L Multiply the last terms: $7(-5) = -35$.

The product $(x+7)(x-5)$ is $x^2 + 2x - 35$.

6. Multiply $(7x-3)(4y+5)$.

   First $\quad 7x(4y) = 28xy$

   Outer $\quad 7x(5) = 35x$

   Inner $\quad -3(4y) = -12y$

   Last $\quad -3(5) = -15$

   The product $(7x-3)(4y+5)$ is
   $28xy + 35x - 12y - 15$.

6. Multiply $(8y-7)(2x+9)$.

   _____

7. Find the product.

   $(3k+7m)(2k+9m)$

   $(3k+7m)(2k+9m)$
   $= 3k(2k) + 3k(9m) + 7m(2k) + 7m(9m)$
   $= 6k^2 + 27km + 14km + 63m^2$
   $= 6k^2 + 41km + 63m^2$

7. Find the product.

   $(5k+8n)(3k+4n)$

   _____

Objective 3 Practice Exercises

For extra help, see Examples 5–7 on page 301 of your text.

Find each product.

7. $(5a-b)(4a+3b)$

   7. _____

8. $(3+4a)(1+2a)$

   8. _____

9. $(2m+3n)(-3m+4n)$

   9. _____

Name: Date:
Instructor: Section:

Chapter 4 EXPONENTS AND POLYNOMIALS

4.6 Special Products

Learning Objectives
1 Square binomials.
2 Find the product of the sum and difference of two terms.
3 Find greater powers of binomials.

Key Terms

Use the vocabulary terms listed below to complete each statement in exercises 1–2.

    conjugate   **binomial**

1. A polynomial with two terms is called a _____.

2. The _____ of $a + b$ is $a - b$.

Objective 1   Square binomials.

Video Examples

Review these examples for Objective 1:
2. Square each binomial.

    a. $(6x+3y)^2$

$$(6x+3y)^2 = (6x)^2 + 2(6x)(3y) + (3y)^2$$
$$= 36x^2 + 36xy + 9y^2$$

    b. $\left(6n+\dfrac{1}{4}\right)^2$

$$\left(6n+\dfrac{1}{4}\right)^2 = (6n)^2 + 2(6n)\left(\dfrac{1}{4}\right) + \left(\dfrac{1}{4}\right)^2$$
$$= 36n^2 + 3n + \dfrac{1}{16}$$

Now Try:
2. Square each binomial.

    a. $(2a+9k)^2$

    _____

    b. $\left(3p+\dfrac{1}{6}\right)^2$

    _____

Objective 1 Practice Exercises

For extra help, see Examples 1–2 on pages 305–306 of your text.

Find each square by using the pattern for the square of a binomial.

1. $(7+x)^2$ 1. _____

Name: Date:
Instructor: Section:

2. $(2m-3p)^2$ 2. _____

3. $(4y-0.7)^2$ 3. _____

Objective 2   Find the product of the sum and difference of two terms.

Video Examples

Review these examples for Objective 2:
3. Find each product.

   a. $(x+5)(x-5)$

   Use the rule for the product of the sum and difference of two terms.
   $$(x+5)(x-5) = x^2 - 5^2$$
   $$= x^2 - 25$$

   b. $\left(\frac{3}{4}-y\right)\left(\frac{3}{4}+y\right)$

   $$\left(\frac{3}{4}-y\right)\left(\frac{3}{4}+y\right) = \left(\frac{3}{4}\right)^2 - y^2$$
   $$= \frac{9}{16} - y^2$$

4. Find each product.

   a. $(6x+w)(6x-w)$

   $$(6x+w)(6x-w) = (6x)^2 - w^2$$
   $$= 36x^2 - w^2$$

   b. $3q(q^2+4)(q^2-4)$

   First, multiply the conjugates.
   $$3q(q^2+4)(q^2-4) = 3q(q^4-16)$$
   $$= 3q^5 - 48q$$

Now Try:
3. Find each product.

   a. $(x+9)(x-9)$

   _____

   b. $\left(\frac{5}{6}+a\right)\left(\frac{5}{6}-a\right)$

   _____

4. Find each product.

   a. $(11x-y)(11x+y)$

   _____

   b. $4p(p^2+6)(p^2-6)$

   _____

Copyright © 2016 Pearson Education, Inc.

Name:
Instructor:
Date:
Section:

Objective 2 Practice Exercises

For extra help, see Examples 3–4 on page 307 of your text.

Find each product by using the pattern for the sum and difference of two terms.

4. $(12+x)(12-x)$ 4. _____

5. $(8k+5p)(8k-5p)$ 5. _____

6. $\left(\frac{4}{7}t+2u\right)\left(\frac{4}{7}t-2u\right)$ 6. _____

Objective 3   Find greater powers of binomials.

Video Examples

Review these examples for Objective 3:

5. Find each product.

    a. $(x+4)^3$

$(x+4)^3$
$=(x+4)^2(x+4)$
$=(x^2+8x+16)(x+4)$
$=x^3+8x^2+16x+4x^2+32x+64$
$=x^3+12x^2+48x+64$

    b. $(5y-4)^4$

$(5y-4)^4$
$=(5y-4)^2(5y-4)^2$
$=(25y^2-40y+16)(25y^2-40y+16)$
$=625y^4-1000y^3+400y^2$
$\quad -1000y^3+1600y^2-640y$
$\quad +400y^2-640y+256$
$=625y^4-2000y^3+2400y^2-1280y+256$

Now Try:

5. Find each product.

    a. $(x+6)^3$

_____

    b. $(3x-5)^4$

_____

Name: Date:
Instructor: Section:

Objective 3 Practice Exercises

For extra help, see Example 5 on page 308 of your text.

Find each product.

7. $(a-3)^3$ 7. _____

8. $(j+3)^4$ 8. _____

9. $(4s+3t)^4$ 9. _____

Name: Date:
Instructor: Section:

Chapter 4 EXPONENTS AND POLYNOMIALS

4.7 Dividing Polynomials

Learning Objectives
1 Divide a polynomial by a monomial.
2 Divide a polynomial by a polynomial.
3 Use division in a geometry application.

Key Terms

Use the vocabulary terms listed below to complete each statement in exercises 1–3.

   quotient   dividend   divisor

1. In the division $\dfrac{5x^5 - 10x^3}{5x^2} = x^3 - 2x$, the expression $5x^5 - 10x^3$ is the
   _____.

2. In the division $\dfrac{5x^5 - 10x^3}{5x^2} = x^3 - 2x$, the expression $x^3 - 2x$ is the _____.

3. In the division $\dfrac{5x^5 - 10x^3}{5x^2} = x^3 - 2x$, the expression $5x^2$ is the _____.

Objective 1 Divide a polynomial by a monomial.

Video Examples

Review these examples for Objective 1:	Now Try:
1. Divide $6x^4 - 18x^3$ by $6x^2$. $$\frac{6x^4 - 18x^3}{6x^2} = \frac{6x^4}{6x^2} - \frac{18x^3}{6x^2}$$ $$= x^2 - 3x$$ Check Multiply. $6x^2(x^2 - 3x) = 6x^4 - 18x^3$	1. Divide $20x^4 - 10x^2$ by $2x$. _____
2. Divide. $\dfrac{25a^6 - 15a^4 + 10a^2}{5a^3}$ Divide each term by $5a^3$. $$\frac{25a^6 - 15a^4 + 10a^2}{5a^3} = \frac{25a^6}{5a^3} - \frac{15a^4}{5a^3} + \frac{10a^2}{5a^3}$$ $$= 5a^3 - 3a + \frac{2}{a}$$	2. Divide. $\dfrac{27n^5 - 36n^4 - 18n^2}{9n^3}$ _____

Name: Date:
Instructor: Section:

3. Divide $-12x^4 + 15x^5 - 5x$ by $-5x$.

   Write the polynomial in descending powers before dividing.
   $$\frac{15x^5 - 12x^4 - 5x}{-5x} = \frac{15x^5}{-5x} - \frac{12x^4}{-5x} - \frac{5x}{-5x}$$
   $$= -3x^4 + \frac{12}{5}x^3 + 1$$

   Check $-5x\left(-3x^4 + \frac{12}{5}x^3 + 1\right)$
   $$= -5x(-3x^4) - 5x\left(\frac{12}{5}x^3\right) - 5x(1)$$
   $$= 15x^5 - 12x^4 - 5x$$

4. Divide
   $225x^5y^9 - 150x^3y^7 + 110x^2y^5 - 80xy^3 + 75y^2$
   by $-25xy^2$.

   $$\frac{225x^5y^9 - 150x^3y^7 + 110x^2y^5 - 80xy^3 + 75y^2}{-25xy^2}$$
   $$= \frac{225x^5y^9}{-25xy^2} - \frac{150x^3y^7}{-25xy^2} + \frac{110x^2y^5}{-25xy^2} - \frac{80xy^3}{-25xy^2} + \frac{75y^2}{-25xy^2}$$
   $$= -9x^4y^7 + 6x^2y^5 - \frac{22xy^3}{5} + \frac{16y}{5} - \frac{3}{x}$$

   Check by multiplying the quotient by the divisor.

3. Divide $-8z^5 + 7z^6 - 10z - 6$ by $2z^2$.

   _____

4. Divide
   $80a^5b^3 + 160a^4b^2 - 120a^2b$ by $-40a^2b$.

   _____

Objective 1 Practice Exercises

For extra help, see Examples 1–4 on pages 311–312 of your text.

Perform each division.

1. $\dfrac{16a^5 - 24a^3}{8a^2}$

   1. _____

2. $\dfrac{12z^5 + 28z^4 - 8z^3 + 3z}{4z^3}$

   2. _____

3. $\dfrac{39m^4 - 12m^3 + 15}{-3m^2}$

   3. _____

Name: Date:
Instructor: Section:

Objective 2 Divide a polynomial by a polynomial.

Video Examples

Review these examples for Objective 2:

5. Divide $\dfrac{2x^2 - 11x + 15}{x - 3}$.

   Step 1 $2x^2$ divided by x is $2x$.
   $2x(x-3) = 2x^2 - 6x$

   Step 2 Subtract. Bring down the next term.

   Step 3 $-5x$ divided by x is -5.
   $-5(x-3) = -5x + 15$

   Step 4 Subtract. The remainder is 0.

   $$\begin{array}{r} 2x - 5 \\ x-3 \overline{\smash{\big)}\, 2x^2 - 11x + 15} \\ \underline{2x^2 - 6x } \\ -5x + 15 \\ \underline{-5x + 15} \\ 0 \end{array}$$

   $\dfrac{2x^2 - 11x + 15}{x - 3} = 2x - 5$

   Check $(x-3)(2x-5) = 2x^2 - 5x - 6x + 15$
   $ = 2x^2 - 11x + 15$

6. Divide $\dfrac{8x + 9x^3 - 7 - 9x^2}{3x - 1}$.

   Write the dividend in descending powers as $9x^3 - 9x^2 + 8x - 7$.

   Step 1 $9x^3$ divided by $3x$ is $3x^2$.
   $3x^2(3x-1) = 9x^3 - 3x^2$

   Step 2 Subtract. Bring down the next term.

   Step 3 $-6x^2$ divided by $3x$ is $-2x$.
   $-2x(3x-1) = -6x^2 + 2x$

   Step 4 Subtract. Bring down the next term.

   Step 5 $6x$ divided by $3x$ is 2.
   $2(3x-1) = 6x - 2$

Now Try:

5. Divide $\dfrac{4x^2 - 5x - 6}{x - 2}$.

   _____

6. Divide $\dfrac{-12x^2 + 10x^3 - 3 - 8x}{5x - 1}$.

   _____

Name: Date:
Instructor: Section:

$$\begin{array}{r} 3x^2 - 2x + 2 \\ 3x-1 \overline{) 9x^3 - 9x^2 + 8x - 7} \\ \underline{9x^3 - 3x^2} \\ -6x^2 + 8x \\ \underline{-6x^2 + 2x} \\ 6x - 7 \\ \underline{6x - 2} \\ -5 \end{array}$$

$$\frac{9x^3 - 9x^2 + 8x - 7}{3x - 1} = 3x^2 - 2x + 2 + \frac{-5}{3x - 1}$$

Step 7 Multiply to check.

Check $(3x-1)\left(3x^2 - 2x + 2 + \frac{-5}{3x-1}\right)$

$= (3x-1)(3x^2) + (3x-1)(-2x)$
$\quad + (3x-1)(2) + (3x-1)\left(\frac{-5}{3x-1}\right)$
$= 9x^3 - 3x^2 - 6x^2 + 2x + 6x - 2 - 5$
$= 9x^3 - 9x^2 + 8x - 7$

7. Divide $x^3 - 64$ by $x - 4$.

    Here the dividend is missing the x^2-term and the x-term. We use 0 as the coefficient for each missing term.

    $$\begin{array}{r} x^2 + 4x + 16 \\ x-4 \overline{) x^3 + 0x^2 + 0x - 64} \\ \underline{x^3 - 4x^2} \\ 4x^2 + 0x \\ \underline{4x^2 - 16x} \\ 16x - 64 \\ \underline{16x - 64} \\ 0 \end{array}$$

    The remainder is 0. The quotient is $x^2 + 4x + 16$.

    Check $(x-4)(x^2 + 4x + 16)$
    $= x^3 + 4x^2 + 16x - 4x^2 - 16x - 64$
    $= x^3 - 64$

7. Divide $x^3 - 1000$ by $x - 10$.

    _____

Copyright © 2016 Pearson Education, Inc.

Name: Date:
Instructor: Section:

8. Divide $x^4 - 3x^3 + 7x^2 - 8x + 14$ by $x^2 + 2$.

Since $x^2 + 2$ is missing the x-term, we write it as $x^2 + 0x + 2$.

$$\begin{array}{r}x^2 - 3x + 5\\x^2 + 0x + 2\overline{\smash{\big)}x^4 - 3x^3 + 7x^2 - 8x + 14}\\\underline{x^4 + 0x^3 + 2x^2}\\-3x^3 + 5x^2 - 8x\\\underline{-3x^3 + 0x^2 - 6x}\\5x^2 - 2x + 14\\\underline{5x^2 + 0x + 10}\\-2x + 4\end{array}$$

The quotient is $x^2 - 3x + 5 + \dfrac{-2x + 4}{x^2 + 2}$

The check shows that the quotient multiplied by the divisor gives the original dividend.

8. Divide $3x^4 + 5x^3 - 7x^2 - 12x + 9$ by $x^2 - 4$.

_____

Objective 2 Practice Exercises

For extra help, see Examples 5–9 on pages 314–316 of your text.

Perform each division.

4. $\dfrac{-6x^2 + 23x - 20}{2x - 5}$

4. _____

5. $\dfrac{6x^4 - 12x^3 + 13x^2 - 5x - 1}{2x^2 + 3}$

5. _____

Name: Date:
Instructor: Section:

6. $\dfrac{2a^4+5a^2+3}{2a^2+3}$

6. _____

Objective 3 Use division in a geometry application.

Video Examples

Review this example for Objective 3:

10. The area of a rectangle is given by $12p^3-7p^2+5p-1$ square units, and the width is $4p-1$ units. What is the length of the rectangle?

   For a rectangle, $A=LW$. Solving for L gives $L=\dfrac{A}{W}$. Divide the area, $12p^3-7p^2+5p-1$ by the width $4p-1$.

$$\begin{array}{r}
3p^2-p+1 \\
4p-1\overline{)12p^3-7p^2+5p-1} \\
\underline{12p^3-3p^2} \\
-4p^2+5p \\
\underline{-4p^2+p} \\
4p-1 \\
\underline{4p-1} \\
0
\end{array}$$

   The length is $3p^2-p+1$ units.

Now Try:

10. The area of a rectangle is given by $6r^3-5r^2+16r-5$ square units, and the width is $3r-1$ units. What is the length of the rectangle?

   _____

Name: Date:
Instructor: Section:

Objective 3 Practice Exercises

For extra help, see Example 10 on page 316 of your text.

Work each problem.

7. The area of a parallelogram is given by $4y^3 - 44y - 600$ square units, and the height is $y - 6$ units. What is the base of the parallelogram?

7. _____

8. The area of a parallelogram is given by $3t^3 + 16t^2 - 32t - 64$ square units, and the base is $t^2 + 4t - 16$ units. What is the height of the parallelogram?

8. _____

Name: Date:
Instructor: Section:

Chapter 5 FACTORING AND APPLICATIONS

5.1 The Greatest Common Factor; Factoring by Grouping

Learning Objectives
1 Find the greatest common factor of a list of terms.
2 Factor out the greatest common factor.
3 Factor by grouping.

Key Terms

Use the vocabulary terms listed below to complete each statement in exercises 1–4.

factor   **factored form**   **greatest common factor (GCF)**

factoring

1. The process of writing a polynomial as a product is called _____.

2. An expression is in _____ when it is written as a product.

3. The _____ is the largest quantity that is a factor of each of a group of quantities.

4. An expression A is a _____ of an expression B if B can be divided by A with 0 remainder.

Objective 1   Find the greatest common factor of a list of terms.

Video Examples

Review these examples for Objective 1:	Now Try:

1. Find the greatest common factor for each list of numbers.
 	
   a. 60, 45
 	
   Write the prime factored form of each number.
    $60 = 2 \cdot 2 \cdot 3 \cdot 5$
 	
    $45 = 3 \cdot 3 \cdot 5$
   Use each prime the least number of times it appears in all the factored forms.
    $\text{GCF} = 3^1 \cdot 5^1 = 15$
	1. Find the greatest common factor for each list of numbers.
	
	   a. 18, 36, 42
	
	   _____

Name: Date:
Instructor: Section:

    b. 90, 36, 108

    Write the prime factored form of each number.
    $90 = 2 \cdot 3 \cdot 3 \cdot 5$
    $36 = 2 \cdot 2 \cdot 3 \cdot 3$
    $108 = 2 \cdot 2 \cdot 3 \cdot 3 \cdot 3$

    There is one factor of 2 and two factors of 3.
    $\text{GCF} = 2^1 \cdot 3^2 = 18$

    c. 17, 18, 24

    Write the prime factored form of each number.
    $17 = 17$
    $18 = 2 \cdot 3 \cdot 3$
    $24 = 2 \cdot 2 \cdot 2 \cdot 3$

    There are no primes common to all three numbers, so the GCF is 1.

2. Find the greatest common factor for the list of terms.

    $28m^4, 35m^6, 49m^9, 70m^5$

    $28m^4 = 2 \cdot 2 \cdot 7 \cdot m^4$
    $35m^6 = 5 \cdot 7 \cdot m^6$
    $49m^9 = 7 \cdot 7 \cdot m^9$
    $70m^5 = 2 \cdot 5 \cdot 7 \cdot m^5$

    Then, $\text{GCF} = 7m^4$.

b. 32, 40, 72 _____

c. 26, 27, 28 _____

2. Find the greatest common factor for the list of terms.

    $54x^5, 48x^7, 42x^9, 30x^4$

    _____

Objective 1 Practice Exercises

For extra help, see Examples 1–2 on pages 331–332 of your text.

Find the greatest common factor for each list of terms.

1. 84, 280, 112

1. _____

2. $6k^2m^4n^5, 8k^3m^7n^4, k^4m^8n^7$

2. _____

3. $9xy^4, 72x^4y^7, 27xy^2, 108x^2y^5$

3. _____

Name: Date:
Instructor: Section:

Objective 2   **Factor out the greatest common factor.**

Video Examples

Review these examples for Objective 2:

3. Write in factored form by factoring out the greatest common factor.

$$12x^5 + 27x^4 - 15x^3$$

$\text{GCF} = 3x^3$
$12x^5 + 27x^4 - 15x^3$
$= 3x^3(4x^2) + 3x^3(9x) + 3x^3(-5)$
$= 3x^3(4x^2 + 9x - 5)$

Check   Multiply the factored form.
$3x^3(4x^2 + 9x - 5)$
$= 3x^3(4x^2) + 3x^3(9x) + 3x^3(-5)$
$= 12x^5 + 27x^4 - 15x^3$

5. Write in factored form by factoring out the greatest common factor.

   a. $x(x+9) + 7(x+9)$

   Factor out $x+9$.
   $x(x+9) + 7(x+9) = (x+9)(x+7)$

   b. $a^2(a+6) - 7(a+6)$

   Factor out $a+6$.
   $a^2(a+6) - 7(a+6) = (a+6)(a^2 - 7)$

Now Try:

3. Write in factored form by factoring out the greatest common factor.
$20y^4 - 12y^3 + 4y^2$

_____

5. Write in factored form by factoring out the greatest common factor.

   a. $y(y+8) + 4(y+8)$

_____

   b. $z^2(z+5) - 11(z+5)$

_____

Objective 2 Practice Exercises

For extra help, see Examples 3–5 on pages 332–334 of your text.

Factor out the greatest common factor or a negative common factor if the coefficient of the term of greatest degree is negative.

4. $20x^2 + 40x^2y - 70xy^2$ 4. _____

5. $2a(x-2y) + 9b(x-2y)$ 5. _____

Name:
Instructor:
Date:
Section:

6. $26x^8 - 13x^{12} + 52x^{10}$ 6. _____

Objective 3   Factor by grouping.

Video Examples

Review these examples for Objective 3:
6. Factor by grouping.

   a. $7by + 28y + b + 4$

   Group the terms, then factor each group.
   $7by + 28y + b + 4$
   $= (7by + 28y) + (b + 4)$
   $= 7y(b + 4) + 1(b + 4)$
   $= (b + 4)(7y + 1)$

   Check   Use the FOIL method.
   $(b + 4)(7y + 1) = 7by + 1b + 4(7y) + 4$
   $= 7by + 28y + b + 4$

   b. $3x^2 - 18x + 5xy - 30y$

   $3x^2 - 18x + 5xy - 30y$
   $= (3x^2 - 18x) + (5xy - 30y)$
   $= 3x(x - 6) + 5y(x - 6)$
   $= (x - 6)(3x + 5y)$

   Check by multiplying using the FOIL method.

   c. $r^3 + 3r^2 - 5r - 15$

   $r^3 + 3r^2 - 5r - 15$
   $= (r^3 + 3r^2) + (-5r - 15)$
   $= r^2(r + 3) - 5(r + 3)$
   $= (r + 3)(r^2 - 5)$

   Check by multiplying using the FOIL method.

Now Try:
6. Factor by grouping.

   a. $36x + 4tx + 9 + t$

   _____

   b. $4x^2 - 28x + 5xy - 35y$

   _____

   c. $x^3 + 7x^2 - 2x - 14$

   _____

Name: Date:
Instructor: Section:

7. Factor by grouping.

   $18x^2 - 20y + 24x - 15xy$

   Group the terms, then factor each group.
   $18x^2 - 20y + 24x - 15xy$
   $= 2(9x^2 - 10y) + 3x(8 - 5y)$

   This does not lead to a common factor, so we try rearranging the terms.
   $18x^2 - 20y + 24x - 15xy$
   $= 18x^2 - 15xy + 24x - 20y$
   $= (18x^2 - 15xy) + (24x - 20y)$
   $= 3x(6x - 5y) + 4(6x - 5y)$
   $= (6x - 5y)(3x + 4)$

   Check Use the FOIL method.
   $(6x - 5y)(3x + 4)$
   $= 18x^2 + 24x - 15xy - 20y$
   $= 18x^2 - 20y + 24x - 15xy$

7. Factor by grouping.

   $56x^2 + 32x - 21xy - 12y$

   _____

Objective 3 Practice Exercises

For extra help, see Examples 6–7 on pages 334–336 of your text.

Factor each polynomial by grouping.

7. $15 - 5x - 3y + xy$

7. _____

8. $2x^2 - 14xy + xy - 7y^2$

8. _____

9. $3r^3 - 2r^2s + 3s^2r - 2s^3$

9. _____

Name: Date:
Instructor: Section:

Chapter 5 FACTORING AND APPLICATIONS

5.2 Factoring Trinomials

Learning Objectives
1 Factor trinomials with coefficient 1 for the second-degree term.
2 Factor such trinomials after factoring out the greatest common factor.

Key Terms

Use the vocabulary terms listed below to complete each statement in exercises 1–3.

   prime polynomial   factoring   greatest common factor

1. _____ is the process of writing a polynomial as a product.

2. The _____ of a polynomial is the greatest term that is a factor of all the terms in the polynomial.

3. A _____ is a polynomial that cannot be factored using only integers.

Objective 1   Factor trinomials with coefficient 1 for the second-degree term.

Video Examples

Review these examples for Objective 1:

1. Factor $m^2 + 8m + 15$.

   Look for integers whose product is 15 and whose sum is 8. Only positive signs are needed.

Factors of 15	Sums of Factors
15, 1	$15 + 1 = 16$
5, 3	$5 + 3 = 8$

   From the table, 5 and 3 are the required integers.
   $m^2 + 8m + 15$ factors as $(m+5)(m+3)$

   Check   Use the FOIL method.
   $(m+5)(m+3) = m^2 + 3m + 5m + 15$
   $= m^2 + 8m + 15$

Now Try:
1. Factor $x^2 + 11x + 24$.

   _____

Name: Date:
Instructor: Section:

2. Factor $x^2 - 11x + 28$.

Look for integers whose product is 28 and whose sum is −11. Since the numbers have a positive product and a negative sum, we consider only pairs of negative integers.

Factors of 28	Sums of Factors
−28, −1	−28 + (−1) = −29
−14, −2	−14 + (−2) = −16
−7, −4	−7 + (−4) = −11

The required integers are −7 and −4.
$x^2 - 11x + 28$ factors as $(x-7)(x-4)$

Check   Use the FOIL method.
$(x-7)(x-4) = x^2 - 4x - 7x + 28$
$= x^2 - 11x + 28$

3. Factor $x^2 + 2x - 15$.

Look for integers whose product is −15 and whose sum is 2. To get a negative product, the pairs of integers must have different signs.

Factors of −15	Sums of Factors
15, −1	15 + (−1) = 14
−15, 1	−15 + 1 = −14
5, −3	5 + (−3) = 2

The required integers are 5 and −3.
$x^2 + 2x - 15$ factors as $(x+5)(x-3)$

Check   Use the FOIL method.
$(x+5)(x-3) = x^2 - 3x + 5x - 15$
$= x^2 + 2x - 15$

5. Factor the trinomial.

$x^2 - 7x + 18$

Look for integers whose product is 18 and whose sum is −7. Since the numbers have a positive product and a negative sum, we consider only pairs of negative integers.

2. Factor $y^2 - 12y + 35$.

_____

3. Factor $p^2 + 6p - 27$.

_____

5. Factor the trinomial.

$m^2 - 7m + 5$

_____

Name: Date:
Instructor: Section:

Factors of 18	Sums of Factors
$-18, -1$	$-18 + (-1) = -19$
$-9, -2$	$-9 + (-2) = -11$
$-6, -3$	$-6 + (-3) = -9$

None of the pairs of integers has a sum of -7.

$x^2 - 7x + 18$ cannot be factored.

It is a prime polynomial.

6. Factor $x^2 - 6xy - 7y^2$.

Here, the coefficient of x in the middle term is $-6y$, so we need to find two expressions whose product is $-7y^2$ and whose sum is $-6y$.

Factors of $-7y^2$	Sums of Factors
$7y, -y$	$7y + (-y) = 6y$
$-7y, y$	$-7y + y = -6y$

$x^2 - 6xy - 7y^2$ factors as $(x - 7y)(x + y)$

Check Use the FOIL method.
$$(x - 7y)(x + y) = x^2 + xy - 7xy - 7y^2$$
$$= x^2 - 6xy - 7y^2$$

6. Factor $p^2 - 5pq - 14q^2$.

6. _____

Objective 1 Practice Exercises

For extra help, see Examples 1–6 on pages 340–342 of your text.

Factor completely. If a polynomial cannot be factored, write prime.

1. $r^2 + r + 3$

1. _____

2. $x^2 - 11x + 28$

2. _____

3. $x^2 - 8x - 33$

3. _____

Name: Date:
Instructor: Section:

Objective 2 Factor such trinomials after factoring out the greatest common factor.

Video Examples

Review this example for Objective 2:

7. Factor $5x^5 - 45x^4 + 90x^3$.

   There is no second-degree term. Look for a common factor.
   $$5x^5 - 45x^4 + 90x^3 = 5x^3(x^2 - 9x + 18)$$
   Now factor $x^2 - 9x + 18$. The integers –3 and –6 have a product of 18 and a sum of –9.
   $5x^5 - 45x^4 + 90x^3$ factors as $5x^3(x-6)(x-3)$

   Check Use the FOIL method.
   $$5x^3(x-6)(x-3)$$
   $$= 5x^3(x^2 - 3x - 6x + 18)$$
   $$= 5x^3(x^2 - 9x + 18)$$
   $$= 5x^5 - 45x^4 + 90x^3$$

Now Try:

7. Factor $7x^6 - 49x^5 + 70x^4$.

   _____

Objective 2 Practice Exercises

For extra help, see Example 7 on page 342 of your text.

Factor completely. If a polynomial cannot be factored, write **prime**.

4. $2n^4 - 16n^3 + 30n^2$

4. _____

5. $2a^3b - 10a^2b^2 + 12ab^3$

5. _____

6. $10k^6 + 70k^5 + 100k^4$

6. _____

Name: Date:
Instructor: Section:

Chapter 5 FACTORING AND APPLICATIONS

5.3 More on Factoring Trinomials

Learning Objectives
1. Factor trinomials by grouping when the coefficient of the second-degree term is not 1.
2. Factor trinomials using the FOIL method.

Key Terms

Use the vocabulary terms listed below to complete each statement in exercises 1–2.

    coefficient   trinomial   FOIL

    outer product   inner product

1. In the term $6x^2y$, 6 is the _____.

2. A polynomial with three terms is a _____.

3. The _____ of $(2y-5)(y+8)$ is $-5y$.

4. _____ is a shortcut method for finding the product of two binomials.

5. The _____ of $(2y-5)(y+8)$ is $16y$.

Objective 1 Factor trinomials by grouping when the coefficient of the second-degree term is not 1.

Video Examples

Review these examples for Objective 1:
1. Factor $3x^2 + 11x + 10$.

    Look for two positive integers whose product is $3 \cdot 10 = 30$ and whose sum is 11.

    The integers are 5 and 6, since $5 \cdot 6 = 30$ and $5 + 6 = 11$.

$$3x^2 + 11x + 10 = 3x^2 + 5x + 6x + 10$$
$$= (3x^2 + 5x) + (6x + 10)$$
$$= x(3x + 5) + 2(3x + 5)$$
$$= (3x + 5)(x + 2)$$

    Check Multiply $(3x+5)(x+2)$ to obtain $3x^2 + 11x + 10$.

Now Try:
1. Factor $5x^2 + 17x + 6$.

    _____

Name: Date:
Instructor: Section:

2. Factor each trinomial.

a. $8x^2 - 2x - 1$

We must find two integers with a product of $8(-1) = -8$ and a sum of -2. The integers are -4 and 2. We write the middle term as $-4x + 2x$.

$$8x^2 - 2x - 1 = 8x^2 - 4x + 2x - 1$$
$$= (8x^2 - 4x) + (2x - 1)$$
$$= 4x(2x - 1) + 1(2x - 1)$$
$$= (2x - 1)(4x + 1)$$

Check Multiply $(2x - 1)(4x + 1)$ to obtain $8x^2 - 2x - 1$.

b. $15z^2 + z - 2$

Look for two integers whose product is $15(-2) = -30$ and whose sum is 1.

The integers are 6 and -5.

$$15z^2 + z - 2 = 15z^2 - 5z + 6z - 2$$
$$= (15z^2 - 5z) + (6z - 2)$$
$$= 5z(3z - 1) + 2(3z - 1)$$
$$= (3z - 1)(5z + 2)$$

Check Multiply $(3z - 1)(5z + 2)$ to obtain $15z^2 + z - 2$.

c. $12r^2 + 5rs - 2s^2$

Two integers whose product is $12(-2) = -24$ and whose sum is 5 are 8 and -3. Rewrite the trinomial with four terms.

$$12r^2 + 5rs - 2s^2 = 12r^2 + 8rs - 3rs - 2s^2$$
$$= (12r^2 + 8rs) + (-3rs - 2s^2)$$
$$= 4r(3r + 2s) - s(3r + 2s)$$
$$= (3r + 2s)(4r - s)$$

Check Multiply $(3r + 2s)(4r - s)$ to obtain $12r^2 + 5rs - 2s^2$.

2. Factor each trinomial.

a. $14x^2 - 3x - 5$

_____

b. $3m^2 - m - 14$

_____

c. $10x^2 + xy - 3y^2$

_____

Name: Date:
Instructor: Section:

3. Factor $100x^5 + 140x^4 - 15x^3$.

Factor out the greatest common factor, $5x^3$.
$100x^5 + 140x^4 - 15x^3 = 5x^3(20x^2 + 28x - 3)$

To factor $20x^2 + 28x - 3$, find two integers whose product is $20(-3) = -60$ and whose sum is 28. Factor 60 into prime factors.
$60 = 2 \cdot 2 \cdot 3 \cdot 5$
Combine the prime factors in pairs using one positive factor and one negative factor to get –60. The factors of 30 and –2 have the correct sum, 28.

$100x^5 + 140x^4 - 15x^3$
$= 5x^3(20x^2 + 28x - 3)$
$= 5x^3(20x^2 + 30x - 2x - 3)$
$= 5x^3[(20x^2 + 30x) + (-2x - 3)]$
$= 5x^3[10x(2x + 3) - 1(2x + 3)]$
$= 5x^3(2x + 3)(10x - 1)$

3. Factor $30x^5 + 87x^4 - 63x^3$.

3. _____

Objective 1 Practice Exercises

For extra help, see Examples 1–3 on page 345–347 of your text.

Factor each trinomial by grouping.

1. $8b^2 + 18b + 9$

1. _____

2. $7a^2b + 18ab + 8b$

2. _____

3. $10c^2 - 29ct + 21t^2$

3. _____

Name: Date:
Instructor: Section:

Objective 2 Factor trinomials using the FOIL method.

Video Examples

Review these examples for Objective 2:

5. Factor $6x^2 + 13x + 7$.

    The number 6 has several possible pairs of factors, but 7 has only 1 and 7, or –1 and –7. We choose positive factors since all coefficients in the trinomial are positive.
    $$(\_\_+7)(\_\_+1)$$
    The possible pairs of $6x^2$ are $6x$ and x, or $3x$ and $2x$.
    $$(3x+7)(2x+1)$$
    gives middle term $3x + 14x = 17x$. Incorrect.
    $$(2x+7)(3x+1)$$
    gives middle term $2x + 21x = 23x$. Incorrect.
    $$(6x+7)(x+1)$$
    gives middle term $6x + 7x = 13x$. Correct.

    $6x^2 + 13x + 7$ factors as $(6x+7)(x+1)$.

    Check. Multiply $(6x+7)(x+1)$ to obtain $6x^2 + 13x + 7$.

6. Factor $10x^2 - 19x + 7$.

    Since 7 has only 1 and 7 or –1 and –7 as factors, it is better to begin by factoring 7. We need two negative factors because the product of two negative factors is positive and their sum is negative, as required.
    We try –1 and –7.
    $$(\_\_-1)(\_\_-7)$$
    The factors of $10x^2$ are $10x$ and x, or $5x$ and $2x$.

    $(10x-1)(x-7)$
    has middle term $-70x - x = -71x$. Incorrect.
    $(5x-1)(2x-7)$
    has middle term $-35x - 2x = -37x$. Incorrect.
    $(2x-1)(5x-7)$
    has middle term $-14x - 5x = -19x$. Correct.

    Thus $10x^2 - 19x + 7$ factors as $(2x-1)(5x-7)$.

Now Try:

5. Factor $15x^2 + 26x + 7$.

    _____

6. Factor $20x^2 - 13x + 2$.

    _____

Name: Date:
Instructor: Section:

7. Factor $6x^2 - x - 15$.

   The integer 6 has several possible pairs of factors, as does –15. Since the constant term is negative, one positive factor and one negative factor of –15 are needed. Since the coefficient of the middle term is relatively small, it is wise to avoid large factors. We try $3x$ and $2x$ as factors of $6x^2$ and 5 and –3 as factors of –15.

   $(3x+5)(2x-3)$
   has middle term $-9x + 10x = x$. Incorrect.
   $(3x-5)(2x+3)$
   has middle term $9x - 10x = -x$. Correct.

   $6x^2 - x - 15$ factors as $(3x-5)(2x+3)$.

7. Factor $8x^2 + 2x - 21$.

   _____

8. Factor $18x^2 - 3xy - 28y^2$.

   There are several factors of $18x^2$, including
   $18x$ and x, $9x$ and $2x$, and $6x$ and $3x$.
   There are many possible pairs of factors of $-28y^2$, including
   $28y$ and $-y$, $-28y$ and y, $14y$ and $-2y$,
   $-14y$ and $2y$, $7y$ and $-4y$, $-7y$ and $4y$.

   Once again, since the coefficient of the middle term is relatively small, avoid the larger factors. Try the factors of $6x$ and $3x$, and $4y$ and $-7y$.
   $(6x+4y)(3x-7y)$ Incorrect
   The first binomial has a common factor of 2.
   $(6x-7y)(3x+4y)$
   has middle term $24xy - 21xy = 3xy$. Incorrect.
   Interchange the signs of the last two terms.
   $(6x+7y)(3x-4y)$
   has middle term $-24xy + 21xy = -3xy$. Correct.

   Thus, $18x^2 - 3xy - 28y^2$ factors as $(6x+7y)(3x-4y)$.

8. Factor $24x^2 - 2xy - 15y^2$.

   _____

Name: Date:
Instructor: Section:

9. Factor the trinomial.

$$-105a^3 + 65a^2 - 10a$$

The common factor is $-5a$. Then use trial and error.

$$-105a^3 + 65a^2 - 10a = -5a(21a^2 - 13a + 2)$$
$$= -5a(3a - 1)(7a - 2)$$

Check.
$$-5a(3a - 1)(7a - 2) = -5a(21a^2 - 13a + 2)$$
$$= -105a^3 + 65a^2 - 10a$$

9. Factor the trinomial.

$$-18a^3 + 66a^2 - 60a$$

_____

Objective 2 Practice Exercises

For extra help, see Examples 4–9 on pages 347–350 of your text.

Factor each trinomial completely.

4. $8q^2 + 10q + 3$

4. _____

5. $3a^2 + 8ab + 4b^2$

5. _____

6. $4c^2 + 14cd - 8d^2$

6. _____

Name: Date:
Instructor: Section:

Chapter 5 FACTORING AND APPLICATIONS

5.4 Special Factoring Techniques

Learning Objectives
1. Factor a difference of squares.
2. Factor a perfect square trinomial.
3. Factor a difference of cubes.
4. Factor a sum of cubes.

Key Terms

Use the vocabulary terms listed below to complete each statement in exercises 1−2.

    perfect square trinomial   **difference**

1. A _____ is the result of a subtraction.

2. A _____ is a trinomial that can be factored as the square of a binomial.

Objective 1   Factor a difference of squares.

Video Examples

Review these examples for Objective 1:
1. Factor the binomial, if possible.

   $x^2 - 64$

   $x^2 - 64 = x^2 - 8^2 = (x+8)(x-8)$

2. Factor each difference of squares.

   a. $36x^2 - 25$

   $36x^2 - 25 = (6x)^2 - 5^2$
   $= (6x+5)(6x-5)$

   b. $81y^2 - 49$

   $81y^2 - 49 = (9y)^2 - 7^2$
   $= (9y+7)(9y-7)$

Now Try:
1. Factor the binomial, if possible.

   $z^2 - 36$

   _____

2. Factor each difference of squares.

   a. $4x^2 - 81$

   _____

   b. $25t^2 - 49$

   _____

Name: Date:
Instructor: Section:

3. Factor completely.

a. $28y^2 - 175$

$$28y^2 - 175 = 7(4y^2 - 25)$$
$$= 7[(2y)^2 - 5^2]$$
$$= 7(2y+5)(2y-5)$$

b. $p^4 - 81$

$$p^4 - 81 = (p^2)^2 - 9^2$$
$$= (p^2+9)(p^2-9)$$
$$= (p^2+9)(p+3)(p-3)$$

3. Factor completely.

a. $90x^2 - 490$

_____

b. $p^4 - 256$

_____

Objective 1 Practice Exercises

For extra help, see Examples 1–3 on pages 353–354 of your text.

Factor each binomial completely. If a binomial cannot be factored, write **prime**.

1. $x^2 - 49$

1. _____

2. $81x^4 - 16$

2. _____

3. $9x^2 + 16$

3. _____

Objective 2   Factor a perfect square trinomial.

Video Examples

Review these examples for Objective 2:

4. Factor $x^2 + 20x + 100$.

The terms x^2 and 100 are perfect squares.
$$x^2 + 20x + 100 = (x+10)^2$$
Check the middle term. $2(x)(10) = 20x$
The trinomial is a perfect square.

Now Try:

4. Factor $p^2 + 16p + 64$.

_____

Name: Date:
Instructor: Section:

5. Factor each trinomial.

   a. $36y^2 + 42y + 49$

   The first and last terms are perfect squares.
   $36y^2 = (6y)^2$ and $49 = 7^2$
   Twice the product of the first and last terms of the binomial is $2 \cdot (6y)(7) = 84y$, which is not the middle term of $36y^2 + 42y + 49$.
   It is a prime polynomial.

   b. $128z^3 + 192z^2 + 72z$

   $128z^3 + 192z^2 + 72z$
   $= 8z(16z^2 + 24z + 9)$
   $= 8z[(4z)^2 + 2(4z)(3) + 3^2]$
   $= 8z(4z + 3)^2$

   c. $49m^2 - 70m + 25$

   $49m^2 - 70m + 25$
   $= (7m)^2 + 2(7m)(-5) + (-5)^2$
   $= (7m - 5)^2$

5. Factor each trinomial.

   a. $100y^2 - 70y + 49$

   _____

   b. $20x^3 + 100x^2 + 125x$

   _____

   c. $64m^2 + 48m + 9$

   _____

Objective 2 Practice Exercises

For extra help, see Examples 4–5 on pages 355–356 of your text.

Factor each trinomial completely. It may be necessary to factor out the greatest common factor first.

4. $z^2 - 26z + 169$

4. _____

5. $9j^2 + 12j + 4$

5. _____

6. $-12a^2 + 60ab - 75b^2$

6. _____

Name: Date:
Instructor: Section:

Objective 3 Factor a difference of cubes.

Video Examples

Review these examples for Objective 3:

6. Factor each difference of cubes.

   a. $m^3 - 1000$

   Use the pattern for a difference of cubes.
   $m^3 - 1000 = m^3 - 10^3$
   $ = (m-10)(m^2 + 10m + 100)$

   b. $64p^3 - 125$

   $64p^3 - 125 = (4p^3) - 5^3$
   $ = (4p - 5)[(4p)^2 + (4p)(5) + 5^2]$
   $ = (4p - 5)(16p^2 + 20p + 25)$

   c. $64y^3 + 1000x^6$

   $64y^3 + 1000x^6$
   $= (4y)^3 + (10x^2)^3$
   $= (4y + 10x^2)[(4y)^2 - (4y)(10x^2) + (10x^2)^2]$
   $= (4y + 10x^2)(16y^2 - 40x^2y + 100x^4)$

Now Try:

6. Factor each difference of cubes.

   a. $t^3 - 216$

   _____

   b. $27k^3 - y^3$

   _____

   c. $27x^3 + 343y^6$

   _____

Objective 3 Practice Exercises

For extra help, see Example 6 on pages 357–358 of your text.

Factor.

7. $8a^3 - 125b^3$ 7. _____

8. $216x^3 - 8y^3$ 8. _____

9. $(m+n)^3 - (m-n)^3$ 9. _____

Copyright © 2016 Pearson Education, Inc.

Name: Date:
Instructor: Section:

Objective 4 Factor a sum of cubes.

Video Examples

Review these examples for Objective 4:
7. Factor each sum of cubes.

   a. $k^3 + 1000$

   $k^3 + 1000 = k^3 + 10^3$
   $= (k+10)(k^2 - 10k + 100)$

   b. $2m^3 + 250n^3$

   $2m^3 + 250n^3 = 2(m^3 + 125n^3)$
   $= 2[m^3 + (5n)^3]$
   $= 2(m+5n)[m^2 - m(5n) + (5n)^2]$
   $= 2(m+5n)(m^2 - 5mn + 25n^2)$

Now Try:
7. Factor each sum of cubes.

   a. $216x^3 + 1$

   _____

   b. $6x^3 + 48y^3$

   _____

Objective 4 Practice Exercises

For extra help, see Example 7 on page 359 of your text.

Factor.

10. $27r^3 + 8s^3$ 10. _____

11. $8a^3 + 64b^3$ 11. _____

12. $64x^3 + 343y^3$ 12. _____

Name: Date:
Instructor: Section:

Chapter 5 FACTORING AND APPLICATIONS

5.5 Solving Quadratic Equations Using the Zero-Factor Property

Learning Objectives
1. Solve quadratic equations using the zero-factor property.
2. Solve other equations using the zero-factor property.

Key Terms

Use the vocabulary terms listed below to complete each statement in exercises 1–3.

    quadratic equation   **standard form**   **double solution**

1. An equation written in the form $ax^2 + bx + c = 0$ is written in the _____ of a quadratic equation.

2. Two factors are identical and both lead to the same solution, called a _____.

3. An equation that can written in the form $ax^2 + bx + c = 0$, with $a \neq 0$, is a _____.

Objective 1 Solve quadratic equations using the zero-factor property.

Video Examples

Review these examples for Objective 1:
1. Solve each equation.

    a. $(x+9)(5x-6) = 0$

By the zero-factor property, either $x + 9 = 0$ or $5x - 6 = 0$.

    $x + 9 = 0$ or $5x - 6 = 0$
      $x = -9$ or $5x = 6$
                             $x = \dfrac{6}{5}$

Check:
Let $x = -9$.
$$(x+9)(5x-6) = 0$$
$$(-9+9)[5(-9)-6] \stackrel{?}{=} 0$$
$$0(-51) \stackrel{?}{=} 0$$
$$0 = 0 \text{ True}$$

Now Try:
1. Solve each equation.

    a. $(x+12)(4x-7) = 0$

_____

Name: Date:
Instructor: Section:

Let $x = \dfrac{6}{5}$.

$$(x+9)(5x-6) = 0$$

$$\left(\dfrac{6}{5}+9\right)\left[5\left(\dfrac{6}{5}\right)-6\right] \stackrel{?}{=} 0$$

$$\left(\dfrac{51}{5}\right)(6-6) \stackrel{?}{=} 0$$

$$0 = 0 \quad \text{True}$$

Both values check, so the solution set is $\left\{-9,\ \dfrac{6}{5}\right\}$.

b. $x(8x-11) = 0$

Use the zero-factor property.
$$x = 0 \quad \text{or} \quad 8x - 11 = 0$$
$$8x = 11$$
$$x = \dfrac{11}{8}$$

Check these solutions by substituting each in the original equation. The solution set is $\left\{0,\ \dfrac{11}{8}\right\}$.

b. $x(6x-11) = 0$

_____

3. Solve $8p^2 + 30 = 46p$.

$$8p^2 + 30 = 46p$$
$$8p^2 - 46p + 30 = 0 \quad \text{Standard form}$$
$$2(4p^2 - 23p + 15) = 0 \quad \text{Factor out 2.}$$
$$4p^2 - 23p + 15 = 0 \quad \text{Divide each side by 2}$$
$$(4p-3)(p-5) = 0 \quad \text{Factor.}$$
$$4p - 3 = 0 \quad \text{or} \quad p - 5 = 0 \quad \text{Zero-factor}$$
$$p = \dfrac{3}{4} \quad \text{or} \quad p = 5 \quad \text{property}$$

The solution set is $\left\{\dfrac{3}{4},\ 5\right\}$.

3. Solve $15p^2 + 36 = 57p$.

_____

Name: Date:
Instructor: Section:

4. Solve the equation.

$$64m^2 - 49 = 0$$

$$64m^2 - 49 = 0$$
$$(8m+7)(8m-7) = 0$$
$$8m+7 = 0 \quad \text{or} \quad 8m-7 = 0$$
$$8m = -7 \quad \text{or} \quad 8m = 7$$
$$m = -\frac{7}{8} \quad \text{or} \quad m = \frac{7}{8}$$

The solution set is $\left\{-\frac{7}{8}, \frac{7}{8}\right\}$.

4. Solve the equation.

$$100x^2 - 9 = 0$$

_____

Objective 1 Practice Exercises

For extra help, see Examples 1–5 on pages 366–370 of your text.

Solve each equation and check your solutions.

1. $2x^2 - 3x - 20 = 0$

1. _____

2. $25x^2 = 20x$

2. _____

3. $c(5c + 17) = 12$

3. _____

Name: Date:
Instructor: Section:

Objective 2 Solve other equations using the zero-factor property.

Video Examples

Review these examples for Objective 2:

6. Solve each equation.

   a. $12z^3 - 3z = 0$

   $$12z^3 - 3z = 0$$
   $$3z(4z^2 - 1) = 0$$
   $$3z(2z+1)(2z-1) = 0$$

   By an extension of the zero-factor property, we have

   $3z = 0$ or $2z+1 = 0$ or $2z-1 = 0$

   $z = 0$ or $z = -\frac{1}{2}$ or $z = \frac{1}{2}$

   Check by substituting each value in the original equation. The solution set is $\left\{-\frac{1}{2},\ 0,\ \frac{1}{2}\right\}$.

   b. $(3x-1)(x^2 - 13x + 36) = 0$

   $$(3x-1)(x^2 - 13x + 36) = 0$$
   $$(3x-1)(x-4)(x-9) = 0$$
   $3x-1 = 0$ or $x-4 = 0$ or $x-9 = 0$
   $x = \frac{1}{3}$ or $x = 4$ or $x = 9$

   Check by substituting each value in the original equation. The solution set is $\left\{\frac{1}{3},\ 4,\ 9\right\}$.

7. Solve $x(3x-7) = (x-1)^2 + 11$.

   $$x(3x-7) = (x-1)^2 + 11$$
   $$3x^2 - 7x = x^2 - 2x + 1 + 11$$
   $$3x^2 - 7x = x^2 - 2x + 12$$
   $$2x^2 - 5x - 12 = 0$$
   $$(2x+3)(x-4) = 0$$
   $2x+3 = 0$ or $x-4 = 0$
   $x = -\frac{3}{2}$ or $x = 4$

   Check by substituting each value in the original equation. The solution set is $\left\{-\frac{3}{2},\ 4\right\}$.

Now Try:

6. Solve each equation.

   a. $3r^3 = 75r$

   _____

   b. $(5x-2)(x^2 - 11x + 18) = 0$

   _____

7. Solve $x(2x+5) = (x+2)^2 + 8$.

   _____

Name: Date:
Instructor: Section:

Objective 2 Practice Exercises

For extra help, see Examples 6–7 on pages 370–371 of your text.

Solve each equation and check your solutions.

4. $x^3 + 2x^2 - 8x = 0$ 4. _____

5. $z^4 + 8z^3 - 9z^2 = 0$ 5. _____

6. $(y^2 - 5y + 6)(y^2 - 36) = 0$ 6. _____

Name: Date:
Instructor: Section:

Chapter 5 FACTORING AND APPLICATIONS

5.6   Applications of Quadratic Equations

Learning Objectives
1   Solve problems involving geometric figures.
2   Solve problems involving consecutive integers.
3   Solve problems by applying the Pythagorean theorem.
4   Solve problems using given quadratic models.

Key Terms

Use the vocabulary terms listed below to complete each statement in exercises 1–2.

  hypotenuse   legs

1. In a right triangle, the sides that form the right angle are the _____.

2. The longest side of a right triangle is the _____.

Objective 1 Solve problems involving geometric figures.

Video Examples

Review this example for Objective 1:
1. The length of a rectangle is three times its width. If the width was increased by 4 and the length remained the same, the resulting rectangle would have an area of 231 square inches. Find the dimensions of the original rectangle.

   Step 1 Read the problem carefully. Find the dimensions of the original rectangle.

   Step 2 Assign a variable.
    Let x = width.
     $3x$ = length
    $x + 4$ = new width
     $3x$ = length

   Step 3 Write an equation. The area of the rectangle is given by $Area = Length \times Width$. Substitute 231 for area, $3x$ for length, and $x + 4$ for width.
     $231 = 3x(x+4)$

Now Try:
1. Mr. Fixxall is building a box which will have a volume of 60 cubic meters. The height of the box will be 4 meters, and the length will be 2 meters more than the width. Find the width and length of the box.

   _____

Name: Date:
Instructor: Section:

Step 4 Solve.
$$231 = 3x(x+4)$$
$$231 = 3x^2 + 12x$$
$$3x^2 + 12x - 231 = 0$$
$$3(x^2 + 4x - 77) = 0$$
$$x^2 + 4x - 77 = 0$$
$$(x+11)(x-7) = 0$$
$$x+11 = 0 \quad \text{or} \quad x-7 = 0$$
$$x = -11 \quad \text{or} \quad x = 7$$

Step 5 State the answer. The solutions are –11 and 7. A rectangle cannot have a side of negative length, so we discard –11. The width is 7 inches. The length is $3(7) = 21$ inches.

Step 6 Check. The new width is 7 + 4 = 11. The new area of the rectangle is $11(21) = 231$ square inches.

Objective 1 Practice Exercises

For extra help, see Example 1 on page 374 of your text.

Solve each problem. Check your answers to be sure they are reasonable.

1. A book is three times as long as it is wide. Find the length and width of the book in inches if its area is numerically 128 more than its perimeter.

    1. width_____

        length _____

2. The area of a triangle is 42 square centimeters. The base is 2 centimeters less than twice the height. Find the base and height of the triangle.

    2. base_____

        height _____

Name: Date:
Instructor: Section:

3. The volume of a box is 192 cubic feet. If the length of the box is 8 feet and the width is 2 feet more than the height, find the height and width of the box.

3. height _____

   width _____

Objective 2   Solve problems involving consecutive integers.

Video Examples

Review this example for Objective 2:
2. The product of the first and second of three consecutive integers is 2 more than 6 times the third integer. Find the integers.

   Step 1 Read carefully. Note that the integers are consecutive integers.

   Step 2 Assign a variable.
   Let x = the first integer.
   Then $x + 1$ = the second integer,
   and $x + 2$ = the third integer.

   Step 3 Write an equation. The product of the first and second integer is 2 more than 6 times the third integer.
   $$x(x+1) = 6(x+2) + 2$$

   Step 4 Solve.
   $$x(x+1) = 6(x+2) + 2$$
   $$x^2 + x = 6x + 14$$
   $$x^2 - 5x - 14 = 0$$
   $$(x+2)(x-7) = 0$$
   $$x+2 = 0 \quad \text{or} \quad x-7 = 0$$
   $$x = -2 \quad \text{or} \quad x = 7$$

   Step 5 State the answer. The value −2 and 7 each lead to a distinct answer.
   If $x = -2$, then $x + 1 = -1$, and $x + 2 = 0$.
      The integers are −2, −1, 0.
   If $x = 7$, then $x + 1 = 8$, and $x + 2 = 9$.
      The integers are 7, 8, 9.

Now Try:
2. The product of the second and third of three consecutive integers is 2 more than 8 times the first integer. Find the integers.

   _____

Name: Date:
Instructor: Section:

Step 6 Check. The product of the first and second integers must equal 2 more than 6 times the third. Because

$-2(-1) = 6(0) + 2$ and $7(8) = 6(9) + 2$

are both true, both sets of consecutive integers satisfy the statement of the problem.

Objective 2 Practice Exercises

For extra help, see Example 2 on page 375 of your text.

Solve each problem.

4. Find all possible pairs of consecutive odd integers whose sum is equal to their product decreased by 47.

    4. _____

5. Find two consecutive positive even integers whose product is six more than three times its sum.

    5. _____

6. Find three consecutive positive odd integers such that four times the sum of all three equals 13 more than the product of the smaller two.

    6. _____

Name: Date:
Instructor: Section:

Objective 3   Solve problems by applying the Pythagorean theorem.

Video Examples

Review this example for Objective 3:

3. Penny and Carla started biking from the same corner. Penny biked east and Carla biked south. When they were 26 miles apart, Carla had biked 14 miles further than Penny. Find the distance each biked.

   Step 1 Read carefully. Find the two distances.

   Step 2 Assign a variable.
   Let x = Penny's distance.
   Then $x + 14$ = Carla's distance.

   Step 3 Write an equation. Substitute into the Pythagorean theorem.
   $$a^2 + b^2 = c^2$$
   $$x^2 + (x+14)^2 = 26^2$$

   Step 4 Solve.
   $$x^2 + x^2 + 28x + 196 = 676$$
   $$2x^2 + 28x - 480 = 0$$
   $$2(x^2 + 14x - 240) = 0$$
   $$x^2 + 14x - 240 = 0$$
   $$(x+24)(x-10) = 0$$
   $x + 24 = 0$ or $x - 10 = 0$
       $x = -24$ or $x = 10$

   Step 5 State the answer. Since –24 cannot be a distance, 10 is the distance for Penny, and $10 + 14 = 24$ is the distance for Carla.

   Step 6 Check. Since $10^2 + 24^2 = 26^2$ is true, the answer is correct.

Now Try:

3. A ladder is leaning against a building. The distance from the bottom of the ladder to the building is 8 feet less than the length of the ladder. How high up the side of the building is the top of the ladder if that distance is 4 feet less than the length of the ladder?

   _____

Objective 3 Practice Exercises

For extra help, see Example 3 on pages 376–377 of your text.

Solve each problem.

7. A field is in the shape of a right triangle. The shorter leg measures 45 meters. The hypotenuse measures 45 meters less than twice the longer the leg. Find the dimensions of the lot.

      7. _____

Name: Date:
Instructor: Section:

8. A train and a car leave a station at the same time, the train traveling due north and the car traveling west. When they are 100 miles apart, the train has traveled 20 miles farther than the car. Find the distance each has traveled.

8. car _____

train _____

9. Two ships left a dock at the same time. When they were 25 miles apart, the ship that sailed due south had gone 10 miles less than twice the distance traveled by the ship that sailed due west. Find the distance traveled by the ship that sailed due south.

9. _____

Objective 4   Solve problems using given quadratic models.

Video Examples

Review this example for Objective 4:

4. Jeff threw a stone straight upward at 46 feet per second from a dock 6 feet above a lake. The height of the stone above the lake t seconds after it is thrown is given by $h = -16t^2 + 46t + 6$. How long will it take for the stone to reach a height of 39 feet?

Substitute 39 for h.
$$39 = -16t^2 + 46t + 6$$

Solve for t.
$$16t^2 - 46t + 33 = 0$$
$$(8t - 11)(2t - 3) = 0$$
$$8t - 11 = 0 \quad \text{or} \quad 2t - 3 = 0$$
$$t = \frac{11}{8} \quad \text{or} \quad t = \frac{3}{2}$$

Since we have found two acceptable answers,

Now Try:

4. A ball is dropped from the roof of a 19.6 meter high building. Its height h (in meters) t seconds later is given by the equation $h = -4.9t^2 + 19.6$. After how many second is the height 14.7 meters?

_____

Name: Date:
Instructor: Section:

the stone will be at height of 39 feet twice (once on its way up and once on its way down) —at $\frac{11}{8}$ sec or $\frac{3}{2}$ sec.

Objective 4 Practice Exercises

For extra help, see Examples 4–5 on pages 377–378 of your text.

Solve each problem.

10. If an object is propelled upward from a height of 16 feet with an initial velocity of 48 feet per second, its height h (in feet) t seconds later is given by the equation $h = -16t^2 + 48t + 16$.

    (a) After how many seconds is the height 52 feet?

    (b) After how many seconds is the height 48 feet?

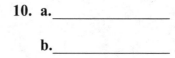

11. A company determines that its daily revenue R (in dollars) for selling x items is modeled by the equation $R = x(150 - x)$. How many items must be sold for its revenue to be $4400?

12. If a ball is batted at an angle of 35°, the distance that the ball travels is given approximately by $D = 0.029v^2 + 0.021v - 1$, where v is the bat speed in miles per hour and D is the distance traveled in feet. Find the distance a batted ball will travel if the ball is batted with a velocity of 90 miles per hour. Round your answer to the nearest whole number.

Name: Date:
Instructor: Section:

Chapter 6 RATIONAL EXPRESSIONS AND APPLICATIONS

6.1 The Fundamental Property of Rational Expressions

Learning Objectives
1. Find the numerical value of a rational expression.
2. Find the values of the variable for which a rational expression is undefined.
3. Write rational expressions in lowest terms.
4. Recognize equivalent forms of rational expressions.

Key Terms

Use the vocabulary terms listed below to complete each statement in exercises 1–2.

  rational expression   lowest terms

1. The quotient of two polynomials with denominator not 0 is called a
   _____.

2. A rational expression is written in _____ if the greatest common factor of its numerator and denominator is 1.

Objective 1 Find the numerical value of a rational expression.

Video Examples

Review this example for Objective 1:

1. Find the numerical value of $\dfrac{4x+8}{3x-6}$ for the value of x.
   $x = -2$

   $\dfrac{4x+8}{3x-6} = \dfrac{4(-2)+8}{3(-2)-6} = \dfrac{0}{-12} = 0$

Now Try:

1. Find the numerical value of $\dfrac{3x-7}{x+5}$ for the value of x.
   $x = -7$

   _____

Objective 1 Practice Exercises

For extra help, see Example 1 on page 394 of your text.

Find the numerical value of each expression when (a) $x = 4$ and (b) $x = -1$.

1. $\dfrac{-3x+1}{2x+1}$

   1. (a) _____

      (b) _____

Name: Date:
Instructor: Section:

2. $\dfrac{2x^2 - 4}{x^2 - 2}$

2. (a) _____

(b) _____

3. $\dfrac{2x - 5}{2 - x - x^2}$

3. (a) _____

(b) _____

Objective 2 Find the values of the variable for which a rational expression is undefined.

Video Examples

Review these examples for Objective 2:
2. Find any values of the variable for which each rational expression is undefined.

a. $\dfrac{3x + 8}{5x + 4}$

Step 1 Set the denominator equal to 0.
$$5x + 4 = 0$$
Step 2 Solve.
$$5x = -4$$
$$x = -\dfrac{4}{5}$$
Step 3 The given expression is undefined for $-\dfrac{4}{5}$, so $x \neq -\dfrac{4}{5}$.

b. $\dfrac{3m^2}{m^2 - 4m - 5}$

Set the denominator equal to 0.
$$m^2 - 4m - 5 = 0$$
$$(m + 1)(m - 5) = 0$$
$$m + 1 = 0 \quad \text{or} \quad m - 5 = 0$$
$$m = -1 \quad \text{or} \quad m = 5$$
The given expression is undefined for -1 and 5, so $m \neq -1$, $m \neq 5$.

Now Try:
2. Find any values of the variable for which each rational expression is undefined.

a. $\dfrac{y + 6}{7y - 1}$

_____

b. $\dfrac{15m^2}{m^2 - m - 20}$

_____

Name: _____ Date: _____
Instructor: _____ Section: _____

c. $\dfrac{6r}{r^2+49}$

This denominator will not equal 0 for any value of *r*. There are no values for which this expression is undefined.

c. $\dfrac{12t^2}{t^2+100}$

_____

Objective 2 Practice Exercises

For extra help, see Example 2 on pages 395–396 of your text.

Find any value(s) of the variable for which each rational expression is undefined. Write answers with $\neq$.

4. $\dfrac{4x^2}{x+7}$

4. _____

5. $\dfrac{2x^2}{x^2+4}$

5. _____

6. $\dfrac{2y-5}{2y^2+4y-16}$

6. _____

Objective 3 Write rational expressions in lowest terms.

Video Examples

Review these examples for Objective 3:

3. Write the expression in lowest terms.

$\dfrac{15k^3}{3k^4}$

Write k^3 as $k \cdot k \cdot k$ and k^4 as $k \cdot k \cdot k \cdot k$.

$\dfrac{15k^3}{3k^4} = \dfrac{3 \cdot 5 \cdot k \cdot k \cdot k}{3 \cdot k \cdot k \cdot k \cdot k}$

$= \dfrac{5 \cdot (3 \cdot k \cdot k \cdot k)}{k \cdot (3 \cdot k \cdot k \cdot k)}$

$= \dfrac{5}{k}$

Now Try:

3. Write the expression in lowest terms.

$\dfrac{12k^5}{4k^8}$

_____

Name: Date:
Instructor: Section:

4. Write each rational expression in lowest terms.

   a. $\dfrac{6x-18}{5x-15}$

   $$\dfrac{6x-18}{5x-15} = \dfrac{6(x-3)}{5(x-3)} = \dfrac{6}{5}$$

   b. $\dfrac{m^2+5m-24}{3m^2-5m-12}$

   $$\dfrac{m^2+5m-24}{3m^2-5m-12} = \dfrac{(m+8)(m-3)}{(3m+4)(m-3)}$$
   $$= \dfrac{m+8}{3m+4}$$

5. Write $\dfrac{3x-2y}{2y-3x}$ in lowest terms.

   Factor -1 from the denominator.
   $$\dfrac{3x-2y}{2y-3x} = \dfrac{3x-2y}{-1(-2y+3x)}$$
   $$= \dfrac{3x-2y}{-1(3x-2y)}$$
   $$= -1$$

4. Write each rational expression in lowest terms.

   a. $\dfrac{7x-35}{9x-45}$

   _____

   b. $\dfrac{m^2-3m-54}{2m^2-15m-27}$

   _____

5. Write $\dfrac{4y-5x}{5x-4y}$ in lowest terms.

   _____

Objective 3 Practice Exercises

For extra help, see Examples 3–6 on pages 396–399 of your text.

Write each rational expression in lowest terms. Assume that no values of any variable make any denominator zero.

7. $\dfrac{15ab^3c^9}{-24ab^2c^{10}}$

7. _____

8. $\dfrac{16-x^2}{2x-8}$

8. _____

Name: Date:
Instructor: Section:

9. $\dfrac{9x^2 - 9x - 108}{2x - 8}$

9. _____

Objective 4   Recognize equivalent forms of rational expressions.

Video Examples

Review this example for Objective 4:
7. Write four equal forms of the following rational expression.

$$-\dfrac{4x+3}{x-8}$$

If we apply the negative sign to the numerator, we obtain the first two equivalent forms.

$$\dfrac{-(4x+3)}{x-8} \text{ and } \dfrac{-4x-3}{x-8}$$

If we apply the negative sign to the denominator, we obtain the last two equivalent forms.

$$\dfrac{4x+3}{-(x-8)} \text{ and } \dfrac{4x+3}{-x+8}$$

Now Try:
7. Write four equal forms of the following rational expression.

$$-\dfrac{10x-7}{4x-3}$$

_____

Objective 4 Practice Exercises

For extra help, see Example 7 on page 400 of your text.

Write four equivalent forms of the following rational expressions. Assume that no values of any variable make any denominator zero.

10. $-\dfrac{4x+5}{3-6x}$

10. _____

11. $\dfrac{2p-1}{1-4p}$

11. _____

12. $-\dfrac{2x-3}{x+2}$

12. _____

Name: Date:
Instructor: Section:

Chapter 6 RATIONAL EXPRESSIONS AND APPLICATIONS

6.2 Multiplying and Dividing Rational Expressions

Learning Objectives
1. Multiply rational expressions.
2. Divide rational expressions.

Key Terms
Use the vocabulary terms listed below to complete each statement in exercises 1–3.

    rational expression   **reciprocal**   **lowest terms**

1. The _____ of the expression $\frac{4x-5}{x+2}$ is $\frac{x+2}{4x-5}$.

2. A _____ is the quotient of two polynomials with denominator not 0.

3. A rational expression is written in _____ when the numerator and denominator have no common terms.

Objective 1   Multiply rational expressions.

Video Examples

Review these examples for Objective 1:
1. Multiply. Write each answer in lowest terms.

    a. $\frac{7}{12} \cdot \frac{3}{14}$

$$\frac{7}{12} \cdot \frac{3}{14} = \frac{7 \cdot 3}{12 \cdot 14}$$
$$= \frac{7 \cdot 3}{2 \cdot 2 \cdot 3 \cdot 2 \cdot 7}$$
$$= \frac{1}{8}$$

    b. $\frac{9}{x^2} \cdot \frac{x^3}{15}$

$$\frac{9}{x^2} \cdot \frac{x^3}{15} = \frac{9 \cdot x^3}{x^2 \cdot 15}$$
$$= \frac{3 \cdot 3 \cdot x \cdot x \cdot x}{3 \cdot 5 \cdot x \cdot x}$$
$$= \frac{3x}{5}$$

Now Try:
1. Multiply. Write each answer in lowest terms.

    a. $\frac{5}{9} \cdot \frac{12}{25}$

    _____

    b. $\frac{8}{x^3} \cdot \frac{x^2}{6}$

    _____

218   Copyright © 2016 Pearson Education, Inc.

Name: Date:
Instructor: Section:

2. Multiply. Write the answer in lowest terms.

$$\frac{3x+2y}{5x} \cdot \frac{x^2}{(3x+2y)^2}$$

Multiply numerators, multiply denominators, factor, and identify the common factors.

$$\frac{3x+2y}{5x} \cdot \frac{x^2}{(3x+2y)^2} = \frac{(3x+2y)x^2}{5x(3x+2y)^2}$$

$$= \frac{(3x+2y) \cdot x \cdot x}{5x(3x+2y)(3x+2y)}$$

$$= \frac{x}{5(3x+2y)}$$

3. Multiply. Write the answer in lowest terms.

$$\frac{x^2+10x+9}{x^2-5x} \cdot \frac{x^2+3x-40}{x^2+9x+8}$$

$$\frac{x^2+10x+9}{x^2-5x} \cdot \frac{x^2+3x-40}{x^2+9x+8}$$

$$= \frac{(x^2+10x+9)(x^2+3x-40)}{(x^2-5x)(x^2+9x+8)}$$

$$= \frac{(x+9)(x+1)(x+8)(x-5)}{x(x-5)(x+1)(x+8)}$$

$$= \frac{x+9}{x}$$

The quotients $\frac{x+1}{x+1}$, $\frac{x+8}{x+8}$, and $\frac{x-5}{x-5}$ are all equal to 1, justifying the final product $\frac{x+9}{x}$.

2. Multiply. Write the answer in lowest terms.

$$\frac{r-s}{6s} \cdot \frac{s^3}{(r-s)^2}$$

_____

3. Multiply. Write the answer in lowest terms.

$$\frac{7x^2-7}{x^2+x-2} \cdot \frac{5x+10}{x^2+x}$$

_____

Objective 1 Practice Exercises

For extra help, see Examples 1–3 on pages 404–405 of your text.

Multiply. Write each answer in lowest terms.

1. $\dfrac{8m^4n^3}{3} \cdot \dfrac{5}{4mn^2}$ **1.** _____

Name: Date:
Instructor: Section:

2. $\dfrac{m^2-16}{m-3} \cdot \dfrac{9-m^2}{4-m}$

2. _____

3. $\dfrac{3x+12}{6x-30} \cdot \dfrac{x^2-x-20}{x^2-16}$

3. _____

Objective 2 Divide rational expressions.

Video Examples

Review these examples for Objective 2:
4. Divide. Write each answer in lowest terms.

  a. $\dfrac{7}{9} \div \dfrac{4}{27}$

  Multiply the first expression by the reciprocal of the second.

  $\dfrac{7}{9} \div \dfrac{4}{27} = \dfrac{7}{9} \cdot \dfrac{27}{4}$

  $= \dfrac{7 \cdot 27}{9 \cdot 4}$

  $= \dfrac{7 \cdot 3 \cdot 9}{9 \cdot 4}$

  $= \dfrac{21}{4}$

  b. $\dfrac{y}{y-5} \div \dfrac{7y}{y+4}$

  $\dfrac{y}{y-5} \div \dfrac{7y}{y+4} = \dfrac{y}{y-5} \cdot \dfrac{y+4}{7y}$

  $= \dfrac{y(y+4)}{(y-5)(7y)}$

  $= \dfrac{y+4}{7(y-5)}$

Now Try:
4. Divide. Write each answer in lowest terms.

  a. $\dfrac{9}{10} \div \dfrac{3}{25}$

  _____

  b. $\dfrac{y}{y+2} \div \dfrac{6y}{y-2}$

  _____

220 Copyright © 2016 Pearson Education, Inc.

Name: Date:
Instructor: Section:

6. Divide. Write the answer in lowest terms.

$$\frac{m^2-16}{(m-5)(m-4)} \div \frac{(m+4)(m-5)}{-6m}$$

Multiply by the reciprocal.

$$\frac{m^2-16}{(m-5)(m-4)} \div \frac{(m+4)(m-5)}{-6m}$$

$$= \frac{m^2-16}{(m-5)(m-4)} \cdot \frac{-6m}{(m+4)(m-5)}$$

$$= \frac{-6m(m^2-16)}{(m-5)(m-4)(m+4)(m-5)}$$

$$= \frac{-6m(m+4)(m-4)}{(m-5)(m-4)(m+4)(m-5)}$$

$$= \frac{-6m}{(m-5)^2}, \text{ or } -\frac{6m}{(m-5)^2}$$

6. Divide. Write the answer in lowest terms.

$$\frac{x^2-49}{(x-7)(x-3)} \div \frac{(x+7)(x-3)}{8x}$$

_____

7. Divide. Write the answer in lowest terms.

$$\frac{x^2-25}{x^2-9} \div \frac{3x^2-15x}{3-x}$$

Multiply by the reciprocal.

$$\frac{x^2-25}{x^2-9} \div \frac{3x^2-15x}{3-x}$$

$$= \frac{x^2-25}{x^2-9} \cdot \frac{3-x}{3x^2-15x}$$

$$= \frac{(x^2-25)(3-x)}{(x^2-9)(3x^2-15x)}$$

$$= \frac{(x+5)(x-5)(3-x)}{(x+3)(x-3)(3x)(x-5)}$$

$$= \frac{-1(x+5)}{3x(x+3)}, \text{ or } \frac{-x-5}{3x(x+3)}$$

7. Divide. Write the answer in lowest terms.

$$\frac{m^2-64}{m^2-81} \div \frac{5m^2+40m}{9-m}$$

_____

Objective 2 Practice Exercises

For extra help, see Examples 4–7 on pages 406–407 of your text.

Divide. Write each answer in lowest terms.

4. $\dfrac{b-7}{16} \div \dfrac{7-b}{8}$

4. _____

Copyright © 2016 Pearson Education, Inc.

Name: Date:
Instructor: Section:

5. $\dfrac{m^2+2mn+n^2}{m^2+m} \div \dfrac{m^2-n^2}{m^2-1}$

5. _____

6. $\dfrac{27-3k^2}{3k^2+8k-3} \div \dfrac{k^2-6k+9}{6k^2-19k+3}$

6. _____

Name: Date:
Instructor: Section:

Chapter 6 RATIONAL EXPRESSIONS AND APPLICATIONS

6.3   Least Common Denominators

Learning Objectives
1   Find the least common denominator for a group of fractions.
2   Write equivalent rational expressions.

Key Terms

Use the vocabulary terms listed below to complete each statement in exercises 1–2.

least common denominator   **equivalent expressions**

1. $\dfrac{24x-8}{9x^2-1}$ and $\dfrac{8}{3x+1}$ are _____.

2. The simplest expression that is divisible by all denominators is called the
    _____.

Objective 1   Find the least common denominator for a list of fractions.

Video Examples

Review these examples for Objective 1:
1. Find the LCD for the pair of fractions.

    $\dfrac{1}{35}, \dfrac{11}{45}$

    Step 1 Factor each denominator into prime factors.
        $35 = 5 \cdot 7$
        $45 = 3 \cdot 3 \cdot 5 = 3^2 \cdot 5$

    Step 2 List each denominator the greatest number of times it appears as a factor in any of the denominators.
        The factor 3 appears two times, and the factors 5 and 7 each appear once.

    Step 3 Multiply to get the LCD.
        $LCD = 3 \cdot 3 \cdot 5 \cdot 7$
        $= 3^2 \cdot 5 \cdot 7$
        $= 315$

Now Try:
1. Find the LCD for the pair of fractions.

    $\dfrac{5}{18}, \dfrac{13}{24}$

    _____

Name: Date:
Instructor: Section:

2. Find the LCD for $\dfrac{9}{28s^3}$ and $\dfrac{5}{42s^2}$.

   Step 1
   $$28s^3 = 2 \cdot 2 \cdot 7 \cdot s^3$$
   $$42s^2 = 2 \cdot 3 \cdot 7 \cdot s^2$$

   Step 2
   Here s appears three times, 2 appears twice, and 3 and 7 each appear once.

   Step 3
   $$\text{LCD} = 2^2 \cdot 3 \cdot 7 \cdot s^3 = 84s^3$$

3. Find the LCD for the fractions in each list.

   a. $\dfrac{5}{7b}$, $\dfrac{8}{b^2 - 5b}$

   $$7b = 7 \cdot b$$
   $$b^2 - 5b = b(b-5)$$
   $$\text{LCD} = 7 \cdot b(b-5) = 7b(b-5)$$

   b. $\dfrac{6}{c^2 - 5c - 6}$, $\dfrac{10}{c^2 + 3c - 54}$, $\dfrac{4}{c^2 - 12c + 36}$

   $$c^2 - 5c - 6 = (c-6)(c+1)$$
   $$c^2 + 3c - 54 = (c-6)(c+9)$$
   $$c^2 - 12c + 36 = (c-6)^2$$

   Use each factor the greatest number of times it appears as a factor.
   $$\text{LCD} = (c+1)(c+9)(c-6)^2$$

   c. $\dfrac{1}{a-8}$, $\dfrac{7}{8-a}$

   $$-(a-8) = -a + 8 = 8 - a$$
   Therefore either $8-a$ or $a-8$ can be used as the LCD.

2. Find the LCD for $\dfrac{15}{40a^2}$ and $\dfrac{13}{24a^4}$.

   _____

3. Find the LCD for the fractions in each list.

   a. $\dfrac{7}{9w}$, $\dfrac{13}{w^2 - 2w}$

   _____

   b. $\dfrac{12}{b^2 - 16}$, $\dfrac{6}{b^2 - 3b - 4}$, $\dfrac{9}{b^2 - 8b + 16}$

   _____

   c. $\dfrac{13}{p-14}$, $\dfrac{12}{14-p}$

   _____

Objective 1 Practice Exercises

For extra help, see Examples 1–3 on pages 411–412 of your text.

Find the least common denominator for each list of rational expressions.

1. $\dfrac{13}{36b^4}$, $\dfrac{17}{27b^2}$

1. _____

Name: Date:
Instructor: Section:

2. $\dfrac{-7}{a^2-2a}$, $\dfrac{3a}{2a^2+a-10}$

2. _____

3. $\dfrac{8}{w^3-9w}$, $\dfrac{4w}{w^2+w-6}$

3. _____

Objective 2 Write equivalent rational expressions.

Video Examples

Review these examples for Objective 2:

4. Write the rational expression as an equivalent expression with the indicated denominator.

$$\dfrac{7}{9} = \dfrac{?}{45}$$

Step 1 Factor both denominators.
$$\dfrac{7}{9} = \dfrac{?}{5 \cdot 9}$$
Step 2 A factor of 5 is missing.
Step 3 Multiply $\dfrac{7}{9}$ by $\dfrac{5}{5}$.
$$\dfrac{7}{9} = \dfrac{7}{9} \cdot \dfrac{5}{5} = \dfrac{35}{45}$$

5. Write each rational expression as an equivalent expression with the indicated denominator.

a. $\dfrac{9}{5x+2} = \dfrac{?}{15x+6}$

Factor the denominator on the right.
$$\dfrac{9}{5x+2} = \dfrac{?}{3(5x+2)}$$
The missing factor is 3, so multiply by $\dfrac{3}{3}$.
$$\dfrac{9}{5x+2} \cdot \dfrac{3}{3} = \dfrac{27}{15x+6}$$

Now Try:

4. Write the rational expression as an equivalent expression with the indicated denominator.

$$\dfrac{13}{6} = \dfrac{?}{30}$$

_____

5. Write each rational expression as an equivalent expression with the indicated denominator.

a. $\dfrac{19}{6c-5} = \dfrac{?}{24c-20}$

_____

Name: Date:
Instructor: Section:

b. $\dfrac{7}{q^2+6q} = \dfrac{?}{q^3+q^2-30q}$

Factor the denominator in each rational expression.

$\dfrac{7}{q(q+6)} = \dfrac{?}{q(q+6)(q-5)}$

The missing factor is $(q-5)$, so multiply by $\dfrac{q-5}{q-5}$.

$\dfrac{7}{q(q+6)} = \dfrac{7}{q(q+6)} \cdot \dfrac{q-5}{q-5}$

$= \dfrac{7(q-5)}{q^3-q^2-30q}$

$= \dfrac{7q-35}{q^3-q^2-30q}$

b. $\dfrac{3}{z^2-7z} = \dfrac{?}{z^3-5z^2-14z}$

_____

Objective 2 Practice Exercises

For extra help, see Examples 4–5 on pages 412–413 of your text.

Rewrite each rational expression with the indicated denominator. Give the numerator of the new fraction.

4. $\dfrac{5a}{8a-3} = \dfrac{?}{6-16a}$

4. _____

5. $\dfrac{3}{5r-10} = \dfrac{?}{50r^2-100r}$

5. _____

6. $\dfrac{3}{k^2+3k} = \dfrac{?}{k^3+10k^2+21k}$

6. _____

Name: Date:
Instructor: Section:

Chapter 6 RATIONAL EXPRESSIONS AND APPLICATIONS

6.4   Adding and Subtracting Rational Expressions

Learning Objectives
1   Add rational expressions having the same denominator.
2   Add rational expressions having different denominators.
3   Subtract rational expressions.

Key Terms

Use the vocabulary terms listed below to complete each statement in exercises 1–2.

least common multiple   **greatest common factor**

1. The _____ of $2m^2 - 5m - 3$ and $2m - 6$ is $m - 3$.

2. The _____ of $2m^2 - 5m - 3$ and $2m - 6$ is $2(m-3)(2m+1)$.

Objective 1   Add rational expressions having the same denominator.

Video Examples

Review these examples for Objective 1:
1. Add. Write each answer in lowest terms.

   a. $\dfrac{1}{8} + \dfrac{3}{8}$

   The denominators are the same, so add the numerators and keep the common denominator.

   $$\dfrac{1}{8} + \dfrac{3}{8} = \dfrac{1+3}{8}$$
   $$= \dfrac{4}{8}$$
   $$= \dfrac{4 \cdot 1}{4 \cdot 2}$$
   $$= \dfrac{1}{2}$$

   b. $\dfrac{4x}{x+2} + \dfrac{8}{x+2}$

   $\dfrac{4x}{x+2} + \dfrac{8}{x+2} = \dfrac{4x+8}{x+2} = \dfrac{4(x+2)}{x+2} = 4$

Now Try:
1. Add. Write each answer in lowest terms.

   a. $\dfrac{9}{20} + \dfrac{7}{20}$

   _____

   b. $\dfrac{2x^2}{x+4} + \dfrac{8x}{x+4}$

   _____

Name: Date:
Instructor: Section:

Objective 1 Practice Exercises

For extra help, see Example 1 on page 416 of your text.

Add. Write each answer in lowest terms.

1. $\dfrac{5}{3w^2} + \dfrac{7}{3w^2}$

1. _____

2. $\dfrac{b}{b^2-4} + \dfrac{2}{b^2-4}$

2. _____

3. $\dfrac{2x+3}{x^2+3x-10} + \dfrac{2-x}{x^2+3x-10}$

3. _____

Objective 2 Add rational expressions having different denominators.

Video Examples

Review these examples for Objective 2:

2. Add. Write each answer in lowest terms.

   a. $\dfrac{5}{18} + \dfrac{7}{24}$

   Step 1 Find the LCD.

   $18 = 2 \cdot 3 \cdot 3 = 2 \cdot 3^2$

   $24 = 2 \cdot 2 \cdot 2 \cdot 3 = 2^3 \cdot 3$

   $\text{LCD} = 2^3 \cdot 3^2 = 72$

   Step 2 Now write each expression as an equivalent expression with the LCD as the denominator.

   $\dfrac{5}{18} + \dfrac{7}{24} = \dfrac{5(4)}{18(4)} + \dfrac{7(3)}{24(3)}$

   $= \dfrac{20}{72} + \dfrac{21}{72}$

   Step 3 Add the numerators.
   Step 4 Write in lowest terms, if necessary.

   $= \dfrac{20+21}{72}$

   $= \dfrac{41}{72}$

Now Try:

2. Add. Write each answer in lowest terms.

   a. $\dfrac{9}{35} + \dfrac{8}{45}$

   _____

Copyright © 2016 Pearson Education, Inc.

Name: Date:
Instructor: Section:

b. $\dfrac{5}{6z}+\dfrac{7}{9z}$ **b.** $\dfrac{4}{9y}+\dfrac{2}{7y}$

Step 1 Find the LCD.
$$6z = 2 \cdot 3 \cdot z$$
$$9z = 3 \cdot 3 \cdot z = 3^2 \cdot z$$
$$\text{LCD} = 2 \cdot 3^2 \cdot z = 18z$$

Step 2 Now write each expression as an equivalent expression with the LCD as the denominator.
$$\dfrac{5}{6z}+\dfrac{7}{9z} = \dfrac{5(3)}{6z(3)}+\dfrac{7(2)}{9z(2)}$$
$$= \dfrac{15}{18z}+\dfrac{14}{18z}$$

Step 3 Add the numerators.
Step 4 Write in lowest terms, if necessary.
$$= \dfrac{15+14}{18z}$$
$$= \dfrac{29}{18z}$$

4. Add. Write the answer in lowest terms.
$$\dfrac{3x}{x^2-x-20}+\dfrac{5}{x^2-2x-15}$$

The LCD is $(x+3)(x+4)(x-5)$.
$$= \dfrac{3x}{(x+4)(x-5)}+\dfrac{5}{(x+3)(x-5)}$$
$$= \dfrac{3x(x+3)}{(x+3)(x+4)(x-5)}+\dfrac{5(x+4)}{(x+3)(x+4)(x-5)}$$
$$= \dfrac{3x(x+3)+5(x+4)}{(x+3)(x+4)(x-5)}$$
$$= \dfrac{3x^2+9x+5x+20}{(x+3)(x+4)(x-5)}$$
$$= \dfrac{3x^2+14x+20}{(x+3)(x+4)(x-5)}$$

4. Add. Write the answer in lowest terms.
$$\dfrac{7x}{x^2-2x-8}+\dfrac{4}{x^2-3x-4}$$

_____

Objective 2 Practice Exercises

For extra help, see Examples 2–5 on pages 417–419 of your text.

Add. Write each answer in lowest terms.

4. $\dfrac{7}{x-5}+\dfrac{4}{x+5}$ **4.** _____

Name: Date:
Instructor: Section:

5. $\dfrac{3z}{z^2-4}+\dfrac{4z-3}{z^2-4z+4}$

5. _____

6. $\dfrac{4z}{z^2+6z+8}+\dfrac{2z-1}{z^2+5z+6}$

6. _____

Objective 3 Subtract rational expressions.

Video Examples

Review these examples for Objective 3:

9. Subtract. Write the answer in lowest terms.

$\dfrac{7x}{x^2-4x+4}-\dfrac{2}{x^2-4}$

$=\dfrac{7x}{(x-2)^2}-\dfrac{2}{(x+2)(x-2)}$

The LCD is $(x+2)(x-2)^2$.

$=\dfrac{7x}{(x-2)^2}-\dfrac{2}{(x+2)(x-2)}$

$=\dfrac{7x(x+2)}{(x+2)(x-2)^2}-\dfrac{2(x-2)}{(x+2)(x-2)^2}$

$=\dfrac{7x(x+2)-2(x-2)}{(x+2)(x-2)^2}$

$=\dfrac{7x^2+14x-2x+4}{(x+2)(x-2)^2}$

$=\dfrac{7x^2+12x+4}{(x+2)(x-2)^2}$

Now Try:

9. Subtract. Write the answer in lowest terms.

$\dfrac{8x}{x^2-10x+25}-\dfrac{3}{x^2-25}$

_____

Name: Date:
Instructor: Section:

8. Subtract. Write the answer in lowest terms.

$$\frac{5x}{x-7} - \frac{x-42}{7-x}$$

The denominators are opposites.

$$\frac{5x}{x-7} - \frac{x-42}{7-x} = \frac{5x}{x-7} - \frac{(x-42)(-1)}{(7-x)(-1)}$$

$$= \frac{5x}{x-7} - \frac{-x+42}{x-7}$$

$$= \frac{5x+x-42}{x-7}$$

$$= \frac{6x-42}{x-7}$$

$$= \frac{6(x-7)}{x-7}$$

$$= 6$$

8. Subtract. Write the answer in lowest terms.

$$\frac{4x}{x-9} - \frac{3x-63}{9-x}$$

_____

Objective 3 Practice Exercises

For extra help, see Examples 6–10 on pages 420–422 of your text.

Subtract. Write each answer in lowest terms.

7. $\dfrac{z+2}{z-2} - \dfrac{z-2}{z+2}$ **7.** _____

8. $\dfrac{-4}{x^2-4} - \dfrac{3}{4-2x}$ **8.** _____

9. $\dfrac{m}{m^2-4} - \dfrac{1-m}{m^2+4m+4}$ **9.** _____

Name: Date:
Instructor: Section:

Chapter 6 RATIONAL EXPRESSIONS AND APPLICATIONS

6.5 Complex Fractions

Learning Objectives
1. Define and recognize a complex fraction.
2. Simplify a complex fraction by writing it as a division problem (Method 1).
3. Simplify a complex fraction by multiplying numerator and denominator by the least common denominator (Method 2).
4. Simplify rational expressions with negative exponents.

Key Terms

Use the vocabulary terms listed below to complete each statement in exercises 1–2.

    complex fraction   **LCD**

1. A _____ is a rational expression with one or more fractions in the numerator, denominator, or both.

2. To simplify a complex fraction, multiply the numerator and denominator by the _____ of all the fractions within the complex fraction.

Objective 1   Define and recognize a complex fraction.

Objective 2   Simplify a complex fraction by writing it as a division problem (Method 1).

Video Examples

Review these examples for Objective 2:
1. Simplify the complex fraction.

$$\frac{8+\frac{4}{x}}{\frac{x}{6}+\frac{1}{12}}$$

Step 1 Write the numerator as a single fraction.

$$8+\frac{4}{x}=\frac{8}{1}+\frac{4}{x}=\frac{8x}{x}+\frac{4}{x}=\frac{8x+4}{x}$$

Do the same with each denominator.

$$\frac{x}{6}+\frac{1}{12}=\frac{x(2)}{6(2)}+\frac{1}{12}=\frac{2x}{12}+\frac{1}{12}=\frac{2x+1}{12}$$

Step 2 Write the equivalent complex fraction as a division problem.

$$\frac{\frac{8x+4}{x}}{\frac{2x+1}{12}}=\frac{8x+4}{x}\div\frac{2x+1}{12}$$

Now Try:
1. Simplify the complex fraction.

$$\frac{9+\frac{3}{x}}{\frac{x}{10}+\frac{1}{30}}$$

_____

232   Copyright © 2016 Pearson Education, Inc.

Name:
Instructor:

Date:
Section:

Step 3 Divide by multiplying by the reciprocal.
$$\frac{8x+4}{x} \div \frac{2x+1}{12} = \frac{8x+4}{x} \cdot \frac{12}{2x+1}$$
$$= \frac{4(2x+1)}{x} \cdot \frac{12}{2x+1}$$
$$= \frac{48}{x}$$

2. Simplify the complex fraction.
$$\frac{\frac{rs^2}{t^3}}{\frac{s^3}{r^3 t}}$$

Use the definition of division and then the fundamental property.
$$= \frac{rs^2}{t^3} \div \frac{s^3}{r^3 t}$$
$$= \frac{rs^2}{t^3} \cdot \frac{r^3 t}{s^3}$$
$$= \frac{r^4}{st^2}$$

2. Simplify the complex fraction.
$$\frac{\frac{a^3 b^2}{c}}{\frac{a^5 b}{c^3}}$$

_____

3. Simplify the complex fraction.
$$\frac{\frac{30}{x+4} - 6}{\frac{4}{x+4} + 1} = \frac{\frac{30}{x+4} - \frac{6(x+4)}{x+4}}{\frac{4}{x+4} + \frac{1(x+4)}{x+4}}$$
$$= \frac{\frac{30 - 6(x+4)}{x+4}}{\frac{4 + 1(x+4)}{x+4}}$$
$$= \frac{\frac{30 - 6x - 24}{x+4}}{\frac{4 + x + 4}{x+4}}$$
$$= \frac{\frac{6 - 6x}{x+4}}{\frac{x+8}{x+4}}$$
$$= \frac{6 - 6x}{x+4} \cdot \frac{x+4}{x+8}$$
$$= \frac{6 - 6x}{x+8}$$

3. Simplify the complex fraction.
$$\frac{\frac{20}{x-5} - 9}{\frac{5}{x-5} + 2}$$

_____

Name: Date:
Instructor: Section:

Objective 2 Practice Exercises

For extra help, see Examples 1–3 on pages 426–427 of your text.

Simplify each complex fraction by writing it as a division problem.

1. $\dfrac{\dfrac{49m^3}{18n^5}}{\dfrac{21m}{27n^2}}$

1. _____

2. $\dfrac{\dfrac{p}{2}-\dfrac{1}{3}}{\dfrac{p}{3}+\dfrac{1}{6}}$

2. _____

3. $\dfrac{3+\dfrac{4}{s}}{2s+\dfrac{2}{3}}$

3. _____

Objective 3 Simplify a complex fraction by multiplying numerator and denominator by the least common denominator (Method 2).

Video Examples

Review these examples for Objective 3:
4. Simplify the complex fraction.

$$\dfrac{12+\dfrac{4}{x}}{\dfrac{x}{5}+\dfrac{1}{15}}$$

 Step 1 Find the LCD for all the denominators.
 The LCD for x, 5, and 15 is $15x$.
 Step 2 Multiply the numerator and denominator of the complex fraction by the LCD.

Now Try:
4. Simplify the complex fraction.

$$\dfrac{4+\dfrac{2}{x}}{\dfrac{x}{3}+\dfrac{1}{6}}$$

_____

Name: Date:
Instructor: Section:

$$\frac{12+\frac{4}{x}}{\frac{x}{5}+\frac{1}{15}} = \frac{15x\left(12+\frac{4}{x}\right)}{15x\left(\frac{x}{5}+\frac{1}{15}\right)}$$

$$= \frac{180x+60}{3x^2+x}$$

$$= \frac{60(3x+1)}{x(3x+1)}$$

$$= \frac{60}{x}$$

5. Simplify the complex fraction.
$$\frac{\frac{9}{7n}-\frac{3}{n^2}}{\frac{8}{3n}+\frac{5}{6n^2}}$$

The LCD is $42n^2$.

$$= \frac{42n^2\left(\frac{9}{7n}-\frac{3}{n^2}\right)}{42n^2\left(\frac{8}{3n}+\frac{5}{6n^2}\right)}$$

$$= \frac{42n^2\left(\frac{9}{7n}\right)-42n^2\left(\frac{3}{n^2}\right)}{42n^2\left(\frac{8}{3n}\right)+42n^2\left(\frac{5}{6n^2}\right)}$$

$$= \frac{54n-126}{112n+35}, \text{ or } \frac{18(3n-7)}{7(16n+5)}$$

5. Simplify the complex fraction.
$$\frac{\frac{2}{9n}-\frac{2}{5n^2}}{\frac{4}{5n}+\frac{2}{3n^2}}$$

_____

Objective 3 Practice Exercises

For extra help, see Examples 4–6 on pages 428–430 of your text.

Simplify each complex fraction by multiplying numerator and denominator by the least common denominator.

4. $\dfrac{\frac{9}{x^2}-1}{\frac{3}{x}-1}$

4. _____

Copyright © 2016 Pearson Education, Inc.

Name: Date:
Instructor: Section:

5. $\dfrac{\dfrac{x-2}{x+2}}{\dfrac{x}{x-2}}$

5. _____

6. $\dfrac{\dfrac{6}{k+1}-\dfrac{5}{k-3}}{\dfrac{3}{k-3}+\dfrac{2}{k+2}}$

6. _____

Objective 4 Simplify rational expressions with negative exponents.

Video Examples

Review this example for Objective 4:
7. Simplify the expression using only positive exponents in the answer.

$$\dfrac{3x^{-2}+5x^{-3}}{10x^{-1}+6}$$

Write with positive exponents. Use Method 2 to simplify.

$$\dfrac{3x^{-2}+5x^{-3}}{10x^{-1}+6} = \dfrac{\dfrac{3}{x^2}+\dfrac{5}{x^3}}{\dfrac{10}{x}+6}$$

$$= \dfrac{x^3\left(\dfrac{3}{x^2}+\dfrac{5}{x^3}\right)}{x^3\left(\dfrac{10}{x}+6\right)}$$

$$= \dfrac{x^3\cdot\dfrac{3}{x^2}+x^3\cdot\dfrac{5}{x^3}}{x^3\cdot\dfrac{10}{x}+x^3\cdot 6}$$

$$= \dfrac{3x+5}{10x^2+6x^3}$$

Now Try:
7. Simplify the expression using only positive exponents in the answer.

$$\dfrac{1+x^{-1}}{(x+y)^{-1}}$$

_____

Name: Date:
Instructor: Section:

Objective 4 Practice Exercises

For extra help, see Example 7 on page 431 of your text.

Simplify each expression using only positive exponents in the answer.

7. $\dfrac{2x^{-1} + y^2}{z^{-3}}$

7. _____

8. $\left(a^{-2} - b^{-2}\right)^{-1}$

8. _____

9. $\dfrac{4x^{-2}}{2 + 6y^{-3}}$

9. _____

Name: Date:
Instructor: Section:

Chapter 6 RATIONAL EXPRESSIONS AND APPLICATIONS

6.6 Solving Equations with Rational Expressions

Learning Objectives
1 Distinguish between operations with rational expressions and equations with terms that are rational expressions.
2 Solve equations with rational expressions.
3 Solve a formula for a specified variable.

Key Terms

Use the vocabulary terms listed below to complete each statement in exercises 1–2.

proposed solution **extraneous solution**

1. A solution that is not an actual solution of a given equation is called a(n) _____.

2. A value of the variable that appears to be a solution after both sides of an equation with rational expressions are multiplied by a variable expression is called a(n) _____.

Objective 1 Distinguish between operations with rational expressions and equations with terms that are rational expressions.

Video Examples

Review these examples for Objective 1:	Now Try:
1. Identify each of the following as an expression or an equation. Then simplify the expression or solve the equation.	1. Identify each of the following as an expression or an equation. Then simplify the expression or solve the equation.
a. $\frac{8}{9}x - \frac{5}{6}x = \frac{2}{3}$	a. $\frac{4}{5}x - \frac{3}{10}x = 7$
Because there is an equality symbol, this is an equation to be solved. The LCD is 18. $$18\left(\frac{8}{9}x - \frac{5}{6}x\right) = 18\left(\frac{2}{3}\right)$$ $$18\left(\frac{8}{9}x\right) - 18\left(\frac{5}{6}x\right) = 18\left(\frac{2}{3}\right)$$ $$16x - 15x = 12$$ $$x = 12$$ A check shows that the solution set is {12}.	

Name: Date:
Instructor: Section:

b. $\dfrac{8}{9}x - \dfrac{5}{6}x$

This is a difference of two terms. It represents an expression since there is no equality symbol.
The LCD is 18.

$\dfrac{8}{9}x - \dfrac{5}{6}x = \dfrac{2 \cdot 8}{2 \cdot 9}x - \dfrac{3 \cdot 5}{3 \cdot 6}x$

$= \dfrac{16}{18}x - \dfrac{15}{18}x$

$= \dfrac{1}{18}x$

b. $\dfrac{4}{5}x - \dfrac{3}{10}x$

_____

Objective 1 Practice Exercises

For extra help, see Example 1 on page 435 of your text.

Identify each of the following as an expression or an equation. Then simplify the expression or solve the equation.

1. $\dfrac{3x}{5} - \dfrac{4x}{3} = \dfrac{22}{15}$

1. _____

2. $\dfrac{4x}{5} - \dfrac{5x}{10}$

2. _____

3. $\dfrac{2x}{5} + \dfrac{7x}{3}$

3. _____

Objective 2 Solve equations with rational expressions.

Video Examples

Review this example for Objective 2:

4. Solve, and check the proposed solution.

$\dfrac{x}{x-3} - \dfrac{3}{x-3} = 3$

Note that x cannot equal 3, since 3 causes both denominators to equal 0.
Multiply by the LCD, $x - 3$.

Now Try:

4. Solve, and check the proposed solution.

$\dfrac{8x}{x+1} + \dfrac{8}{x+1} = 4$

_____

$$(x-3)\left(\frac{x}{x-3}-\frac{3}{x-3}\right)=(x-3)(3)$$

$$(x-3)\left(\frac{x}{x-3}\right)-(x-3)\left(\frac{3}{x-3}\right)=(x-3)(3)$$

$$x-3=3x-9$$
$$-2x-3=-9$$
$$-2x=-6$$
$$x=3$$

Check $\dfrac{x}{x-3}-\dfrac{3}{x-3}=3$

$$\frac{3}{3-3}-\frac{3}{3-3}\overset{?}{=}3$$

$$\frac{3}{0}-\frac{3}{0}\overset{?}{=}3$$

Division by 0 is undefined.

Thus, the proposed solution 3 must be rejected, and the solution set is ∅.

7. Solve, and check the proposed solution(s).

$$\frac{6}{x^2-1}=1-\frac{3}{x+1}$$

$x \neq 1, -1$ or a denominator is 0.
Factor the denominator.

$$\frac{6}{(x+1)(x-1)}=1-\frac{3}{x+1}$$

The LCD is $(x+1)(x-1)$.

$$(x+1)(x-1)\frac{6}{(x+1)(x-1)}$$
$$=(x+1)(x-1)\left(1-\frac{3}{x+1}\right)$$

$$(x+1)(x-1)\frac{6}{(x+1)(x-1)}$$
$$=(x+1)(x-1)-(x+1)(x-1)\frac{3}{x+1}$$

$$6=(x+1)(x-1)-3(x-1)$$
$$6=x^2-1-3x+3$$
$$0=x^2-3x-4$$
$$0=(x+1)(x-4)$$
$$x+1=0 \quad \text{or} \quad x-4=0$$
$$x=-1 \quad \text{or} \quad x=4$$

Since −1 makes the original denominator equal

7. Solve, and check the proposed solution(s).

$$\frac{x}{x+1}+\frac{4}{x}=\frac{4}{x^2+x}$$

———————

Name: Date:
Instructor: Section:

0, the proposed solution −1 is an extraneous value.

Check $\dfrac{6}{x^2-1} = 1 - \dfrac{3}{x+1}$

$\dfrac{6}{4^2-1} \stackrel{?}{=} 1 - \dfrac{3}{4+1}$

$\dfrac{2}{5} = \dfrac{2}{5}$ True

A check shows that {4} is the solution set.

Objective 2 Practice Exercises

For extra help, see Examples 2–8 on pages 436–441 of your text.

Solve each equation and check your solutions.

4. $\dfrac{4}{n+2} - \dfrac{2}{n} = \dfrac{1}{6}$ 4. _____

5. $\dfrac{x}{3x+16} = \dfrac{4}{x}$ 5. _____

6. $\dfrac{-16}{n^2-8n+12} = \dfrac{3}{n-2} + \dfrac{n}{n-6}$ 6. _____

Name: Date:
Instructor: Section:

Objective 3 Solve a formula for a specified variable.

Video Examples

Review these examples for Objective 3:
9. Solve each formula for the specified variable.

   a. $r = \dfrac{s+w}{v}$ for w

   Isolate w. Multiply by v.
   $$r = \dfrac{s+w}{v}$$
   $$rv = s + w$$
   $$rv - s = w$$

   Check $r = \dfrac{s+w}{v}$
   $$r \stackrel{?}{=} \dfrac{s + rv - s}{v}$$
   $$r \stackrel{?}{=} \dfrac{rv}{v}$$
   $$r = r \quad \text{True}$$

   b. $S = \dfrac{a_1}{1-r}$ for r

   Isolate r. Multiply by $1 - r$.
   $$S = \dfrac{a_1}{1-r}$$
   $$S(1-r) = a_1$$
   $$S - rS = a_1$$
   $$-rS = a_1 - S$$
   $$r = \dfrac{a_1 - S}{-S}, \text{ or } \dfrac{S - a_1}{S}$$

Now Try:
9. Solve each formula for the specified variable.

   a. $a = \dfrac{b-c}{q}$ for b

   _____

   b. $w = \dfrac{x}{y+z}$ for y

   _____

Name: Date:
Instructor: Section:

10. Solve the formula $\frac{1}{p}+\frac{1}{q}=\frac{1}{r}$ for p.

Isolate p, the specified variable. Multiply by the LCD, pqr.

$$pqr\left(\frac{1}{p}+\frac{1}{q}\right)=pqr\left(\frac{1}{r}\right)$$

$$pqr\left(\frac{1}{p}\right)+pqr\left(\frac{1}{q}\right)=pqr\left(\frac{1}{r}\right)$$

$$qr+pr=pq$$

$$qr=pq-pr$$

$$qr=p(q-r)$$

$$\frac{qr}{q-r}=p$$

10. Solve the formula $\frac{1}{x}=\frac{1}{y}+\frac{1}{z}$ for x.

_____

Objective 3 Practice Exercises

For extra help, see Examples 9–10 on pages 441–442 of your text.

Solve each formula for the specified variable.

7. $\frac{1}{f}=\frac{1}{d_0}+\frac{1}{d_1}$ for f

7. _____

8. $m=\frac{y_2-y_1}{x_2-x_1}$ for y_1

8. _____

9. $A=\frac{2pf}{b(q+1)}$ for q

9. _____

Name: Date:
Instructor: Section:

Chapter 6 RATIONAL EXPRESSIONS AND APPLICATIONS

6.7 Applications of Rational Expressions

Learning Objectives
1 Solve problems about numbers.
2 Solve problems about distance, rate, and time.
3 Solve problems about work.

Key Terms

Use the vocabulary terms listed below to complete each statement in exercises 1–3.

    reciprocal   **numerator**   **denominator**

1. In the fraction $\frac{x+5}{x-2}$, $x+5$ is the _____.

2. In the fraction $\frac{x+5}{x-2}$, $x-2$ is the _____.

3. The fraction $\frac{x+5}{x-2}$ is the _____ of the fraction $\frac{x-2}{x+5}$.

Objective 1   Solve problems about numbers.

Video Examples

Review this example for Objective 1:
1. If a certain number is added to the numerator and twice that number is subtracted from the denominator of the fraction $\frac{3}{5}$, the result is equal to 5. Find the number.

   Step 1 Read the problem carefully. We are trying to find a number.

   Step 2 Assign a variable.
       Let x = the number.

   Step 3 Write an equation. The fraction $\frac{3+x}{5-2x}$ represents adding the number to the numerator and twice that number is subtracted from the denominator of the fraction $\frac{3}{5}$. The result is equal to 5.
   $$\frac{3+x}{5-2x} = 5$$

Now Try:
1. If the same number is added to the numerator and denominator of the fraction $\frac{5}{9}$, the value of the resulting fraction is $\frac{2}{3}$. Find the number.

   _____

Name: Date:
Instructor: Section:

Step 4 Solve. Multiply by the LCD, $5-2x$.

$$(5-2x)\frac{3+x}{5-2x} = (5-2x)5$$
$$3+x = 25-10x$$
$$11x = 22$$
$$x = 2$$

Step 5 State the answer. The number is 2.

Step 6 Check the solution in the original problem. If 2 is added to the numerator, and twice 2 is subtracted from the denominator of $\frac{3}{5}$, the result is $\frac{3+2}{5-2(2)} = \frac{5}{1}$, or 5, as required.

Objective 1 Practice Exercises

For extra help, see Example 1 on page 448 of your text.

Solve each problem. Check your answers to be sure they are reasonable.

1. If three times a number is subtracted from twice its reciprocal, the result is –1. Find the number.

   1. _____

2. If two times a number is added to one-half of its reciprocal, the result is $\frac{13}{6}$. Find the number.

   2. _____

3. The denominator of a fraction is 1 less than twice the numerator. If the numerator and the denominator are each increased by 3, the resulting fraction simplifies to $\frac{3}{4}$. Find the original fraction.

   3. _____

Name: Date:
Instructor: Section:

Objective 2 Solve problems about distance, rate, and time.

Video Examples

Review this example for Objective 2:

2. A boat goes 6 miles per hour in still water. It takes as long to go 40 miles upstream as 80 miles downstream. Find the speed of the current.

   Step 1 Read the problem carefully. Find the speed of the current.

   Step 2 Assign a variable.
   Let x = the speed of the current.
   The rate of traveling upstream is $6 - x$.
   The rate of traveling downstream is $6 + x$.

	d	r	t
Upstream	40	$6-x$	$\dfrac{40}{6-x}$
Downstream	80	$6+x$	$\dfrac{80}{6+x}$

   The times are equal.

   Step 3 Write an equation.
   $$\frac{40}{6-x} = \frac{80}{6+x}$$

   Step 4 Solve. The LCD is $(6+x)(6-x)$.
   $$(6+x)(6-x)\frac{40}{6-x} = (6+x)(6-x)\frac{80}{6+x}$$
   $$40(6+x) = 80(6-x)$$
   $$240 + 40x = 480 - 80x$$
   $$240 + 120x = 480$$
   $$120x = 240$$
   $$x = 2$$

   Step 5 State the answer. The speed of the current is 2 miles per hour.

   Step 6 Check.
   Upstream: $\dfrac{40}{6-2} = 10$ hr
   Downstream: $\dfrac{80}{6+2} = 10$ hr
   The time upstream is the same as the time downstream, as required.

Now Try:

2. The Cuyahoga River has a current of 2 miles per hour. Ali can paddle 10 miles downstream in the time it takes her to paddle 2 miles upstream. How fast can Ali paddle?

   _____

Name: Date:
Instructor: Section:

Objective 2 Practice Exercises

For extra help, see Example 2 on pages 449–450 of your text.

Solve each problem.

4. A boat travels 15 miles per hour in still water. The boat travels 20 miles downstream in the same time it takes the boat to travel 10 miles upstream. How fast is the current?

4. _____

5. A ship goes 120 miles downriver in $2\frac{2}{3}$ hours less than it takes to go the same distance upriver. If the speed of the current is 6 miles per hour, find the speed of the ship.

5. _____

6. On Saturday, Pablo jogged 6 miles. On Monday, jogging at the same speed, it took him 30 minutes longer to cover 10 miles. How fast did Pablo jog?

6. _____

Name: Date:
Instructor: Section:

Objective 3 Solve problems about work.

Video Examples

Review this example for Objective 3:

3. Skip can paint a house in 8 hours and Phil can paint a house in 12 hours. How long will it take to paint a house if they work together?

Step 1 Read the problem carefully. Find the time working together.

Step 2 Assign a variable.
Let x = the number of hours working together.

The rate for Skip is $\frac{1}{8}$.

The rate for Phil is $\frac{1}{12}$.

Step 3 Write an equation. The sum of the fractional part for each multiplied by the time working together is the whole job.

$$\frac{1}{8}x + \frac{1}{12}x = 1$$

Step 4 Solve. The LCD is 24.

$$24\left(\frac{1}{8}x + \frac{1}{12}x\right) = 24(1)$$

$$24\left(\frac{1}{8}x\right) + 24\left(\frac{1}{12}x\right) = 24$$

$$3x + 2x = 24$$

$$5x = 24$$

$$x = \frac{24}{5} = 4\frac{4}{5}$$

Step 5 State the answer. Working together, it takes $4\frac{4}{5}$ hours to paint the house.

Step 6 Check. Substitute $\frac{24}{5}$ for x in the equation from Step 3.

$$\frac{1}{8}x + \frac{1}{12}x = 1$$

$$\frac{1}{8}\left(\frac{24}{5}\right) + \frac{1}{12}\left(\frac{24}{5}\right) = 1$$

$$\frac{3}{5} + \frac{2}{5} = 1 \quad \text{True}$$

Now Try:

3. Chuck can weed the garden in $\frac{1}{2}$ hour, but David takes 2 hours. How long does it take them to weed the garden if they work together?

_____

Name: Date:
Instructor: Section:

Objective 3 Practice Exercises

For extra help, see Example 3 on pages 450–451 of your text.

Solve each problem.

7. Kelly can clean the house in 6 hours, but it takes Linda 4 hours. How long would it take them to clean the house if they worked together?

7. _____

8. Michael can type twice as fast as Sharon. Together they can type a certain job in 2 hours. How long would it take Michael to type the entire job by himself?

8. _____

Name: Date:
Instructor: Section:

Chapter 7 GRAPHS, LINEAR EQUATIONS, AND SYSTEMS

7.1   Review of Graphs and Slopes of Lines

Learning Objectives
1. Plot ordered pairs.
2. Graph lines and find intercepts.
3. Recognize equations of horizontal and vertical lines.
4. Use the midpoint formula.
5. Find the slope of a line.
6. Graph a line given its slope and a point on the line.
7. Use slopes to determine whether two lines are parallel, perpendicular, or neither.
8. Solve problems involving average rate of change.

Key Terms

Use the vocabulary terms listed below to complete each statement in exercises 1−17.

>   ordered pair   origin   x-axis   y-axis
>
>   rectangular (Cartesian) coordinate system   plot
>
>   components   coordinate   quadrant   graph of an equation
>
>   first-degree equation   linear equation in two variables
>
>   y-intercept   x-intercept   rise   run   slope

1. If a graph intersects the y-axis at k, then the _____ is (0, k).

2. An equation that can be written in the form $Ax + By = C$, where A, B, and C are real numbers and A, B ≠ 0, is called a _____.

3. Each number in an ordered pair represents a _____ of the corresponding point.

4. The axis lines in a coordinate system intersect at the _____.

5. If a graph intersects the x-axis at k, then the _____ is (k, 0).

6. In a rectangular coordinate system, the horizontal number line is called the _____.

7. A _____ is one of the four regions in the plane determined by a rectangular coordinate system.

8. A pair of numbers written between parentheses in which order is important is called a(n) _____.

9. In a rectangular coordinate system, the vertical number line is called the _____.

Name: Date:
Instructor: Section:

10. The two numbers in an ordered pair are the _____ of the ordered pair.

11. To _____ an ordered pair is to locate the corresponding point on a coordinate system.

12. Together, the *x*-axis and the *y*-axis form a _____.

13. The _____ of a line is the ratio of the change in *y* compared to the change in *x* when moving along the line from one point to another.

14. The _____ is the set of points corresponding to all ordered pairs that satisfy the equation.

15. A _____ has no term with a variable to a power greater than one.

16. The vertical change between two different points on a line is called the _____.

17. The horizontal change between two different points on a line is called the _____.

Objective 1 Plot ordered pairs.

For extra help, see page 468 of your text.

Objective 2 Graph lines and find intercepts.

Video Examples

Review these examples for Objective 2:
1. Find the *x*- and *y*-intercepts of $3x + y = 6$ and graph the equation.

    To find the *y*-intercept, let $x = 0$.
    To find the *x*-intercept, let $y = 0$.

    $$\begin{array}{c|c} 3(0)+y=6 & 3x+0=6 \\ 0+y=6 & 3x=6 \\ y=6 & x=2 \end{array}$$

    The intercepts are (0, 6) and (2, 0). To find a third point, as a check, we let $x = 1$.

    $$3(1)+y=6$$
    $$3+y=6$$
    $$y=3$$

    This gives the ordered pair (1, 3).

Now Try:
1. Find the *x*- and *y*-intercepts of $5x - 2y = -10$ and graph the equation.

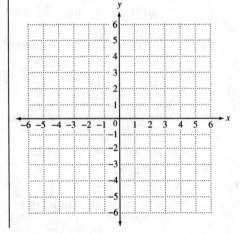

Name: Date:
Instructor: Section:

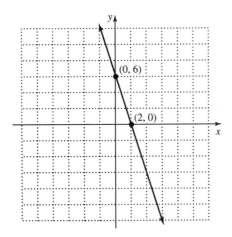

2. Graph $x + 5y = 0$.

To find the y-intercept, let $x = 0$.
To find the x-intercept, let $y = 0$.

$$\begin{array}{c|c} 0 + 5y = 0 & x + 5(0) = 0 \\ 5y = 0 & x + 0 = 0 \\ y = 0 & x = 0 \end{array}$$

The x- and y-intercepts are the same point $(0, 0)$. We must select two other values for x or y to find two other points. We choose $y = 1$ and $y = -1$.

$$\begin{array}{c|c} x + 5(1) = 0 & x + 5(-1) = 0 \\ x + 5 = 0 & x - 5 = 0 \\ x = -5 & x = 5 \end{array}$$

We use $(-5, 1)$, $(0, 0)$, and $(5, -1)$ to draw the graph.

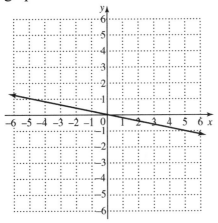

2. Graph $3x - y = 0$.

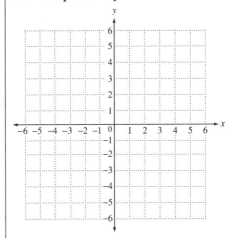

Copyright © 2016 Pearson Education, Inc.

Name: Date:
Instructor: Section:

Objective 2 Practice Exercises

For extra help, see Examples 1–2 on pages 470–471 of your text.

Find the intercepts, then graph the equation.

1. $4x - y = 4$

1. _____

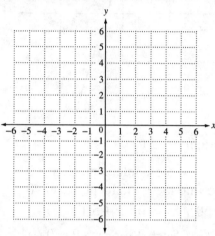

2. $2x - 3y = 6$

2. _____

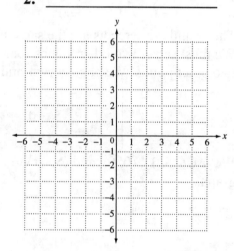

254   Copyright © 2016 Pearson Education, Inc.

Name: Date:
Instructor: Section:

Objective 3 Recognize equations of horizontal and vertical lines.

Video Examples

Review these examples for Objective 3:
3. Graph each equation.

   a. $y = -2$

   For any value of x, y is always –2. Three ordered pairs that satisfy the equation are (–4, –2), (0, –2) and (2, –2). Drawing a line through these points gives the horizontal line. The y-intercept is (0, –2). There is no x-intercept.

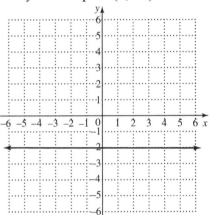

Now Try:
3. Graph each equation.

   a. $y = 4$

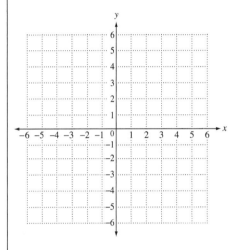

b. $x + 4 = 0$

First we subtract 4 from each side of the equation to get the equivalent equation $x = -4$. All ordered-pair solutions of this equation have x-coordinate –4.
Three ordered pairs that satisfy the equation are (–4, –1), (–4, 0), and (–4, 3). The graph is a vertical line. The x-intercept is (–4, 0). There is no y-intercept.

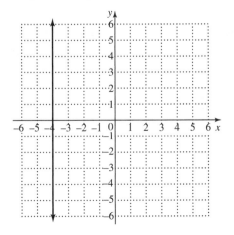

b. $x = 0$

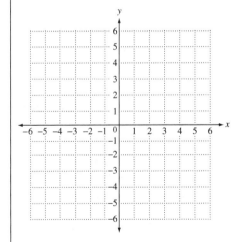

Copyright © 2016 Pearson Education, Inc.

Name: Date:
Instructor: Section:

Objective 3 Practice Exercises

For extra help, see Example 3 on page 471 of your text.

Find the intercepts, and graph the line.

3. $x - 1 = 0$

3.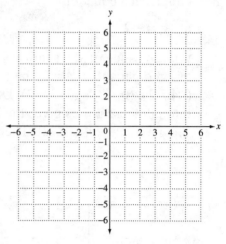

4. $y + 3 = 0$

4.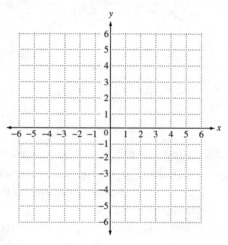

Objective 4 Use the midpoint formula.

Video Examples

Review this example for Objective 4:

4. Find the coordinates of the midpoint of the line segment PQ with endpoints $P(8, -5)$ and $Q(4, -3)$.

$P(8, -5) = (x_1, y_1)$ and $Q(4, -3) = (x_2, y_2)$

$$\left(\frac{x_1 + x_2}{2}, \frac{y_1 + y_2}{2}\right) = \left(\frac{8+4}{2}, \frac{-5+(-3)}{2}\right)$$

$$= \left(\frac{12}{2}, \frac{-8}{2}\right)$$

$$= (6, -4)$$

The midpoint of PQ is $(6, -4)$.

Now Try:

4. Find the coordinates of the midpoint of the line segment PQ with endpoints $P(7, -6)$ and $Q(3, 2)$.

_____

Name: Date:
Instructor: Section:

Objective 4 Practice Exercises

For extra help, see Example 4 on page 472 of your text.

Find the midpoint of each segment with the given endpoints.

5. (–4, 8) and (8, –4) 5. _____

6. (8, 5) and (–3, –11) 6. _____

7. (–2.2, –9.3) and (–8.4, 5.7) 7. _____

Objective 5 Find the slope of a line.

Video Examples

Review this example for Objective 5:

5. Find the slope of the line passing through (–5, 4) and (2, –6)

Apply the slope formula.
$(x_1, y_1) = (-5, 4)$ and $(x_2, y_2) = (2, -6)$

$$\text{slope } m = \frac{y_2 - y_1}{x_2 - x_1} = \frac{-6 - 4}{2 - (-5)}$$

$$= \frac{-10}{7}, \text{ or } -\frac{10}{7}$$

Now Try:

5. Find the slope of the line passing through (–6, 7) and (3, –9)

_____

Objective 5 Practice Exercises

For extra help, see Examples 5–7 on pages 474–475 of your text.

Find the slope of the line through the given points.

8. (4, 3) and (3, 5) 8. _____

9. (5, –2) and (2, 7) 9. _____

10. (7, 2) and (–7, 3) 10. _____

Name: Date:
Instructor: Section:

Objective 6   Graph a line, given its slope and a point on the line.

Video Examples

Review this example for Objective 6:

8. Graph the line passing through the point $(1,-3)$, with slope $-\frac{5}{2}$.

   First, locate the point $(1, -3)$. Then write the slope $-\frac{5}{2}$ as

   $$\text{slope } m = \frac{\text{change in } y \text{ (rise)}}{\text{change in } x \text{ (run)}} = \frac{5}{-2}.$$

   Locate another point on the line by counting up 5 units from $(1,-3)$, and then to the left 2 units. Finally, draw the line through this new point, $(-1, 2)$.

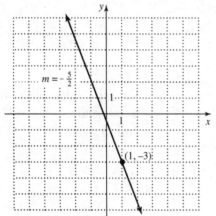

Now Try:

8. Graph the line passing through the point $(2, 2)$, with slope $\frac{1}{3}$.

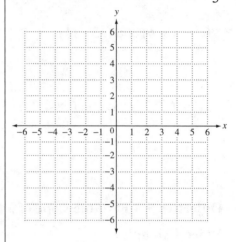

258   Copyright © 2016 Pearson Education, Inc.

Name: Date:
Instructor: Section:

Objective 6 Practice Exercises

For extra help, see Example 8 on page 476 of your text.

Graph the line passing through the given point and having the given slope.

10. $(4, -2)$; $m = -1$ 10.

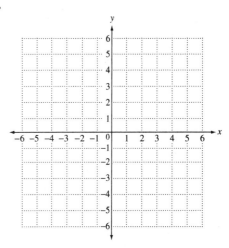

11. $(-3, -2)$; $m = \frac{2}{3}$ 11.

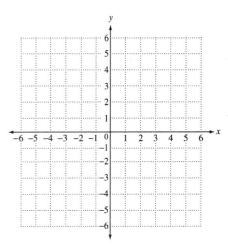

12. $(-3, -1)$; undefined slope   12.

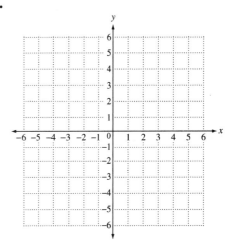

Name: Date:
Instructor: Section:

Objective 7 Use slopes to determine whether two lines are parallel, perpendicular, or neither.

Video Examples

Review these examples for Objective 7:

9a. Determine whether the lines L_1 passing through (7, 6) and (–4, 2) and L_2 passing through (0, –4) and (11, 0), are parallel.

Slope of L_1: $m_1 = \dfrac{2-6}{-4-7} = \dfrac{-4}{-11} = \dfrac{4}{11}$

Slope of L_2: $m_2 = \dfrac{0-(-4)}{11-0} = \dfrac{4}{11}$

The slopes are equal, the two lines are parallel.

9b. Are the lines with equations $x + 3y = 8$ and $-3x + y = 5$ perpendicular?

Find the slope of each line by first solving each equation for y.

$3y = -x + 8$ | $y = 3x + 5$
$y = -\dfrac{1}{3}x + \dfrac{8}{3}$ |
The slope is $-\dfrac{1}{3}$. | The slope is 3.

Check the product of the slopes: $-\dfrac{1}{3}(3) = -1$.

The two lines are perpendicular because the product of their slopes is –1.

Now Try:

9a. Determine whether the lines L_1 passing through (6, 3) and (–4, 5) and L_2 passing through (0, 3) and (15, 0), are parallel.

_____

9b. Are the lines with equations $9x - y = 7$ and $x + 9y = 11$ perpendicular?

_____

Objective 7 Practice Exercises

For extra help, see Example 9 on pages 477–478 of your text.

*Decide whether the lines in each pair are **parallel**, **perpendicular**, or **neither**.*

13. $y = -5x - 2$
    $y = 5x + 11$

13. _____

14. $-x + y = -7$
    $x - y = -3$

14. _____

Name: Date:
Instructor: Section:

15. $2x + 2y = 7$
     $2x - 2y = 5$

15. _____

Objective 8 Solve problems involving average rate of change.

Video Examples

Review these examples for Objective 8:

10. A small company had the following sales during their first three years of operation.

Year	Sales
2005	$82,250
2006	$89,790
2007	$96,100

    a. What was the rate of change in 2005–2006?
    b. What was the rate of change in 2006–2007?
    c. What was the rate of change in 2005–2007?

    a. We use the ordered pairs (2005, 82,250) and (2006, 89,790).

    $$\text{average rate of change} = \frac{89,790 - 82,250}{2006 - 2005}$$
    $$= \frac{7540}{1} = 7540$$

    This means sales increased by an average of $7540 from 2005 to 2006.

    b. We use the ordered pairs (2006, 89,790) and (2007, 96,100).

    $$\text{average rate of change} = \frac{96,100 - 89,790}{2007 - 2006}$$
    $$= \frac{6310}{1} = 6310$$

    This means sales increased by an average of $6310 from 2006 to 2007.

    c. We use the ordered pairs (2005, 82,250) and (2007, 96,100).

    $$\text{average rate of change} = \frac{96,100 - 82,250}{2007 - 2005}$$
    $$= \frac{13,850}{2} = 6925$$

    This means sales increased by an average of $6925 from 2005 to 2007.

Now Try:

10. A plane had an altitude of 8500 feet at 4:02 P.M. and 12,700 feet at 4:39 P.M. What was the average rate of change in the altitude in feet per minute?

_____

Name: Date:
Instructor: Section:

11. Enrollment in a college was 11,500 two years ago, 10,975 last year, and 10,800 this year. What is the average rate of change in enrollment per year for this 3-year period?

We use the ordered pairs (1, 11,500) and (3, 10,800).

$$\text{average rate of change} = \frac{10,800 - 11,500}{3 - 1}$$
$$= \frac{-700}{2}$$
$$= -350$$

The enrollment decreases at a rate of 350 students per year.

11. A company had 44 employees during the first year of operation. During their eighth year, the company had 79 employees. What was the average rate of change in the number of employees per year?

_____

Objective 8 Practice Exercises

For extra help, see Examples 10–11 on pages 478–479 of your text.

Solve each problem.

16. Suppose in 2005, the sales of a company were $1,625,000. In 2010, the company had sales of $2,250,000. Find the average rate of change in the sales per year.

16. _____

17. A state had a population of 755,000 in 2000 and a population of 809,000 in 2012. Find the average rate of change in population per year.

17. _____

18. Suppose a man's salary was $45,750 in 1995 and $60,000 in 2010. Find the average rate of change in the salary per year.

18. _____

Name: Date:
Instructor: Section:

Chapter 7 GRAPHS, LINEAR EQUATIONS, AND SYSTEMS

7.2   Review of Equations of Lines; Linear Models

Learning Objectives
1   Write an equation of a line, given its slope and *y*-intercept.
2   Graph a line, using its slope and *y*-intercept.
3   Write an equation of a line, given its slope and a point on the line.
4   Write an equation of a line, given two points on the line.
5   Write equations of horizontal and vertical lines.
6   Write an equation of a line parallel or perpendicular to a given line.
7   Write an equation of a line that models real data.

Key Terms

Use the vocabulary terms listed below to complete each statement in exercises 1–3.

   slope-intercept form   **point-slope form**   **standard form**

1. A linear equation in the form $y - y_1 = m(x - x_1)$ is written in
   _____.

2. A linear equation in the form $Ax + By = C$ is written in
   _____.

3. A linear equation in the form $y = mx + b$ is written in
   _____.

Objective 1   Write an equation of a line given its slope and *y*-intercept.

Video Examples

Review this example for Objective 1:	Now Try:
1. Write an equation of the line with slope $\frac{5}{7}$ and *y*-intercept $(0, -6)$. Here, $m = \frac{5}{7}$ and $b = -6$, so we can write the following equation. $y = mx + b$ $y = \frac{5}{7}x + (-6)$, or $y = \frac{5}{7}x - 6$	1. Write an equation of the line with slope $\frac{7}{9}$ and *y*-intercept $(0, 8)$. _____

Name: Date:
Instructor: Section:

Objective 1 Practice Exercises

For extra help, see Example 1 on page 485 of your text.

Write the slope-intercept form equation of the line with the given slope and y-intercept.

1. $m = \dfrac{3}{2};\ b = -\dfrac{2}{3}$

   1. _____

2. $m = -7;\ b = -2$

   2. _____

3. Slope: $-\dfrac{6}{5}$; y-intercept $\left(0,\ \dfrac{2}{5}\right)$

   3. _____

Objective 2 **Graph a line, using its slope and y-intercept.**

Video Examples

Review this example for Objective 2:

2. Graph the equation by using the slope and y-intercept.

   $2x - 3y = 6$

   Solve for y to write the equation in slope-intercept form.
   $$2x - 3y = 6$$
   $$-3y = -2x + 6$$
   $$y = \dfrac{2}{3}x - 2$$

   The y-intercept is (0,–2). Graph this point.

   The slope is $\dfrac{2}{3}$. By definition,

   $$\text{slope } m = \dfrac{\text{change in } y \text{ (rise)}}{\text{change in } x \text{ (run)}} = \dfrac{2}{3}$$

   From the y-intercept, count up 2 units and to the right 3 units to obtain the point (3, 0).

   Draw the line through the points (0,–2) and (3, 0) to obtain the graph.

Now Try:

2. Graph the equation by using the slope and y-intercept.

   $2x - 3y = 0$

   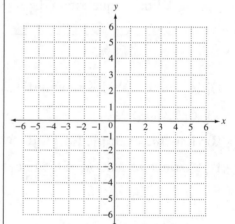

Name: Date:
Instructor: Section:

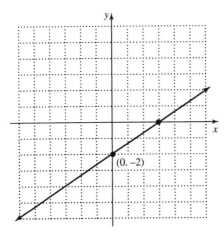

Objective 2 Practice Exercises

For extra help, see Example 2 on page 486 of your text.

Graph each equation by using the slope and y-intercept.

4. $4x - y = 4$

4.
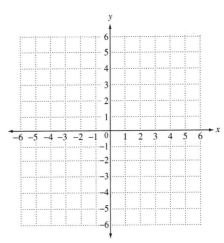

5. $y = -3x + 6$

5.
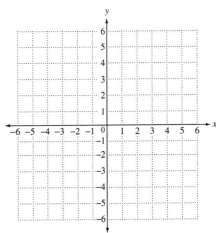

Name:
Instructor:
Date:
Section:

Graph the line passing through the given point and having the given slope.

6. $(-2,-2);\ m=0$

6.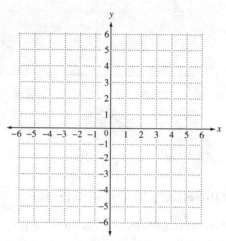

Objective 3   Write an equation of a line, given its slope and a point on the line.

Video Examples

Review this example for Objective 3:

3. Write an equation of the line with slope $\frac{2}{3}$ passing through the point $(4,-7)$.

 Method 1 Use point-slope form, with $(x_1, y_1)=(4,-7)$ and $m=\frac{2}{3}$.

$$y - y_1 = m(x - x_1)$$
$$y-(-7)=\frac{2}{3}(x-4)$$
$$y+7=\frac{2}{3}(x-4)$$
$$3y+21=2x-8$$
$$3y=2x-29$$
$$y=\frac{2}{3}x-\frac{29}{3}$$

 Method 2 Use slope-intercept form, with $(x_1, y_1)=(4,-7)$ and $m=\frac{2}{3}$.

$$y=mx+b$$
$$-7=\frac{2}{3}(4)+b$$
$$-7=\frac{8}{3}+b$$
$$-\frac{29}{3}=b,\ \text{or}\ b=-\frac{29}{3}$$

Now Try:

3. Write an equation of the line with slope $\frac{4}{5}$ passing through the point $(6,-4)$.

_____

Name: Date:
Instructor: Section:

Knowing $m = \frac{2}{3}$ and $b = -\frac{29}{3}$ gives the equation $y = \frac{2}{3}x - \frac{29}{3}$, same as Method 1.

Objective 3 Practice Exercises

For extra help, see Example 3 on page 487 of your text.

Write the equation in standard form of the line satisfying the given conditions.

7. $(-3, 4)$; $m = -\frac{3}{5}$

7. _____

8. $(-4, -7)$; $m = \frac{4}{3}$

8. _____

9. $(-1, 2)$; $m = \frac{2}{3}$

9. _____

Objective 4 Write an equation of a line, given two points on the line.

Video Examples

Review this example for Objective 4:
4. Write the equation of the line passing through the point (6, 8) and (−3, 5). Give the final answer in slope-intercept form and then in standard form.

First, find the slope of the line.
$(x_1, y_1) = (6, 8)$ and $(x_2, y_2) = (-3, 5)$

slope $m = \frac{y_2 - y_1}{x_2 - x_1} = \frac{5 - 8}{-3 - 6} = \frac{-3}{-9} = \frac{1}{3}$

Now Try:
4. Write the equation of the line passing through the point (7, 15) and (15, 9). Give the final answer in slope-intercept form and then in standard form.

_____

Name: Date:
Instructor: Section:

Now use (x_1, y_1), here $(6, 8)$ and point-slope form.

$$y - y_1 = m(x - x_1)$$
$$y - 8 = \frac{1}{3}(x - 6)$$
$$y - 8 = \frac{1}{3}x - 2$$
$$y = \frac{1}{3}x + 6 \quad \text{Slope-intercept form}$$
$$3y = x + 18$$
$$-x + 3y = 18$$
$$x - 3y = -18 \quad \text{Standard form}$$

Objective 4 Practice Exercises

For extra help, see Example 4 on page 488 of your text.

Write the equation in standard form of the line through the given points.

10. $(3, 7), (5, 4)$ 10. _____

11. $(2, -1), (5, -2)$ 11. _____

12. $(-1, -4), (-2, -3)$ 12. _____

Objective 5 Write equations of horizontal and vertical lines.

Video Examples

Review these examples for Objective 5:
5. Write an equation of the line passing through the point $(2, -2)$ that satisfies the given condition.

  a. The line has slope 0.

  Since the slope is 0, this is a horizontal line.
  $y = -2$.

Now Try:
5. Write an equation of the line passing through the point $(-5, 5)$ that satisfies the given condition.

  a. The line has slope 0.

  _____

Name: Date:
Instructor: Section:

b. The line has undefined slope.

This is a vertical line, since the slope is undefined.
$x = 2$

b. The line has undefined slope.

_____

Objective 5 Practice Exercises

For extra help, see Example 5 on page 489 of your text.

Write the equation in standard form of the line through the given points.

13. $(-1, -7), (-1, 8)$ 13. _____

14. $(0, 2), (0, -6)$ 14. _____

15. $(4, -5), (8, -5)$ 15. _____

Objective 6 Write an equation of a line parallel or perpendicular to a given line.

Video Examples

Review these examples for Objective 6:

6. Write an equation in slope-intercept form of the line passing through the point $(-4, 5)$ that satisfies the given condition.

   a. The line is parallel to $5x + 2y = 10$.

   First, find the slope of the given line.
   $$5x + 2y = 10$$
   $$2y = -5x + 10$$
   $$y = -\frac{5}{2}x + 5$$
   The slope is $-\frac{5}{2}$.

   Use point-slope form with $(x_1, y_1) = (-4, 5)$ and $m = -\frac{5}{2}$.

Now Try:

6. Write an equation in slope-intercept form of the line passing through the point $(-6, 8)$ that satisfies the given condition.

   a. The line is parallel to $3x + 4y = 12$.

   _____

Name: Date:
Instructor: Section:

$$y - y_1 = m(x - x_1)$$
$$y - 5 = -\frac{5}{2}(x - (-4))$$
$$y - 5 = -\frac{5}{2}(x + 4)$$
$$y - 5 = -\frac{5}{2}x - 10$$
$$y = -\frac{5}{2}x - 5$$

b. The line is perpendicular to $5x + 2y = 10$.

From part (a), the line in slope-intercept form is $y = -\frac{5}{2}x + 5$.

The line perpendicular to this line must have slope $\frac{2}{5}$, the negative reciprocal of $-\frac{5}{2}$.

Use point-slope form with $(x_1, y_1) = (-4, 5)$ and $m = \frac{2}{5}$.

$$y - y_1 = m(x - x_1)$$
$$y - 5 = \frac{2}{5}(x - (-4))$$
$$y - 5 = \frac{2}{5}(x + 4)$$
$$y - 5 = \frac{2}{5}x + \frac{8}{5}$$
$$y = \frac{2}{5}x + \frac{33}{5}$$

b. The line is perpendicular to $3x + 4y = 12$.

_____

Objective 6 Practice Exercises

For extra help, see Example 6 on pages 489–490 of your text.

Write the equation in standard form of the line satisfying the given conditions.

16. parallel to $2x + 3y = -12$, through $(9, -3)$ 16. _____

17. parallel to $4x - 3y = 8$, through $(-2, 3)$.   17. _____

18. perpendicular to $x - 3y = 0$, through $(-10, 2)$ 18. _____

Name: Date:
Instructor: Section:

Objective 7 Write an equation of a line that models real data.

Video Examples

Review these examples for Objective 7:

7. The table and scatter graph shows the number of internet users in the world from 1998 to 2005, where year 0 represents 1998.

Year	Number of Internet Users (millions)
0	147
2	361
4	587
6	817
8	1093

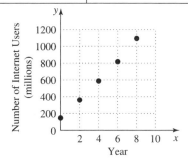

a. Find an equation that models the data.

The points appear to lie approximately in a straight line. y represents the number of internet users in year x. To find an equation of the line, we choose the ordered pairs (0, 147) and (8, 1093) from the table and find the slope of the line through these points.

$(x_1, y_1) = (0, 147)$ and $(x_2, y_2) = (8, 1093)$

$$\text{slope } m = \frac{y_2 - y_1}{x_2 - x_1} = \frac{1093 - 147}{8 - 0} = \frac{946}{8}$$

$$= 118.25$$

Now Try:

7. The table and scatter graph shows the average annual telephone expenditures for residential and pay telephones from 2001 to 2006, where year 0 represents 2001.

Year	Annual Telephone Expenditures
0	$686
2	$620
3	$592
4	$570
5	$542

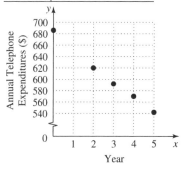

a. Find an equation that models the data.

_____

Copyright © 2016 Pearson Education, Inc.

Name: Date:
Instructor: Section:

Use the slope, 118.25, and the point (0, 147) in slope-intercept form.
$$y = mx + b$$
$$147 = 118.25(0) + b$$
$$147 = b$$
Thus, $m = 118.25$ and $b = 147$, so the equation of the line is $y = 118.25x + 147$.

b. Find and interpret the ordered pair associated with the equation for $x = 5$.

If $x = 5$, then
$$y = 118.25(5) + 147$$
$$= 591.25 + 147$$
$$= 738.25$$
In 2003, there were 738.25 million internet users.

b. Find the ordered pair associated with the equation for $x = 1$.

_____

Objective 7 Practice Exercises

For extra help, see Examples 7–8 on pages 491–492 of your text.

Solve each problem.

19. To run a newspaper ad, there is a $25 set up fee plus a charge of $1.25 per line of type in the ad. Let x represent the number of lines in the ad so that y represents the total cost of the ad (in dollars).
    a. Write an equation in the form $y = mx + b$.
    b. Give three ordered pairs associated with the equation for x-values 0, 5, and 10.

19.

a. _____

b. _____

Name: _____ Date: _____
Instructor: _____ Section: _____

20. The table and scatter graph shows the U.S. municipal solid waste recycling percent since 1985, where year 0 represents 1985.
   a. Find an equation that models the data.
   b. Use the equation from part (a) to predict the percent of municipal solid waste recycling in the year 2015.

20.

a. _____

b. _____

Year	Recycling Percent
0	10.1
5	16.2
10	26.0
15	29.1
20	32.5

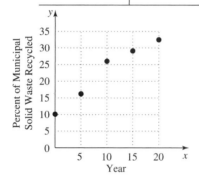

Name: Date:
Instructor: Section:

Chapter 7 GRAPHS, LINEAR EQUATIONS, AND SYSTEMS

7.3 Solving Systems of Linear Equations by Graphing

Learning Objectives
1. Decide whether a given ordered pair is a solution of a system.
2. Solve linear systems by graphing.
3. Solve special systems by graphing.
4. Identify special systems without graphing.

Key Terms

Use the vocabulary terms listed below to complete each statement in exercises 1–7.

system of linear equations   **solution of the system**

solution set of the system   **consistent system**

inconsistent system   **independent equations**

dependent equations

1. Equations of a system that have different graphs are called _____.

2. A system of equations with at least one solution is a _____.

3. The set of all ordered pairs that are solutions of a system is the _____.

4. The _____ of linear equations is an ordered pair that makes all the equations of the system true at the same time.

5. Equations of a system that have the same graph (because they are different forms of the same equation) are called _____.

6. A system with no solution is called a(n) _____.

7. A(n) _____ consists of two or more linear equations with the same variables.

Name: Date:
Instructor: Section:

Objective 1   Decide whether a given ordered pair is a solution of a system.

Video Examples

Review this example for Objective 1:

1. Determine whether the ordered pair (5, –2) is a solution of the system.

$$4x + 5y = 10$$
$$3x + 8y = 6$$

Again, substitute 5 for x and –2 for y in each equation.

$$4x + 5y = 10 \qquad\qquad 3x + 8y = 6$$
$$4(5) + 5(-2) \stackrel{?}{=} 10 \qquad 3(5) + 8(-2) \stackrel{?}{=} 6$$
$$20 - 10 \stackrel{?}{=} 10 \qquad\qquad 15 - 16 \stackrel{?}{=} 6$$
$$10 = 10 \text{ True} \qquad \text{False} \;\; -1 = 6$$

The ordered pair (5, –2) is not a solution of this system because it does not satisfy the second equation.

Now Try:

1. Determine whether the ordered pair (6, 5) is a solution of the system.
$$5x - 6y = 0$$
$$6x + 5y = 50$$

1. _____

Objective 1 Practice Exercises

For extra help, see Example 1 on page 498 of your text.

Decide whether the given ordered pair is a solution of the given system.

1. $(2, -4)$
   $$2x + 3y = 6$$
   $$3x - 2y = 14$$

   1. _____

2. $(-3, -1)$
   $$5x - 3y = -12$$
   $$2x + 3y = -9$$

   2. _____

3. $(4, 0)$
   $$4x + 3y = 16$$
   $$x - 4y = -4$$

   3. _____

Name: Date:
Instructor: Section:

Objective 2 Solve linear systems by graphing.

Video Examples

Review this example for Objective 2:

2. Solve the system of equations by graphing both equations on the same axes.

$$6x - 5y = 4$$
$$2x - 5y = 8$$

Graph these equations by plotting several points for each line. To find the *x*-intercept, let $y = 0$. To find the *y*-intercept, let $x = 0$.
The tables show the intercepts and a check point for each graph.

$6x - 5y = 4$

x	y
0	$-\frac{4}{5}$
$\frac{2}{3}$	0
4	4

$2x - 5y = 8$

x	y
0	$\frac{8}{5}$
4	0
-6	-4

Now Try:

2. Solve the system of equations by graphing both equations on the same axes.

$$3x - y = -7$$
$$2x + y = -3$$

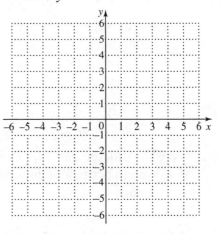

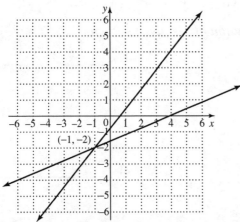

The lines suggest that the graphs intersect at the point $(-1, -2)$. We check by substituting -1 for x and -2 for y in both equations.

$$6x - 5y = 4 \qquad\qquad 2x - 5y = 8$$
$$6(-1) - 5(-2) \stackrel{?}{=} 4 \qquad 2(-1) - 5(-2) \stackrel{?}{=} 8$$
$$-6 + 10 \stackrel{?}{=} 4 \qquad\qquad -2 + 10 \stackrel{?}{=} 8$$
$$4 = 4 \text{ True} \qquad\qquad \text{True } 8 = 8$$

Because $(-1, -2)$ satisfies both equations, the solution set of this system is $\{(-1, -2)\}$.

Name: Date:
Instructor: Section:

Objective 2 Practice Exercises

For extra help, see Example 2 on page 499 of your text.

Solve each system by graphing both equations on the same axes.

4. $x - 2y = 6$
   $2x + y = 2$

4. _____

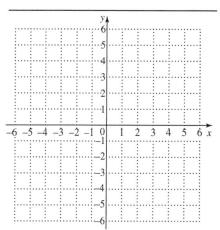

5. $2x = y$
   $5x + 3y = 0$

5. _____

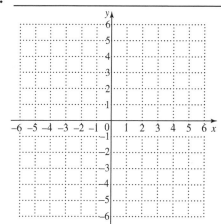

6. $3x + 2 = y$
   $2x - y = 0$

6. _____

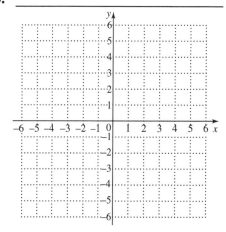

Name:
Instructor:
Date:
Section:

Objective 3   Solve special systems by graphing.

Video Examples

Review these examples for Objective 3:

3. Solve each system by graphing.

   a. $x - y = 1$

   $x - y = -1$

   The graphs of these two equations are parallel and have no points in common. There is no solution for this system. It is inconsistent. The solution set is $\varnothing$.

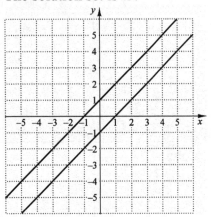

Now Try:

3. Solve each system by graphing.

   a. $x - 3y = 6$

   $x - 3y = 4$

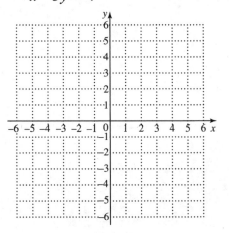

_____

b. $3x - y = 0$

$2y = 6x$

The graphs of these two equations are the same line.

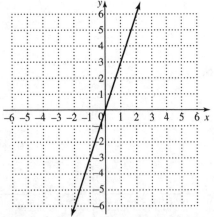

In this case, every point on the line is a solution of the system, and the solution set contains an infinite number of ordered pairs, each of which satisfies both equations of the system. We write the solution set as

$\{(x, y) \mid 3x - y = 0\}$.

b. $4x - 2y = 8$

$6x - 3y = 12$

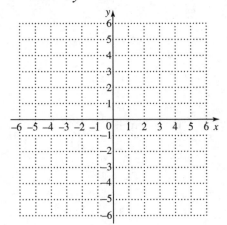

_____

278   Copyright © 2016 Pearson Education, Inc.

Name: Date:
Instructor: Section:

Objective 3 Practice Exercises

For extra help, see Example 3 on page 500 of your text.

*Solve each system of equations by graphing both equations on the same axes. If the two equations produce parallel lines, write **no solution**. If the two equations produce the same line, write **infinite number of solutions**.*

7. $8x + 4y = -1$
   $4x + 2y = 3$

7. _____

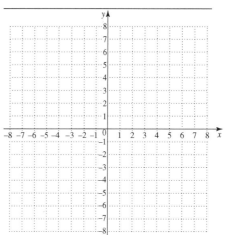

8. $-3x + 2y = 6$
   $-6x + 4y = 12$

8. _____

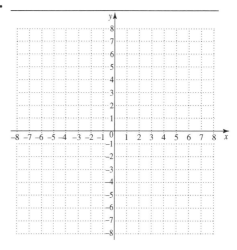

Name: Date:
Instructor: Section:

Objective 4   Identify special systems without graphing.

Video Examples

Review these examples for Objective 4:

4. Given the linear system,
$$3x - 4y = 12$$
$$2x - 3y = 6$$
answer the following questions without graphing.

   a. Is the system inconsistent, are the equations dependent, or neither?

   Write each equation in slope-intercept form.

   $3x - 4y = 12$ | $2x - 3y = 6$
   $-4y = -3x + 12$ | $-3y = -2x + 6$
   $y = \frac{3}{4}x - 3$ | $y = \frac{2}{3}x - 2$

   The slopes are different. This system is not inconsistent nor is it dependent. The system is neither.

   b. Is the graph a pair of intersecting lines, a pair of parallel lines, or one line?

   From part (a), we have written the system in slope-intercept form.
   $$y = \frac{3}{4}x - 3 \qquad y = \frac{2}{3}x - 2$$
   The graphs are neither parallel nor the same line, since the slopes are different. The system is a pair of intersecting lines.

   c. Does the system have one solution, no solution, or an infinite number of solutions?

   From part (a), we have written the system in slope-intercept form.
   $$y = \frac{3}{4}x - 3 \qquad y = \frac{2}{3}x - 2$$
   Since the slopes are different, the graph is neither parallel nor the same line, and therefore cannot have no solution nor an infinite number of solutions. This system has exactly one solution.

Now Try:

4. Given the linear system,
$$x - 6y = 4$$
$$6x - y = 9$$
answer the following questions without graphing.

   a. Is the system inconsistent, are the equations dependent, or neither?

   _____

   b. Is the graph a pair of intersecting lines, a pair of parallel lines, or one line?

   _____

   c. Does the system have one solution, no solution, or an infinite number of solutions?

   _____

Name: Date:
Instructor: Section:

Objective 4 Practice Exercises

For extra help, see Example 4 on page 402 of your text.

Without graphing, answer the following equations for each linear system.
(a) Is the system inconsistent, are the equations dependent, or neither?
(b) Is the graph a pair of intersecting lines, a pair of parallel lines, or one line?
(c) Does the system have one solution, no solution, or an infinite number of solutions?

9. $y = 2x + 1$
   $3x - y = 7$

   9. (a)_____
   (b)_____
   (c)_____

10. $-2x + y = 4$
    $-4x + 2y = -2$

    10. (a)_____
    (b)_____
    (c)_____

11. $4x + 3y = 12$
    $-12x = -36 + 9y$

    11. (a)_____
    (b)_____
    (c)_____

Name:
Instructor:
Date:
Section:

Chapter 7 GRAPHS, LINEAR EQUATIONS, AND SYSTEMS

7.4 Solving Systems of Linear Equations by Substitution

Learning Objectives
1. Solve linear systems by substitution.
2. Solve special systems by substitution.
3. Solve linear systems with fractions and decimals.

Key Terms

Use the vocabulary terms listed below to complete each statement in exercises 1–4.

substitution **ordered pair** **inconsistent system**

dependent system

1. The solution of a linear system of equations is written as a(n) _____.

2. When one expression is replaced by another, _____ is being used.

3. A system of equations in which all solutions of the first equation are also solutions of the second equation is a(n) _____.

4. A system of equations that has no common solution is called a(n) _____.

Objective 1 Solve linear systems by substitution.

Video Examples

Review these examples for Objective 1:
1. Solve the system by the substitution method.
$$2x + 5y = 22 \quad (1)$$
$$y = 4x \quad (2)$$

Equation (2) is already solved for y. We substitute $4x$ for y in equation (1).
$$2x + 5y = 22$$
$$2x + 5(4x) = 22$$
$$2x + 20x = 22$$
$$22x = 22$$
$$x = 1$$

Find the value of y by substituting 1 for x in either equation. We use equation (2).
$$y = 4x$$
$$y = 4(1) = 4$$

We check the solution (1, 4) by substituting 1 for

Now Try:
1. Solve the system by the substitution method.
$$x + y = 7$$
$$y = 6x$$

_____

282 Copyright © 2016 Pearson Education, Inc.

Name: Date:
Instructor: Section:

x and 4 for y in both equations.

$2x+5y=22$ | $y=4x$
$2(1)+5(4)\stackrel{?}{=}22$ | $4\stackrel{?}{=}4(1)$
$2+20\stackrel{?}{=}22$ | True $4=4$
$22=22$ True

Since (1, 4) satisfies both equations, the solution set of the system is {(1, 4)}.

2. Solve the system by the substitution method.
$$4x+5y=13 \quad (1)$$
$$x=-y+2 \quad (2)$$

Equation (2) gives x in terms of y. We substitute $-y+2$ for x in equation (1).
$$4x+5y=13$$
$$4(-y+2)+5y=13$$
$$-4y+8+5y=13$$
$$y+8=13$$
$$y=5$$

Find the value of x by substituting 5 for y in either equation. We use equation (2).
$$x=-y+2$$
$$x=-5+2=-3$$

We check the solution (–3, 5) by substituting –3 for x and 5 for y in both equations.

$4x+5y=13$ | $x=-y+2$
$4(-3)+5(5)\stackrel{?}{=}13$ | $-3\stackrel{?}{=}-5+2$
$-12+25\stackrel{?}{=}13$ | True $-3=-3$
$13=13$ True

Both results are true, so the solution set of the system is {(–3, 5)}.

3. Solve the system by the substitution method.
$$4x=5-y \quad (1)$$
$$7x+3y=15 \quad (2)$$

Step 1 Solve one of the equations for x or y. Solve equation (1) for y to avoid fractions.
$$4x=5-y$$
$$y+4x=5$$
$$y=-4x+5$$

2. Solve the system by the substitution method.
$$2x+3y=6$$
$$x=5-y$$

_____

3. Solve the system by the substitution method.
$$2x+7y=2$$
$$3y=2-x$$

_____

Step 2 Now substitute $-4x+5$ for y in equation (2).
$$7x+3y=15$$
$$7x+3(-4x+5)=15$$

Step 3 Solve the equation from Step 2.
$$7x-12x+15=15$$
$$-5x+15=15$$
$$-5x=0$$
$$x=0$$

Step 4 Equation (1) solved for y is $y=-4x+5$. Substitute 0 for x.
$$y=-4(0)+5=5$$

Step 5 Check that (0, 5) is the solution.

$4x=5-y$	$7x+3y=15$
$4(0)\stackrel{?}{=}5-5$	$7(0)+3(5)\stackrel{?}{=}15$
$0=0$ True	True $15=15$

Since both results are true, the solution set of the system is $\{(0, 5)\}$.

Objective 1 Practice Exercises

For extra help, see Examples 1–3 on pages 506–509 of your text.

Solve each system by the substitution method. Check each solution.

1. $3x+2y=14$
   $y=x+2$

   1. _____

2. $x+y=9$
   $5x-2y=-4$

   2. _____

3. $3x-21=y$
   $y+2x=-1$

   3. _____

Name: Date:
Instructor: Section:

Objective 2   Solve special systems by substitution.

Video Examples

Review these examples for Objective 2: | **Now Try:**

4. Use substitution to solve the system.
    $$x = 7 - 3y \quad (1)$$
    $$5x + 15y = 1 \quad (2)$$

    Because equation (1) is already solved for x, we substitute $7 - 3y$ for x in equation (2).
    $$5x + 15y = 1$$
    $$5(7 - 3y) + 15y = 1$$
    $$35 - 15y + 15y = 1$$
    $$35 = 1 \quad \text{False}$$
    A false result, here $35 = 1$, means that the equations in the system have graphs that are parallel lines. The system is inconsistent and has no solution, so the solution set is $\emptyset$.

4. Use substitution to solve the system.
    $$5x - 10y = 8$$
    $$x = 2y + 5$$

    _____

5. Use substitution to solve the system.
    $$14x - 7y = 21 \quad (1)$$
    $$-2x + y = -3 \quad (2)$$

    Begin by solving equation (2) for y to get $y = 2x - 3$. Substitute $2x - 3$ for y in equation (1).
    $$14x - 7y = 21$$
    $$14x - 7(2x - 3) = 21$$
    $$14x - 14x + 21 = 21$$
    $$0 = 0$$
    This true result means that every solution of one equation is also a solution of the other, so the system has an infinite number of solutions. The solution set is $\{(x, y) | 14x - 7y = 21\}$.

5. Use substitution to solve the system.
    $$5x + 4y = 20$$
    $$-10x + 40 = 8y$$

    _____

Objective 2 Practice Exercises

For extra help, see Examples 4–5 on pages 509–510 of your text.

Solve each system by the substitution method. Use set-builder notation for dependent equations.

4. $$y = -\frac{1}{3}x + 5$$
    $$3y + x = -9$$

4. _____

Name:
Instructor:
Date:
Section:

5. $\frac{1}{2}x+3=y$
   $6=-x+2y$

   5. _____

6. $4x+3y=2$
   $8x+6y=6$

   6. _____

Objective 3   Solve linear systems with fractions and decimals.

Video Examples

Review these examples for Objective 3:
6. Solve the system by the substitution method.

$$\frac{1}{2}x-y=3 \quad (1)$$
$$\frac{1}{5}x+\frac{1}{2}y=\frac{3}{10} \quad (2)$$

Clear equation (1) of fractions by multiplying each side by 2.

$$2\left(\frac{1}{2}x-y\right)=2(3)$$
$$2\left(\frac{1}{2}x\right)-2y=2(3)$$
$$x-2y=6$$

Clear equation (2) of fractions by multiplying each side by 10.

$$10\left(\frac{1}{5}x+\frac{1}{2}y\right)=10\left(\frac{3}{10}\right)$$
$$10\left(\frac{1}{5}x\right)+10\left(\frac{1}{2}y\right)=10\left(\frac{3}{10}\right)$$
$$2x+5y=3$$

The given system of equations has been simplified to an equivalent system.

$$x-2y=6 \quad (3)$$
$$2x+5y=3 \quad (4)$$

To solve the system by substitution, solve equation (3) for x.

$$x-2y=6$$
$$x=2y+6$$

Now substitute the result for x in equation (4).

Now Try:
6. Solve the system by the substitution method.

$$x+\frac{1}{2}y=\frac{1}{2}$$
$$\frac{1}{2}x+\frac{1}{5}y=0$$

_____

Name: Date:
Instructor: Section:

$$2x+5y=3$$
$$2(2y+6)+5y=3$$
$$4y+12+5y=3$$
$$9y+12=3$$
$$9y=-9$$
$$y=-1$$

Substitute -1 for y in $x=2y+6$ (equation (3) solved for x).
$$x=2(-1)+6=4$$
Check $(4,-1)$ in both of the original equations.
The solution set is $\{(4,-1)\}$.

7. Solve the system by the substitution method.
$$0.4x+2.5y=8 \quad (1)$$
$$-0.1x+3.5y=14.5 \quad (2)$$

Clear each equation of decimals, by multiplying by 10.
$$0.4x+2.5y=8$$
$$10(0.4x+2.5y)=10(8)$$
$$10(0.4x)+10(2.5y)=10(8)$$
$$4x+25y=80$$

$$-0.1x+3.5y=14.5$$
$$10(-0.1x+3.5y)=10(14.5)$$
$$10(-0.1x)+10(3.5y)=10(14.5)$$
$$-x+35y=145$$

Now solve the equivalent system of equations by substitution.
$$4x+25y=80 \quad (3)$$
$$-x+35y=145 \quad (4)$$

Equation (4) can be solved for x.
$$x=35y-145$$

Substitute this result for x in equation (3).
$$4x+25y=80$$
$$4(35y-145)+25y=80$$
$$140y-580+25y=80$$
$$165y-580=80$$
$$165y=660$$
$$y=4$$

7. Solve the system by the substitution method.
$$0.6x+0.8y=-2.2$$
$$0.7x-0.1y=2.6$$

_____

Copyright © 2016 Pearson Education, Inc.

Name: Date:
Instructor: Section:

Since equation (4) solved for x is $x = 35y - 145$, substitute 4 for y.
$$x = 35(4) - 145 = -5$$
Check $(-5, 4)$ in both of the original equations.
The solution set is $\{(-5, 4)\}$.

Objective 3 Practice Exercises

For extra help, see Examples 6–7 on pages 510–512 of your text.

Solve each system by the substitution method. Check each solution.

7. $\frac{5}{4}x - y = -\frac{1}{4}$

    $-\frac{7}{8}x + \frac{5}{8}y = 1$

    7. _____

8. $\frac{1}{4}x + \frac{3}{8}y = -3$

    $\frac{5}{6}x - \frac{3}{7}y = -10$

    8. _____

9. $0.6x + 0.8y = 1$

    $0.4y = 0.5 - 0.3x$

    9. _____

Name: Date:
Instructor: Section:

Chapter 7 GRAPHS, LINEAR EQUATIONS, AND SYSTEMS

7.5 Solving Systems of Linear Equations by Elimination

Learning Objectives	
1	Solve linear systems by elimination.
2	Multiply when using the elimination method.
3	Use an alternative method to find the second value in a solution.
4	Solve special systems by elimination.

Key Terms

Use the vocabulary terms listed below to complete each statement in exercises 1–3.

addition property of equality   **elimination method**   **substitution**

1. Using the addition property to solve a system of equations is called the _____.

2. The _____ states that the same added quantity to each side of an equation results in equal sums.

3. _____ is being used when one expression is replaced by another.

Objective 1 Solve linear systems by elimination.

Video Examples

Review this example for Objective 1:
1. Use the elimination method to solve the system.
$$x + y = 6 \quad (1)$$
$$-x + y = 4 \quad (2)$$

Add the equations vertically.
$$x + y = 6 \quad (1)$$
$$\underline{-x + y = 4} \quad (2)$$
$$2y = 10$$
$$y = 5$$

To find the x-value, substitute 5 for y in either of the two equations of the system. We choose equation (1).
$$x + y = 6$$
$$x + 5 = 6$$
$$x = 1$$

Now Try:
1. Use the elimination method to solve the system.
$$x + y = 11$$
$$x - y = 5$$

_____

Name: Date:
Instructor: Section:

Check the solution (1, 5), by substituting 1 for x and 5 for y in both equations of the given system.

$$x + y = 6 \qquad\qquad -x + y = 4$$
$$1 + 5 \stackrel{?}{=} 6 \qquad\qquad -1 + 5 \stackrel{?}{=} 4$$
$$6 = 6 \text{ True} \qquad\qquad \text{True } 4 = 4$$

Since both results are true, the solution set of the system is $\{(1, 5)\}$.

Objective 1 Practice Exercises

For extra help, see Examples 1–2 on pages 514–515 of your text.

Solve each system by the elimination method. Check your answers.

1. $x - 4y = -4$
   $-x + y = -5$

   1. _____

2. $2x - y = 10$
   $3x + y = 10$

   2. _____

3. $x - 3y = 5$
   $-x + 4y = -5$

   3. _____

Name: Date:
Instructor: Section:

Objective 2 Multiply when using the elimination method.

Video Examples

Review this example for Objective 2:

3. Solve the system.
$$3x + 8y = -2 \quad (1)$$
$$2x + 7y = 2 \quad (2)$$

To eliminate x, multiply equation (1) by 2 and multiply equation (2) by -3. Then add.
$$6x + 16y = -4$$
$$\underline{-6x - 21y = -6}$$
$$-5y = -10$$
$$y = 2$$

Find the value of x by substituting 2 for y in either equation (1) or (2).
$$2x + 7y = 2$$
$$2x + 7(2) = 2$$
$$2x + 14 = 2$$
$$2x = -12$$
$$x = -6$$

Check that the solution set of the system is $\{(-6, 2)\}$.

Now Try:

3. Solve the system.
$$3x + 4y = 24$$
$$4x + 3y = 11$$

3. _____

Objective 2 Practice Exercises

For extra help, see Example 3 on page 516 of your text.

Solve each system by the elimination method. Check your answers.

4. $6x + 7y = 10$
   $2x - 3y = 14$

4. _____

5. $8x + 6y = 10$
   $4x - y = 1$

5. _____

6. $6x + y = 1$
   $3x - 4y = 23$

6. _____

Name: Date:
Instructor: Section:

Objective 3 Use an alternative method to find the second value in a solution.

Video Examples

Review this example for Objective 3:

4. Solve the system.
$$6x = 7 - 3y \quad (1)$$
$$8x - 5y = 5 \quad (2)$$

Write equation (1) in standard form.
$$6x + 3y = 7 \quad (3)$$
$$8x - 5y = 5 \quad (4)$$

One way to proceed is to eliminate y by multiplying each side of equation (3) by 5 and each side of equation (4) by 3, and then adding.

$$30x + 15y = 35$$
$$24x - 15y = 15$$
$$\overline{54x = 50}$$
$$x = \frac{50}{54} \text{ or } \frac{25}{27}$$

Substituting $\frac{25}{27}$ for x in one of the given equations would give y, but the arithmetic would be complicated. Instead, solve for y by starting again with the original equations in standard form, and eliminating x.

Multiply equation (3) by 4 and equation (4) by –3.

$$24x + 12y = 28$$
$$-24x + 15y = -15$$
$$\overline{27y = 13}$$
$$y = \frac{13}{27}$$

The solution set is $\left\{\left(\frac{25}{27}, \frac{13}{27}\right)\right\}$.

Now Try:

4. Solve the system.
$$8x = 5y + 1$$
$$6x - 8y = -2$$

4. _____

Objective 3 Practice Exercises

For extra help, see Example 4 on page 517 of your text.

Solve each system by the elimination method. Check your answers.

7. $4x - 3y - 20 = 0$
$6x + 5y + 8 = 0$

7. _____

292   Copyright © 2016 Pearson Education, Inc.

Name: Date:
Instructor: Section:

8. $6x = 16 - 7y$
    $4x = 3y + 26$

    8. _____

9. $2x = 14 + 4y$
    $6y = -5x + 3$

    9. _____

Objective 4 Solve special systems by elimination.

Video Examples

Review these examples for Objective 4:

5. Solve each system by the elimination method.

   a. $7x + y = 9$ (1)
       $-14x - 2y = -18$ (2)

   Multiply each side of equation (1) by 2.
   $14x + 2y = 18$
   $-14x - 2y = -18$
   _____
   $0 = 0$ True

   A true statement occurs when the equations are equivalent. This indicates that every solution of one equation is also a solution of the other. The solution set is $\{(x, y) \mid 7x + y = 9\}$.

   b. $5x + 10y = 9$ (1)
       $3x + 6y = 8$ (2)

   Multiply each side of equation (1) by 3 and each side of equation (2) by –5.
   $15x + 30y = 27$
   $-15x - 30y = -40$
   _____
   $0 = -13$ False

   The false statement $0 = -13$ indicates that the system has solution set $\varnothing$.

Now Try:

5. Solve each system by the elimination method.

   a. $9x - 7y = 5$
      $18x = 14y + 10$

   _____

   b. $2x + 6y = 5$
      $5x + 15y = 8$

   _____

Name: Date:
Instructor: Section:

Objective 4 Practice Exercises

For extra help, see Example 5 on pages 517–518 of your text.

Solve each system by the elimination method. Use set-builder notation for dependent equations. Check your answers.

10. $12x - 8y = 3$
    $6x - 4y = 6$

10. _____

11. $2x + 4y = -6$
    $-x - 2y = 3$

11. _____

12. $15x + 6y = 9$
    $10x + 4y = 18$

12. _____

Name: Date:
Instructor: Section:

Chapter 7 GRAPHS, LINEAR EQUATIONS, AND SYSTEMS

7.6 Systems of Linear Equations in Three Variables

Learning Objectives
1. Understand the geometry of systems of three equations in three variables.
2. Solve linear systems (with three equations and three variables) by elimination.
3. Solve linear systems (with three equations and three variables) in which some of the equations have missing terms.
4. Solve special systems.

Key Terms

Use the vocabulary terms listed below to complete each statement in exercises 1–3.

    ordered triple   **inconsistent system**   **dependent system**

1. The solution of a linear system of equations in three variables is written as a(n) _____.

2. A system of equations in which all solutions of the first equation are also solutions of the second equation is a(n) _____.

3. A system of equations that has no common solution is called a(n) _____.

Objective 1   Understand the geometry of systems of three equations in three variables.

Objective 1 Practice Exercises

For extra help, see pages 521–522 of your text.

Answer each question.

1. If a system of linear equations in three variables has a single solution, how do the planes that are the graphs of the equations intersect?

    1. _____

2. If a system of linear equations in three variables has no solution, how do the planes that are the graphs of the equations intersect?

    2. _____

Name: Date:
Instructor: Section:

Objective 2 Solve linear systems (with three equations and three variables) by elimination.

Video Examples

Review this example for Objective 2:
1. Solve the system.
$$x - 2y + 5z = -7 \quad (1)$$
$$2x + 3y - 4z = 14 \quad (2)$$
$$3x - 5y + z = 7 \quad (3)$$

Step 1: Since x in equation (1) has coefficient 1, choose x as the focus variable and (1) as the working equation.

Step 2: Multiply working equation (1) by -2 and add the result to equation (2) to eliminate focus variable x.
$$-2x + 4y - 10z = 14 \quad \text{Multiply (1) by } -2$$
$$\underline{2x + 3y - 4z = 14 \quad (2)}$$
$$7y - 14z = 28 \quad (4)$$

Step 3: Multiply working equation (1) by -3 and add the result to equation (3) to eliminate focus variable x.
$$-3x + 6y - 15z = 21 \quad \text{Multiply (1) by } -3$$
$$\underline{3x - 5y + z = 7 \quad (3)}$$
$$y - 14z = 28 \quad (5)$$

Step 4: Write equations (4) and (5) as a system, then solve the system.
$$7y - 14z = 28 \quad (4)$$
$$y - 14z = 28 \quad (5)$$
We will eliminate z.
$$-7y + 14z = -28 \quad \text{Multiply (4) by } -1$$
$$\underline{y - 14z = 28 \quad (5)}$$
$$-6y = 0 \quad \text{Add.}$$
$$y = 0 \quad \text{Divide by } -6.$$
Substitute 0 for y in either equation to find z.
$$y - 14z = 28 \quad (5)$$
$$0 - 14z = 28 \quad \text{Let } y = 0.$$
$$-14z = 28$$
$$z = -2$$

Now Try:
1. Solve the system.
$$2x + y + 2z = -1$$
$$3x - y + 2z = -6$$
$$3x + y - z = -10$$

_____

Name: Date:
Instructor: Section:

Step 5: Now substitute $y = 0$ and $z = -2$ in working equation (1) to find the value of the remaining variable, focus variable x.

$$x - 2y + 5z = -7 \quad (1)$$
$$x - 2(0) + 5(-2) = -7 \quad \text{Let } y = 0 \text{ and } z = -2.$$
$$x - 10 = -7$$
$$x = 3$$

Step 6: It appears that the ordered triple $(3, 0, -2)$ is the solution of the system. We must check that the solution satisfies all three original equations of the system.

$$x - 2y + 5z = -7 \quad (1)$$
$$3 - 2(0) + 5(-2) \stackrel{?}{=} -7$$
$$3 - 0 - 10 \stackrel{?}{=} -7$$
$$-7 = -7$$

Because $(3, 0, -2)$ also satisfies equations (2) and (3), the solution set is $\{(3, 0, -2)\}$.

Objective 2 Practice Exercises

For extra help, see Example 1 on pages 523–524 of your text.

Solve each system of equations.

3. $x + y + z = 2$
   $x - y + z = -2$
   $x - y - z = -4$

   3. _____

4. $2x + y - z = 9$
   $x + 2y + z = 3$
   $3x + 3y - z = 14$

   4. _____

Name: Date:
Instructor: Section:

5. $2x - 5y + 2z = 30$
    $x + 4y + 5z = -7$
    $\frac{1}{2}x - \frac{1}{4}y + z = 4$

5. _____

Objective 3 Solve linear systems (with three equations and three variables) in which some of the equations have missing terms.

Video Examples

Review this example for Objective 3:
2. Solve the system.
   $3x \quad - 4z = -23$ (1)
   $\quad y + 5z = 24$ (2)
   $x - 3y \quad = 2$ (3)

   Since equation (1) is missing the variable y, one way to begin is to eliminate y again, using equations (2) and (3).
   $\quad 3y + 15z = 72$ Multiply (2) by 3
   $\underline{x - 3y \quad = 2}$ (3)
   $x \quad + 15z = 74$ (4)

   Now solve the system composed of equations (1) and (4).
   $3x - 4z = -23$ (1)
   $\underline{-3x - 45z = -222}$ Multiply (4) by -3
   $-49z = -245$
   $z = 5$

   Substitute 5 for z in (1) and solve for x.
   $3x - 4z = -23$ (1)
   $3x - 4(5) = -23$ Let $z = 5$.
   $3x - 20 = -23$
   $3x = -3$
   $x = -1$

Now Try:
2. Solve the system.
   $x + 5y \quad = -23$
   $\quad 4y - 3z = -29$
   $2x \quad + 5z = 19$

   _____

298 Copyright © 2016 Pearson Education, Inc.

Name: Date:
Instructor: Section:

Substitute 5 for z in (2) and solve for y.
$$y + 5z = 24 \quad (2)$$
$$y + 5(5) = 24 \quad \text{Let } z = 5.$$
$$y + 25 = 24$$
$$y = -1$$

A check verifies that the solution set is $\{(-1, -1, 5)\}$.

Objective 3 Practice Exercises

For extra help, see Example 2 on pages 524–525 of your text.

Solve each system of equations.

6. $7x \quad\quad + z = -1$
   $\quad\quad 3y - 2z = 8$
   $5x + y \quad\quad = 2$

   6. _____

7. $2x + 5y \quad\quad = 18$
   $\quad\quad 3y + 2z = 4$
   $\frac{1}{4}x - y \quad\quad = -1$

   7. _____

8. $5x \quad\quad - 2z = 8$
   $\quad\quad 4y + 3z = -9$
   $\frac{1}{2}x + \frac{2}{3}y \quad\quad = -1$

   8. _____

Name: Date:
Instructor: Section:

Objective 4 Solve special systems.

Video Examples

Review these examples for Objective 4:

4. Solve the system.
$$3x - 2y + 5z = 4 \quad (1)$$
$$-6x + 4y - 10z = -8 \quad (2)$$
$$\tfrac{3}{2}x - y + \tfrac{5}{2}z = 2 \quad (3)$$

Multiplying each side of equation (1) by -2 gives equation (2). Multiplying each side of equation (1) by $\tfrac{1}{2}$ gives equation (3). Thus, the equations are dependent, and all three equations have the same graph as shown below. The solution set is written
$\{(x,y,z) \mid 3x - 2y + 5z = 4\}$.

3. Solve the system.
$$x - y + z = 7 \quad (1)$$
$$2x + 5y - 4z = 2 \quad (2)$$
$$-x + y - z = 4 \quad (3)$$

Since x in equation (1) has coefficient 1, choose x as the focus variable and (1) as the working equation. Using equations (1) and (3), we have
$$x - y + z = 7 \quad (1)$$
$$\underline{-x + y - z = 4} \quad (3)$$
$$0 = 11 \quad \text{False}$$

The resulting false statement indicates that equations (1) and (3) have no common solution. Thus, the system is inconsistent and the solution set is ∅. The graph of this system would show that three planes are parallel to each other as shown below.

Now Try:

4. Solve the system.
$$x - 5y + 2z = 0$$
$$-x + 5y - 2z = 0$$
$$\tfrac{1}{2}x - \tfrac{5}{2}y + z = 0$$

_____

3. Solve the system.
$$-4x - 2y + z = -19$$
$$-6x + 2y - 6z = -8$$
$$-4x + 2y - 5z = -6$$

_____

Name: Date:
Instructor: Section:

Objective 4 Practice Exercises

For extra help, see Examples 3–5 on pages 525–526 of your text.

Solve each system of equations.

9. $8x - 7y + 2z = 1$
    $3x + 4y - z = 6$
    $-8x + 7y - 2z = 5$

9. _____

10. $3x - 2y + 4z = 5$
    $-3x + 2y - 4z = -5$
    $\frac{3}{2}x - y + 2z = \frac{5}{2}$

10. _____

11. $-x + 5y - 2z = 3$
    $2x - 10y + 4z = -6$
    $-3x + 15y - 6z = 9$

11. _____

Name: Date:
Instructor: Section:

Chapter 7 GRAPHS, LINEAR EQUATIONS, AND SYSTEMS

7.7   Applications of Systems of Linear Equations

Learning Objectives
1   Solve geometry problems using two variables.
2   Solve money problems using two variables.
3   Solve mixture problems using two variables
4   Solve distance-rate-time problems using two variables.
5   Solve problems with three variables using a system of three equations.

Key Terms

Use the vocabulary terms listed below to complete each statement in exercises 1–2.

    elimination method   **substitution**

1. Using the addition property to solve a system of equations is called the _____.

2. _____ is being used when one expression is replaced by another.

Objective 1   Solve geometry problems using two variables.

Video Examples

Review this example for Objective 1:
1. The length of a rectangular field is 5 m more than the width. Find the length and width if the perimeter is 70 m.

   Step 1 Read the problem. We must find the dimensions of the field.

   Step 2 Assign variables. Let L = length and W = width.

   Step 3 Write a system of equations. We use the equation for perimeter.
   $2L + 2W = 70$
   A second equation uses the information given.
   $L = W + 5$
   The system of equations is
   $2L + 2W = 70$ (1)
   $L = W + 5$ (2)

   Step 4 Solve the system. Since equation (2) is solved for L, we use substitution.

Now Try:
1. The length of a rectangle is 7 ft more than the width. The perimeter is 54 ft. Find the dimensions of the rectangle.

   _____

Name: Date:
Instructor: Section:

$$2L + 2W = 70$$
$$2(W+5) + 2W = 70$$
$$2W + 10 + 2W = 70$$
$$4W + 10 = 70$$
$$4W = 60$$
$$W = 15$$

Let $W = 15$ in equation (2) to find L.
$$L = 15 + 5 = 20$$

Step 5 State the answer. The length is 20 m and the width is 15 m.

Step 6 Check.
$$2(20) + 2(15) = 70$$
$$20 = 15 + 5$$
The answer is correct.

Objective 1 Practice Exercises

For extra help, see Example 1 on pages 529–530 of your text.

Solve each problem.

1. The side of a square is 5 centimeters shorter than the side of an equilateral triangle. The perimeter of the square is 7 centimeters less than the perimeter of the triangle. Find the lengths of a side of the square and of a side of the triangle.

   1. square: _____

   triangle: _____

2. The perimeter of a rectangle is 96 inches. If the width were tripled, the width would be 36 inches more than the length. Find the length and width of the rectangle.

   2. length: _____

   width: _____

Name: Date:
Instructor: Section:

3. The perimeter of a triangle is 70 centimeters. Two sides of the triangle have the same length. The third side is 7 centimeters longer than either of the equal sides. Find the length of the equal sides of the triangle.

3. _____

Objective 2 Solve money problems using two variables.

Video Examples

Review this example for Objective 2:

2. The total receipts for a basketball game were $4690.50. There were 723 tickets sold, some for children and some for adults. If the adult tickets cost $9.50 and the children's tickets cost $4, how many of each type were there?

Step 1 Read the problem. There are two unknowns.

Step 2 Assign variables.
Let a = the number of adult tickets sold.
Let c = the number of child tickets sold.

Step 3 Write a system of equations. We write one equation using the total number of tickets.
$a + c = 723$
We write another equation using the cost.
$9.50a + 4c = 4690.50$
The system of equations is
$a + c = 723$ (1)
$9.50a + 4c = 4690.50$ (2)

Step 4 Solve the system. To eliminate c, multiply equation (1) by -4, and add.

$-4a - 4c = -2892$ Multiply (1) by -4.
$\underline{9.5a + 4c = 4690.5}$ (2)
$5.5a = 1798.5$ Add.
$a = 327$ Divide by 5.5.

Now Try:

2. The Garden Center ordered 6 ounces of marigold seed and 8 ounces of carnation seed, paying $214.54. They later ordered another 12 ounces of marigold seed and 18 ounces of carnation seed, paying $464.28. Find the price per ounce for each type of seed.

marigold _____

carnation _____

304 Copyright © 2016 Pearson Education, Inc.

Name: Date:
Instructor: Section:

To find the value of c, let $a = 327$ in equation (1).
$$327 + c = 723$$
$$c = 396$$

Step 5 State the answer. The number of adult tickets sold is 327 and the number of child tickets sold is 396.

Step 6 Check.
$$327 + 396 = 723$$
$$9.50(327) + 4(396) = 4690.50$$
The answer is correct.

Objective 2 Practice Exercises

For extra help, see Example 2 on page 531 of your text.

Solve each problem.

4. Pablo has some $10-bills and some $20-bills. The total value of the money is $650, with a total of 40 bills. How many of each are there?

    4. $10-bills _____

        $20-bills _____

5. Big Giant Super Market will sell 5 large jars and 2 small jars of their peanut butter for $36. They will also sell 2 large jars and 5 small jars for $27. What is the price of each jar?

    5. small _____

        large _____

6. A taxi charges a flat rate plus a certain charge per mile. A trip of 7 miles costs $5.30, while a trip of 3 miles costs $3.70. Find the flat rate and the charge per mile.

    6. flat rate _____

        per mile _____

Name: Date:
Instructor: Section:

Objective 3   Solve mixture problems using two variables.

Video Examples

Review this example for Objective 3:

3. A 75% solution will be mixed with a 55% solution to get 70 liters of 63% solution? How many liters of the 55% and 75% solutions should be used?

   Step 1 Read the problem. There are two solution strengths. We are looking for an "in between" strength.

   Step 2 Assign variables.
       Let x = the number of liters of 75% solution.
       Let y = the number of liters of 55% solution.

   Step 3 Write a system of equations.
   Write one equation using the total amount.
       $x + y = 70$

   Write each percent as a decimal and multiply each solution by its concentration.
       $0.75x + 0.55y = 0.63(70)$

   The system of equations is
       $x + y = 70$ (1)
       $0.75x + 0.55y = 44.1$ (2)

   Step 4 Solve the system. Multiply equation (2) by 100. Multiply equation (1) by –55 to eliminate y.

   $-55x - 55y = -3850$ Multiply (1) by -55.
   $\underline{75x + 55y = 4410}$ Multiply (2) by 100.
      $20x = 560$ Add.
              $x = 28$ Divide by 20.

   Substitute the 28 for x in equation (1) to find the value of y.
       $28 + y = 70$
             $y = 42$

   Step 5 State the answer. The desired mixture will contain 28 liters of 75% solution and 42 liters of 55% solution.

   Step 6 Check.
   Total amount: $28 + 42 = 70$
   Total concentration: $0.75(28) + 0.55(42) = 44.1$
   The answer is correct.

Now Try:

3. How many liters of water should be added to 25% antifreeze solution to get 30 liters of a 20% solution? How many liters of 25% solution are needed?

   water   _____

   25% solution   _____

Name: Date:
Instructor: Section:

Objective 3 Practice Exercises

For extra help, see Example 3 on pages 532–533 of your text.

Solve each problem.

7. Jorge wishes to make 150 pounds of coffee blend that can be sold for $8 per pound. The blend will be a mixture of coffee worth $6 per pound and coffee worth $12 per pound. How many pounds of each kind of coffee should be used in the mixture?

   7.
   $6 coffee_____
   $12 coffee_____

8. Bags of coffee worth $90 a bag must be mixed with coffee worth $75 a bag to get 50 bags worth $87 a bag. How many bags of each are needed?

   8.
   $90 coffee_____
   $75 coffee_____

9. A pharmacist wants to add water to a solution that contains 80% medicine. She wants to obtain 12 oz. of a solution that is 20% medicine. How much water and how much of the 80% solution should she use?

   9.
   water_____
   80% solution _____

Copyright © 2016 Pearson Education, Inc.

Name: Date:
Instructor: Section:

Objective 4   Solve distance-rate-time problems using two variables.

Video Examples

Review this example for Objective 4:

4. A train travels 600 kilometers in the same time that a truck travels 520 kilometers. Find the speed of the train and the truck if the train's average speed is 8 kilometers per hour faster than the truck's.

Step 1 Read the problem. We need to find the rate of each vehicle.

Step 2 Assign variables.
Let x = the rate of the train.
Let y = the rate of the truck.

	d	r	t
Train	600	x	$\frac{600}{x}$
Truck	520	y	$\frac{520}{y}$

Step 3 Write a system of equations. From comparing the two speeds we have an equation.
$x = y + 8$

Since both vehicles travel for the same time, we have a second equation.
$$\frac{600}{x} = \frac{520}{y}$$

Multiplying both sides by xy, we have
$600y = 520x$.

The system of equations is
$x = y + 8$ (1)
$600y = 520x$ (2)

Step 4 Solve the system. We solve the system by substitution. Replace x with $y + 8$ in equation (2).
$600y = 520(y + 8)$
$600y = 520y + 4160$
$80y = 4160$
$y = 52$

Because $x = y + 8$,
$x = 52 + 8 = 60$.

Step 5 State the answer. The train's rate is 60 km per hr. The truck's rate is 52 km per hr.

Now Try:

4. Ashley walks 10 miles in the same time that Taylor walks 6 miles. If Ashley walks 1 mile per hour less than twice Taylor's rate, what is the rate at which each walks?

Ashley _____

Taylor _____

Name: Date:
Instructor: Section:

Step 6 Check.

Train: $\dfrac{600}{60} = 10$ hr   Truck: $\dfrac{520}{52} = 10$ hr

The rate of the train is 8 km more than the rate of the truck.

Objective 4 Practice Exercises

For extra help, see Examples 4–5 on pages 533–535 of your text.

Solve each problem.

10. Two cars start together and travel in the same direction, one going twice as fast as the other. At the end of 3 hours, they are 96 miles apart. How fast is each traveling?

    10. slower_____

    faster_____

11. Travis and his sister Kate jog to school daily. Travis jogs at 9 miles per hour, and Kate jogs at 5 miles per hour. When Travis reaches school, Kate is $\frac{1}{2}$ mile from the school. How far do Travis and Kate live from their school? How long does it take Travis to jog to school?

    11. distance_____

    time_____

Name: Date:
Instructor: Section:

Objective 5 Solve problems with three variables using a system of three equations.

Video Examples

Review this example for Objective 5:

6. Lee has some $5, $10, and $20-bills. He has a total of 51 bills, worth $795. The number of $5-bills is 25 less than the number of $20-bills. Find the number of each type of bill he has.

Step 1: Read the problem again. There are three unknowns.

Step 2: Assign variables.
Let x = the number of $5 bills,
let y = the number of $10 bills,
let z = the number of $20 bills.

Step 3: Write a system of three equations.
There are a total of 51 bills, so
$x + y + z = 51$ (1)
The bills amounted to $795, so
$5x + 10y + 20z = 795$ (2)
The number of $5-bills is 25 less than the number of $20-bills, so $x = z - 25$ or
$x - z = -25$ (3)
The system is
$x + y + z = 51$ (1)
$5x + 10y + 20z = 795$ (2)
$x \quad - z = -25$ (3)

Step 4: Solve the system.
Eliminate y.
$-10x - 10y - 10z = -510$ Multiply (1) by -10
$\underline{5x + 10y + 20z = 795}$ (2)
$-5x \qquad + 10z = 285$ (4)

Solve the system consisting of equations (3) and (4).
$5x - 5z = -125$ Multiply (3) by 5
$\underline{-5x + 10z = 285}$ (4)
$5z = 160$
$z = 32$

Substitute 32 for z in equation (3) and solve for x.
$x - z = -25$ (3)
$x - 32 = -25$ Let $z = 32$.
$x = 7$

Now Try:

6. The manager of the Sweet Candy Shop wishes to mix candy worth $4 per pound, $6 per pound, and $10 per pound to get 100 pounds of a mixture worth $7.60 per pound. The amount of $10 candy must equal the total amounts of the $4 and the $6 candy. How many pounds of each must be used?

$4 candy _____

$6 candy _____

$10 candy _____

Name: Date:
Instructor: Section:

Substitute 7 for *x* and 32 for *z* in equation (1) and solve for *y*.

$x + y + z = 51$ (1)

$7 + y + 32 = 51$ Let $x = 7$, $z = 32$.

$y + 39 = 51$

$y = 12$

Step 5: State the answer. Lee has 7 $5-bills, 12 $10-bills, and 32 $20-bills.

Step 6: Check that the total value of the bills is $795 and that the number of $5-bills is 25 less than the number of $20-bills.

Objective 5 Practice Exercises

For extra help, see Examples 6–7 on pages 536–539 of your text.

Solve each problem involving three unknowns.

12. Julie has $80,000 to invest. She invests part at 5%, one fourth this amount at 6%, and the balance 7%. Her total annual income from interest is $4700. Find the amount invested at each rate.

    12. 5% _____

    6% _____

    7% _____

13. A merchant wishes to mix gourmet coffee selling for $8 per pound, $10 per pound, and $15 per pound to get 50 pounds of a mixture that can be sold for $11.70 per pound. The amount of the $8 coffee must be 3 pounds more than the amount of the $10 coffee. Find the number of pounds of each that must be used.

    13. $8/lb _____

    $10/lb _____

    $15/lb _____

14. A boy scout troop is selling popcorn. There are three different kinds of popcorn in three different arrangements. Arrangement I contains 1 bag of cheddar cheese popcorn, 2 bags of caramel popcorn, and 3 bags of microwave popcorn. Arrangement II contains 3 bags of cheddar cheese popcorn, 1 bag of caramel popcorn, and 2 bags of microwave popcorn. Arrangement III contains 2 bags of cheddar cheese popcorn, 3 bags of caramel popcorn, and 1 bag of microwave popcorn. Jim needs 28 bags of cheddar cheese popcorn, 22 bags of caramel popcorn, and 22 bags of microwave popcorn to give as stocking stuffers for Christmas. How many of each arrangement should he buy?

14. I   2

II   6

III   4

Name: Date:
Instructor: Section:

Chapter 8 INEQUALITIES AND ABSOLUTE VALUE

8.1 Review of Linear Inequalities in One Variable

Learning Objectives
1   Review inequalities and interval notation.
2   Solve linear inequalities using the addition property.
3   Solve linear inequalities using the multiplication property.
4   Solve linear inequalities with three parts.

Key Terms

Use the vocabulary terms listed below to complete each statement in exercises 1–4.

    interval   interval notation   inequality

    linear inequality in one variable

1. The _____ for $a \leq x < b$ is $[a, b)$.

2. A(n) _____ can be written in the form $Ax + B > C$, $Ax + B \geq C$, $Ax + B < C$, or $Ax + B \leq C$, where A, B, and C are real numbers with $A \neq 0$.

3. An algebraic expression related by $>$, $\geq$, $<$, or $\leq$ is called a(n) _____.

4. An _____ is a portion of a number line.

Objective 1   Review inequalities and interval notation.

For extra help, see pages 560–561 of your text.

Objective 2   Solve linear inequalities using the addition property.

Video Examples

Review this example for Objective 2:

1. Solve $x - 6 < -11$ and graph the solution set.

$$x - 6 < -11$$
$$x - 6 + 6 < -11 + 6$$
$$x < -5$$

A check confirms that $(-\infty, -5)$, graphed below, is the solution set.

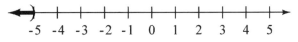

Now Try:

1. Solve $x + 3 < -1$ and graph the solution set.

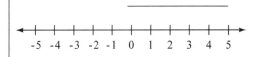

Name: Date:
Instructor: Section:

Objective 2 Practice Exercises

For extra help, see Example 1 on page 561 of your text.

Solve each inequality, giving its solution set in both interval and graph forms. Check your answers.

1. $5a + 3 \leq 6a$

    1. _____

    <-+-+-+-+-+-+-+-+-+-+-+->

2. $6 + 3x < 4x + 4$

    2. _____

    <-+-+-+-+-+-+-+-+-+-+-+->

3. $3 + 5p \leq 4p + 3$

    3. _____

    <-+-+-+-+-+-+-+-+-+-+-+->

Objective 3 Solve linear inequalities using the multiplication property.

Video Examples

Review these examples for Objective 3:	**Now Try:**
2. Solve the inequality, and graph the solution set. $6x < -24$ We divide each side by 6. $6x < -24$ $\dfrac{6x}{6} < \dfrac{-24}{6}$ $x < -4$ The graph of the solution set $(-\infty, -4)$, is shown below. 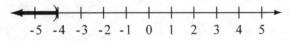	2. Solve the inequality, and graph the solution set. $8x \leq -40$ 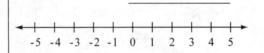

Name: Date:
Instructor: Section:

3. Solve $-5(x+3)+3 \geq 8-3x$, and graph the solution set.

   Step 1 $-5(x+3)+3 \geq 8-3x$
   $-5x-15+3 \geq 8-3x$
   $-5x-12 \geq 8-3x$

   Step 2 $-5x-12+3x \geq 8-3x+3x$
   $-2x-12 \geq 8$
   $-2x-12+12 \geq 8+12$
   $-2x \geq 20$

   Step 3 $\dfrac{-2x}{-2} \leq \dfrac{20}{-2}$
   $x \leq -10$

   The solution set is $(-\infty, -10]$. The graph is shown below.

   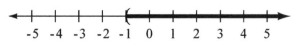

3. Solve $-6(x+4)+1 \geq 9-2x$, and graph the solution set.

   _____

4. Solve $-\dfrac{2}{5}(r-4)-\dfrac{1}{3} < \dfrac{1}{3}(4-r)$, and graph the solution set.

   Step 1 $15\left[-\dfrac{2}{5}(r-4)-\dfrac{1}{3}\right] < 15\left[\dfrac{1}{3}(4-r)\right]$
   $15\left[-\dfrac{2}{5}(r-4)\right]-15\left(\dfrac{1}{3}\right) < 15\left[\dfrac{1}{3}(4-r)\right]$
   $-6(r-4)-5 < 5(4-r)$
   $-6r+24-5 < 20-5r$
   $-6r+19 < 20-5r$

   Step 2 $-6r+19+5r < 20-5r+5r$
   $-r+19 < 20$
   $-r+19-19 < 20-19$
   $-r < 1$

   Step 3 $-1(-r) > -1(1)$
   $r > -1$

   The solution set is $(-1, \infty)$. The graph is shown below.

4. Solve $-\dfrac{1}{2}(r-5)+\dfrac{1}{3} < \dfrac{1}{3}(5-r)$, and graph the solution set.

   _____

Name: Date:
Instructor: Section:

Objective 3 Practice Exercises

For extra help, see Examples 2–4 on pages 562–564 of your text.

Solve each inequality, giving its solution set in both interval and graph forms. Check your answers.

4. $-2s < 4$

4. _____

5. $4k \geq -16$

5. _____

6. $-9m \geq -36$

6. _____

Objective 4 Solve linear inequalities with three parts.

Video Examples

Review this example for Objective 4:
7. Solve $3 \leq 4x - 5 < 7$ and graph the solution set.

$$3 \leq -4x - 5 < 7$$
$$3 + 5 \leq -4x < 7 + 5$$
$$8 \leq -4x < 12$$
$$-\frac{8}{4} \geq -\frac{4x}{4} > -\frac{12}{4}$$
$$-2 \geq x > -3$$
$$-3 < x \leq -2$$

The solution set is $(-3, -2]$. The graph is shown below.

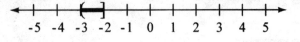

Now Try:
7. Solve $8 \leq -6x - 4 < 20$ and graph the solution set.

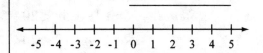

Name: Date:
Instructor: Section:

Objective 4 Practice Exercises

For extra help, see Examples 5–6 on page 565 of your text.

Solve each inequality, giving its solution set in both interval and graph forms. Check your answers.

7. $7 < 2x + 3 \leq 13$

7. _____

<++++++++++++++>

8. $-17 \leq 3x - 2 < -11$

8. _____

<++++++++++++++>

9. $1 < 3z + 4 < 19$

9. _____

<++++++++++++++>

Name: Date:
Instructor: Section:

Chapter 8 INEQUALITIES AND ABSOLUTE VALUE

8.2 Set Operations and Compound Inequalities

Learning Objectives
1. Recognize set intersection and union.
2. Find the intersection of two sets.
3. Solve compound inequalities with the word *and*.
4. Find the union of two sets.
5. Solve compound inequalities with the word *or*.

Key Terms

Use the vocabulary terms listed below to complete each statement in exercises 1–3.

    intersection   compound inequality   union

1. The _____ of two sets, A and B, is the set of elements that belong to either A or B or both.

2. A _____ is formed by joining two inequalities with a connective word such as *and* or *or*.

3. The _____ of two sets, A and B, is the set of elements that belong to both A and B.

Objective 2 Find the intersection of two sets.

Video Examples

Review this example for Objective 2:
1. Let $A = \{6, 7, 8, 9\}$ and $B = \{6, 8, 10\}$. Find $A \cap B$.

   The set $A \cap B$ contains those elements that belong to both A and B.
   $A \cap B = \{6, 7, 8, 9\} \cap \{6, 8, 10\}$
   $ = \{6, 8\}$

Now Try:
1. Let $A = \{20, 30, 40, 50\}$ and $B = \{30, 50, 70\}$. Find $A \cap B$.

   1. _____

Objective 2 Practice Exercises

For extra help, see Example 1 on page 568 of your text.

Let $A = \{0, 1, 2, 3, 4, 5\}$, $B = \{2, 4, 6, 8, 10\}$, $C = \{1, 3, 5, 7, 9\}$, and $D = \{0, 2, 4\}$. Specify each set.

1. $A \cap D$ 1. _____

Name: Date:
Instructor: Section:

2. $B \cap C$ 2. _____

3. $A \cap C$ 3. _____

Objective 3 Solve compound inequalities with the word *and*.

Video Examples

Review these examples for Objective 3:

2. Solve $x+6 \leq 15$ and $x+3 \geq 7$, and graph the solution set.

Step 1 Solve each inequality individually.
$x+6 \leq 15$ and $x+3 \geq 7$
$x+6-6 \leq 15-6$ and $x+3-3 \geq 7-3$
$x \leq 9$ and $x \geq 4$

Step 2 The solution set of the compound inequality includes all numbers that satisfy both inequalities from Step 1.
The graphs of $x \leq 9$ and $x \geq 4$ are shown below.

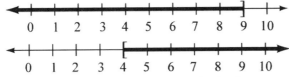

The intersection of these two graphs is the solution set [4, 9].

3. Solve $-5x-6>8$ and $6x-3 \leq -21$, and graph the solution set.

Step 1 Solve each inequality individually.
$-5x-6>8$ and $6x-3 \leq -21$
$-5x>14$ and $6x \leq -18$
$x<-\dfrac{14}{5}$ and $x \leq -3$

The graphs of $x<-\dfrac{14}{5}$ and $x \leq -3$ are shown below. $-\tfrac{14}{5}$

Now Try:

2. Solve $x+4 \leq 20$ and $x-3 \geq 9$ and graph the solution set.

_____

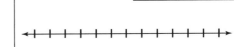

3. Solve $-8x+6>30$ and $4x-7 \leq 11$, and graph the solution set.

_____

Copyright © 2016 Pearson Education, Inc.

Name: Date:
Instructor: Section:

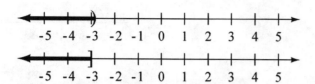

Step 2 Now find all the values of *x* that are less than $-\frac{14}{5}$ and also less than or equal to -3. The solution set is $(-\infty, -3]$.

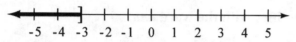

Objective 3 Practice Exercises

For extra help, see Examples 2–4 on pages 569–570 of your text.

For each compound inequality, give the solution set in both interval and graph forms.

4. $1 - 2s \le 7$ and $2s + 7 \ge 11$ 4. _____

5. $3x + 2 < 11$ and $2 - 3x \le 14$ 5. _____

6. $5t > 0$ and $5t + 4 \le 9$ 6. _____

Objective 4 Find the union of two sets.

Video Examples

Review this example for Objective 4:
5. Let $A = \{6, 7, 8, 9\}$ and $B = \{6, 8, 10\}$. Find $A \cup B$.

The set $A \cup B$ contains those elements that belong to either *A* or *B* (or both).
$A \cup B = \{6, 7, 8, 9\} \cup \{6, 8, 10\}$
$ = \{6, 7, 8, 9, 10\}$

Now Try:
5. Let $A = \{20, 30, 40, 50\}$ and $B = \{30, 50, 70\}$. Find $A \cup B$.

_____

Name: Date:
Instructor: Section:

Objective 4 Practice Exercises

For extra help, see Example 5 on page 570 of your text.

Let $A = \{0, 1, 2, 3, 4, 5\}$, $B = \{2, 4, 6, 8, 10\}$, $C = \{1, 3, 5, 7, 9\}$, $D = \{0, 2, 4\}$, and $E = \{0\}$. Specify each set.

7. $A \cup D$ 7. _____

8. $B \cup C$ 8. _____

9. $A \cup E$ 9. _____

Objective 5   Solve compound inequalities with the word *or*.

Video Examples

Review these examples for Objective 5:

6. Solve the compound inequality, and graph the solution set.
$$7x - 6 < 4x \quad \text{or} \quad -4x \leq -16$$

Step 1 Solve each inequality individually.
$$7x - 6 < 4x \quad \text{or} \quad -4x \leq -16$$
$$3x < 6$$
$$x < 2 \quad \text{or} \quad x \geq 4$$

The graphs of $x < 2$ and $x \geq 4$ are shown below.

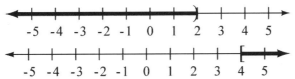

Step 2 Since the inequalities are joined with *or*, find the union of the two solution sets. The union is shown below. $(-\infty, 2) \cup [4, \infty)$

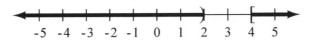

Now Try:

6. Solve the compound inequality, and graph the solution set.
$$5x - 6 < 2x \quad \text{or} \quad -7x \leq -35$$

_____

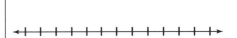

Copyright © 2016 Pearson Education, Inc.

Name: Date:
Instructor: Section:

8. Solve the compound inequality, and graph the solution set.
$-5x+4 \geq -16$ or $6x-11 \geq -29$

Solve each inequality.
$-5x+4 \geq -16$ or $6x-11 \geq -29$
$-5x \geq -20$ or $6x \geq -18$
$x \leq 4$ or $x \geq -3$

Graph each inequality.

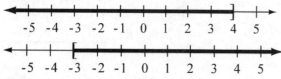

The solution is the union of the two sets, $(-\infty, \infty)$.

8. Solve the compound inequality, and graph the solution set.
$-9x+3 \geq -15$ or $8x-7 \geq -15$

_____

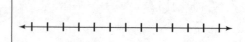

Objective 5 Practice Exercises

For extra help, see Examples 6–9 on pages 571–573 of your text.

For each compound inequality, give the solution set in both interval and graph forms.

10. $q+3 > 7$ or $q+1 \leq -3$ 10. _____

11. $3 > 4m+2$ or $4m-3 \geq -2$ 11. _____

12. $2r+4 \geq 8$ or $4r-3 < 1$ 12. _____

Name: Date:
Instructor: Section:

Chapter 8   INEQUALITIES AND ABSOLUTE VALUE

8.3   Absolute Value Equations and Inequalities

Learning Objectives
1. Use the distance definition of absolute value.
2. Solve equations of the form $|ax + b| = k$, for $k > 0$.
3. Solve inequalities of the form $|ax + b| < k$ and of the form $|ax + b| > k$, for $k > 0$.
4. Solve absolute value equations that involve rewriting.
5. Solve equations of the form $|ax + b| = |cx + d|$.
6. Solve special cases of absolute value equations and inequalities.
7. Solve an application involving relative error.

Key Terms

Use the vocabulary terms listed below to complete each statement in exercises 1–2.

    absolute value equation   **absolute value inequality**

1. An _____ is an equation that involves the absolute value of a variable expression.

2. An _____ is an inequality that involves the absolute value of a variable expression.

Objective 2   Solve equations of the form $|ax + b| = k$, for $k > 0$.

Video Examples

Review this example for Objective 2:

1. Solve $|3x + 2| = 14$. Graph the solution set.

     This is Case 1.
$$3x + 2 = 14 \quad \text{or} \quad 3x + 2 = -14$$
$$3x = 12 \quad \text{or} \quad 3x = -16$$
$$x = 4 \quad \text{or} \quad x = -\frac{16}{3}$$

     Check

         Let $x = 4$ Let $x = -\dfrac{16}{3}$

$$3(4) + 2 \overset{?}{=} 14 \quad \Big| \quad 3\left(-\frac{16}{3}\right) + 2 \overset{?}{=} -14$$
$$12 + 2 \overset{?}{=} 14 \quad \Big| \quad -16 + 2 \overset{?}{=} -14$$
$$14 = 14 \quad \Big| \quad -14 = -14$$

Now Try:

1. Solve $|5x + 4| = 11$. Graph the solution set.

    _____

Copyright © 2016 Pearson Education, Inc.

Name: Date:
Instructor: Section:

The check confirms that the solution set is $\left\{-\frac{16}{3}, 4\right\}$.

Objective 2 Practice Exercises

For extra help, see Example 1 on pages 577–578 of your text.

Solve each equation.

1. $|2x+3|=10$ 1. _____

2. $|5r-15|=0$ 2. _____

3. $\left|\frac{1}{2}x-3\right|=4$ 3. _____

Objective 3 Solve inequalities of the form $|ax+b|<k$ and of the form $|ax+b|>k$, for $k>0$.

Video Examples

Review these examples for Objective 3:

2. Solve $|4x+2|>10$. Graph the solution set.

   This is Case 2.
   $4x+2>10$ or $4x+2<-10$
   $4x>8$ or $4x<-12$
   $x>2$ or $x<-3$

   A check confirms that the solution set is $(-\infty,-3)\cup(2,\infty)$.

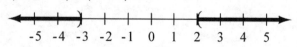

Now Try:

2. Solve $|8x+4|>12$. Graph the solution set.

   _____

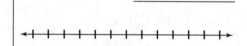

Name: Date:
Instructor: Section:

3. Solve $|10x+5|<25$. Graph the solution set.

This is Case 3.
$$-25 < 10x+5 < 25$$
$$-30 < 10x < 20$$
$$-3 < x < 2$$
A check confirms that the solution set is $(-3, 2)$.

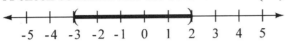

3. Solve $|6x+3|<15$. Graph the solution set.

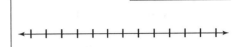

Objective 3 Practice Exercises

For extra help, see Examples 2–4 on pages 578–579 of your text.

Solve each inequality and graph the solution set.

4. $|x-2|>8$

4. _____

5. $|2r-9| \geq 23$

5. _____
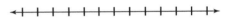

6. $|5r+2|<18$

6. _____

Objective 4 Solve absolute value equations that involve rewriting.

Video Examples

Review this example for Objective 4:

5. Solve $|x-7|+3=15$.

Isolate the absolute value.
$$|x-7|+3=15$$
$$|x-7|+3-3=15-3$$
$$|x-7|=12$$

Now Try:

5. Solve $|x-3|+4=11$.

_____

Copyright © 2016 Pearson Education, Inc.

Name: Date:
Instructor: Section:

Now solve, using Case 1.
$$x - 7 = 12 \quad \text{or} \quad x - 7 = -12$$
$$x = 19 \quad \text{or} \quad x = -5$$

Check
Let $x = 19$ Let $x = -5$
$$|19 - 7| + 3 \stackrel{?}{=} 15 \quad | \quad |-5 - 7| + 3 \stackrel{?}{=} 15$$
$$|12| + 3 \stackrel{?}{=} 15 \quad | \quad |-12| + 3 \stackrel{?}{=} 15$$
$$15 = 15 \quad\quad | \quad\quad 15 = 15$$

The check confirms that the solution set is $\{-5, 19\}$.

Objective 4 Practice Exercises

For extra help, see Examples 5–6 on page 580 of your text.

Solve each equation.

7. $|2w - 1| + 7 = 12$ 7. _____

8. $\left|2 - \frac{1}{2}x\right| - 5 = 18$ 8. _____

9. $|4t + 3| + 8 = 10$ 9. _____

Name: Date:
Instructor: Section:

Objective 5 Solve equations of the form $|ax + b| = |cx + d|$.

Video Examples

Review this example for Objective 5:

7. Solve $|z+5|=|3z-9|$.

$$z+5=3z-9 \quad \text{or} \quad z+5=-(3z-9)$$
$$z+14=3z \quad \text{or} \quad z+5=-3z+9$$
$$14=2z \quad \text{or} \quad 4z=4$$
$$z=7 \quad \text{or} \quad z=1$$

Check
   Let $z = 7$ Let $z = 1$
   $|7+5|\stackrel{?}{=}|3(7)-9|$ $|1+5|\stackrel{?}{=}|3(1)-9|$
   $|12|\stackrel{?}{=}|21-9|$ $|6|\stackrel{?}{=}|-6|$
   $12 = 12$ $6 = 6$

The check confirms that the solution set is {1, 7}.

Now Try:

7. Solve $|z+7|=|2z+8|$.

_____

Objective 5 Practice Exercises

For extra help, see Example 7 on page 581 of your text.

Solve each problem.

10. $|y+5|=|3y+1|$

10. _____

11. $|2p-4|=|7-p|$

11. _____

12. $|3x-2|=|5x+8|$

12. _____

Copyright © 2016 Pearson Education, Inc.   327

Name: Date:
Instructor: Section:

Objective 6 Solve special cases of absolute value equations and inequalities.

Video Examples

Review these examples for Objective 6:

8. Solve each equation.

   a. $|9x-1|=-13$

   The absolute value of an expression can never be negative, so there are no solutions for this equation. The solution set is $\varnothing$.

   b. $|8x-12|=0$

   $$8x-12=0$$
   $$8x=12$$
   $$x=\frac{3}{2}$$

   The solution set is $\left\{\frac{3}{2}\right\}$.

9. Solve each inequality.

   a. $|x|\geq -12$

   The absolute value of a number is always greater than or equal to 0. Thus, $|x|\geq -12$ is always true, and the solution set is $(-\infty, \infty)$.

   b. $|x-11|+7<2$

   $$|x-11|+7<2$$
   $$|x-11|<-5$$

   There is no number whose absolute value is less than -5, so this inequality has no solution. The solution set is $\varnothing$.

   c. $|2x-8|-5\leq -5$

   $$|2x-8|-5\leq -5$$
   $$|2x-8|\leq 0$$

   The value of $|2x-8|$ will never be less than zero. However, $|2x-8|$ will equal 0.
   $$2x-8=0$$
   $$2x=8$$
   $$x=4$$
   The solution set is $\{4\}$.

Now Try:

8. Solve each equation.

   a. $|6x+13|=-5$

   _____

   b. $|7x+35|=0$

   _____

9. Solve each inequality.

   a. $|x|\geq -2$

   _____

   b. $|x+12|+6<3$

   _____

   c. $|5x-20|+7\leq 7$

   _____

Name: Date:
Instructor: Section:

Objective 6 Practice Exercises

For extra help, see Examples 8–9 on pages 581–582 of your text.

Solve each problem.

13. $\left|7+\dfrac{1}{2}x\right|=0$ 13. _____

14. $|m-2|\geq -1$ 14. _____

15. $|k+5|\leq -2$ 15. _____

Objective 7   Solve an application involving relative error.

Video Examples

Review this example for Objective 7:

10. Suppose a machine filling 16.9 oz water bottles is set for a relative error that is no greater than 0.025 oz. How many ounces may a filled water bottle contain?

$$\left|\dfrac{16.9-x}{16.9}\right|\leq 0.025$$

$$-0.025\leq \dfrac{16.9-x}{16.9}\leq 0.025$$

$$-0.4225\leq 16.9-x\leq 0.4225$$

$$-17.3225\leq -x\leq -16.4775$$

$$17.3225\geq x\geq 16.4775$$

$$16.4775\leq x\leq 17.3225$$

The bottle may contain between 16.4775 and 17.3225 oz, inclusive.

Now Try:

10. Suppose a machine filling 16.9 oz water bottles is set for a relative error that is no greater than 0.05 oz. How many ounces may a filled water bottle contain?

_____

Name: Date:
Instructor: Section:

For extra help, see Example 10 on page 582 of your text.

Determine the number of ounces a filled 16.9 oz water bottle may contain for the given relative error.

16.   no greater than 0.04 oz

16. _____

17.   no greater than 0.015 oz

17. _____

18.   no greater than 0.03 oz

17. _____

Name: Date:
Instructor: Section:

Chapter 8 INEQUALITIES AND ABSOLUTE VALUE

8.4 Linear Inequalities and Systems in Two Variables

Learning Objectives
1. Graph linear inequalities in two variables.
2. Solve systems of linear inequalities by graphing.

Key Terms

Use the vocabulary terms listed below to complete each statement in exercises 1–4.

linear inequality in two variables   **boundary line**

system of linear inequalities

solution set of a system of linear inequalities

1. In the graph of a linear inequality, the _____ separates the region that satisfies the inequality from the region that does not satisfy the inequality.

2. All ordered pairs that make all inequalities of the system true at the same time is called the _____.

3. A _____ contains two or more linear inequalities (and no other kinds of inequalities).

4. An inequality that can be written in the form $Ax + By < C$, $Ax + By > C$, $Ax + By \leq C$, or $Ax + By \geq C$ is called a _____.

Objective 1 Graph linear inequalities in two variables.

Video Examples

Review these examples for Objective 1:
1. Graph $3x - 2y \leq 6$.

   The inequality $3x - 2y \leq 6$ means that
   $3x - 2y < 6$ or $3x - 2y = 6$.
   We begin by graphing the line $3x - 2y = 6$ with intercepts (0, –3) and (2, 0). This boundary line divides the plane into two regions, one of which satisfies the inequality. We use the test point (0, 0) to see whether the resulting statement is true or false, thereby determining whether the point is in the shaded region or not.

Now Try:
1. Graph $2x + 5y \leq -8$.

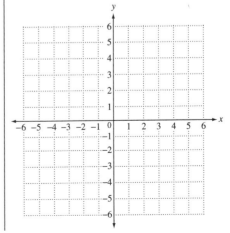

Copyright © 2016 Pearson Education, Inc.

Name: Date:
Instructor: Section:

$$3x - 2y \leq 6$$
$$3(0) - 2(0) \stackrel{?}{\leq} 6$$
$$0 - 0 \stackrel{?}{\leq} 6$$
$$0 \leq 6 \quad \text{True}$$

Since the last statement is true, we shade the region that includes the test point (0, 0). The shaded region, along with the boundary line, is the desired graph.

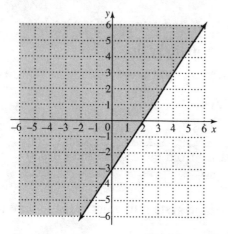

2. Graph $y \geq 3x$.

We graph $y = 3x$ using a solid line through (0, 0), (1, 3) and (2, 6). Because (0, 0) is on the line $y \geq 3x$, it cannot be used as a test point. Instead we choose a test point off the line, say (3, 0).

$$0 \stackrel{?}{\geq} 3(3)$$
$$0 \geq 9 \quad \text{False}$$

Because $0 \geq 9$ is false, shade the other region.

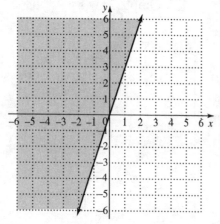

2. Graph $y \geq x$.

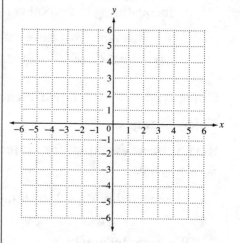

Name: Date:
Instructor: Section:

3. Graph $x - 4 \leq -1$.

First, solve the inequality for x.
$x \leq 3$
Now graph the line $x = 3$, a vertical line through the point $(3, 0)$. Use a solid line, and choose $(0, 0)$ as a test point.
$0 \leq 3$ True
Because $0 \leq 3$ is true, we shade the region containing $(0, 0)$.

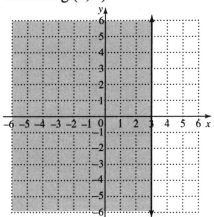

3. Graph $y \geq -1$.

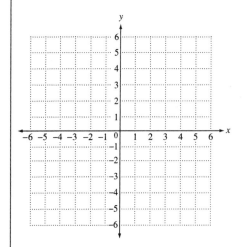

Objective 1 Practice Exercises

For extra help, see Examples 1–3 on pages 589–591 of your text.

Graph each linear inequality.

1. $x - y < 5$

1.
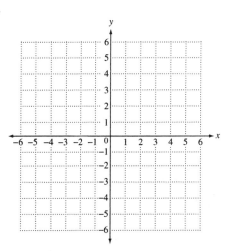

Name:
Instructor:

Date:
Section:

2. $2x + 3y \geq 6$

2.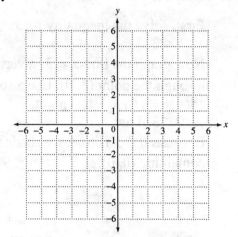

3. $x \leq 4y$

3.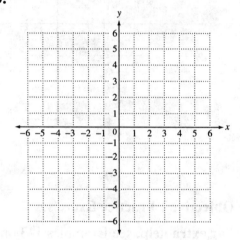

Name: Date:
Instructor: Section:

Objective 2 Solve systems of linear inequalities by graphing.

Video Examples

Review this examples for Objective 2:
4. Graph the solution set of the system.
$$x + y \leq 3$$
$$5x - y \geq 5$$

To graph $x + y \leq 3$, graph the solid boundary line $x + y = 3$ using the intercepts (0, 3) and (3, 0). Determine the region to shade using (0, 0) as a test point.
$$x + y \leq 3$$
$$0 + 0 \stackrel{?}{\leq} 3$$
$$0 \leq 3 \quad \text{True}$$
Shade the region containing (0, 0).

To graph $5x - y \geq 5$, graph the solid boundary line $5x - y = 5$ using the intercepts (0, –5) and (1, 0). Determine the region to shade using (0, 0) as a test point.
$$5x - y \geq 5$$
$$5(0) - 0 \stackrel{?}{\geq} 5$$
$$0 \geq 5 \quad \text{False}$$
Shade the region that does not contain (0, 0).

The solution set of this system includes all points in the intersection (overlap) of the graph of the two inequalities. This intersection is the gray shaded region and portions of the two boundary lines that surround it.

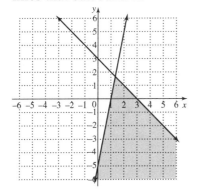

Now Try:
4. Graph the solution set of the system.
$$3x - y \leq 3$$
$$x + y \leq 0$$

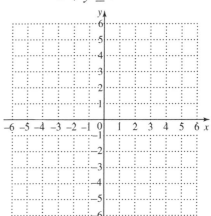

Name: Date:
Instructor: Section:

Objective 2 Practice Exercises

For extra help, see Examples 4–5 on pages 591–592 of your text.

Graph the solution of each system of linear inequalities.

4. $4x + 5y \leq 20$
    $y \leq x + 3$

4.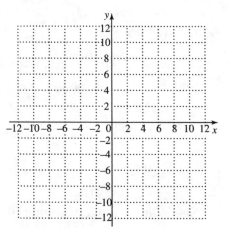

5. $x < 2y + 3$
    $0 < x + y$

5.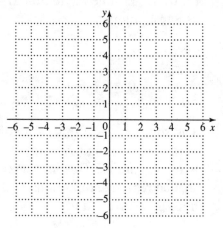

6. $y < 4$
    $x \geq -3$

6.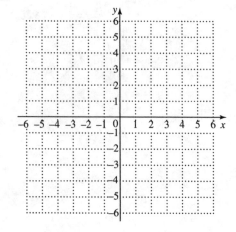

Name: Date:
Instructor: Section:

Chapter 9 RELATIONS AND FUNCTIONS

9.1 Introduction to Relations and Functions

Learning Objectives
1 Define and identify relations and functions.
2 Find domain and range.
3 Identify functions defined by graphs and equations.

Key Terms

Use the vocabulary terms listed below to complete each statement in exercises 1−6.

dependent variable   independent variable   relation

function   domain   range

1. The _____ of a relation is the set of second components (*y*-values) of the ordered pairs of the relation.

2. A _____ is a set of ordered pairs of real numbers.

3. If the quantity *y* depends on *x*, then *y* is called the _____ in a relation between *x* and *y*.

4. The _____ of a relation is the set of first components (*x*-values) of the ordered pairs of the relation.

5. A _____ is a set of ordered pairs in which each value of the first component, *x*, corresponds to exactly one value of the second component, *y*.

6. If the quantity *y* depends on *x*, then *x* is called the _____ in a relation between *x* and *y*.

Objective 1   Define and identify relations and functions.

Video Examples

Review these examples for Objective 1:

1. Write the relation as a set of ordered pairs.

Number of Hours Worked	Paycheck Amount (in dollars)
8	96
16	192
24	288
32	384

{(8, 96), (16, 192), (24, 288), (32, 384)}

Now Try:

1. Write the relation as a set of ordered pairs.

Number of Hours Fishing	Number of Fish Caught
1	3
2	4
3	6
5	7

_____

Copyright © 2016 Pearson Education, Inc.   337

Name: Date:
Instructor: Section:

2. Determine whether each relation defines a function.

   a. $G = \{(-5,-2), (-2, 6), (6, 8), (8, 11), (11, 11)\}$

   Relation G is a function. Although the last two ordered pairs have the same y-value, this does not violate the definition of a function.

   b. $H = \{(-9, 2), (-6, 2), (-6, 9)\}$

   In relation H, the last two ordered pairs have the same x-value pair with different y-values. H is a relation, but not a function.

2. Determine whether each relation defines a function.

   a. $\{(10, 1), (100, 2), (70, 2)\}$

   _____

   b. $\{(0, 2), (2, 6), (6, 3), (0, 7)\}$

   _____

Objective 1 Practice Exercises

For extra help, see Examples 1–2 on pages 604–605 of your text.

Write the relation as a set of ordered pairs.

1. | x | y |
    |---|---|
    | 1 | 3 |
    | 1 | 4 |
    | 2 | -1 |
    | 3 | 7 |

1. _____

Decide whether each relation is a function.

2. $\{(2,-2,), (3,-3), (4,-4)\}$

2. _____

3. $\{(3, 4), (5, 2), (4, 3), (5, 3), (-2, 2)\}$

3. _____

Objective 2   Find domain and range.

Video Examples

Review these examples for Objective 2:

3. Give the domain and range of each relation. Tell whether the relation defines a function.

   $\{(15, 2), (20, 3), (6, 10), (-1, 2)\}$

   The domain is the set of x-values $\{-1, 6, 15, 20\}$.
   The range is the set of y-values $\{2, 3, 10\}$.
   The relation is a function, because each x-value corresponds to exactly one y-value.

Now Try:

3. Give the domain and range of each relation. Tell whether the relation defines a function.
   $\{(13, -1), (13, -2), (13, 4)\}$

   domain: _____

   range: _____

Name: Date:
Instructor: Section:

4. Give the domain and range of each relation.

a.
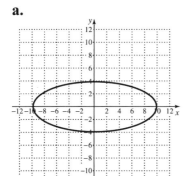

The x-values of the points on the graph include all numbers between –10 and 10, inclusive. The y-values include all numbers between –4 and 4, inclusive.
The domain is [–10, 10]. The range is [–4, 4].

b.
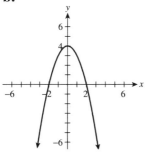

The graph extends indefinitely left and right, as well as downward. The domain is (–∞, ∞). Because there is a greatest y-value, 4, the range includes all numbers less than or equal to 4, written (–∞, 4].

Objective 2 Practice Exercises

For extra help, see Examples 3–4 on pages 607–608 of your text.

Decide whether the relation is a function, and give the domain and range of the relation.

4. $\{(5, 2), (3, -1), (1, -3), (-1, -5)\}$

4. Give the domain and range of each relation.

a.
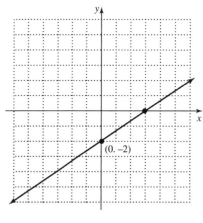

_____

_____

b.
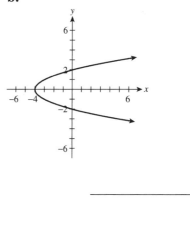

_____

_____

4. _____

domain: _____

range: _____

Name: Date:
Instructor: Section:

5.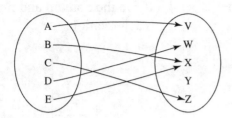

5. _____

domain:_____

range: _____

6.

x	y
1	3
2	−1
−1	4
1	4

6. _____

domain:_____

range: _____

Objective 3   Identify functions defined by graphs and equations.

Video Examples

Review these examples for Objective 3:

5. Use the vertical line test to determine whether the relation graphed is a function.

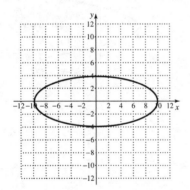

The graph is not a function.

6. Decide whether the relation defines y as a function of x. Give the domain.

$y = 2x - 4$

Each x value corresponds to just one y-value and the relation defines a function. Since x can be any real number, the domain is $(-\infty, \infty)$.

Now Try:

5. Use the vertical line test to determine whether the relation graphed is a function.

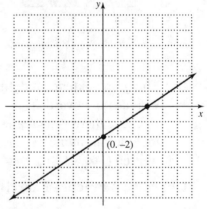

_____

6. Decide whether the relation defines y as a function of x. Give the domain.

$y = 3x - 1$

_____

_____

Name: Date:
Instructor: Section:

Objective 3 Practice Exercises

For extra help, see Examples 5–6 on pages 609–610 of your text.

Use the vertical line test to determine whether the relation graphed is a function.

7.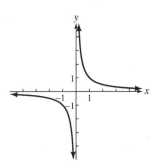

7. _____

Decide whether each equation defines y as a function of x. Give the domain.

8. $y^2 = x+1$

8. _____

9. $y = \dfrac{3}{x+6}$

9. _____

Name: Date:
Instructor: Section:

Chapter 9 RELATIONS AND FUNCTIONS

9.2 Function Notation and Linear Functions

Learning Objectives
1 Use function notation.
2 Graph linear and constant functions.

Key Terms

Use the vocabulary terms listed below to complete each statement in exercises 1–3.

   function notation   **linear function**   **constant function**

1. A function defined by an equation of the form $f(x) = ax + b$, for real numbers a and b, is a _____.

2. _____ $f(x)$ represents the value of the function at x, that is, the y-value that corresponds to x.

3. A _____ is a linear function of the form $f(x) = b$, for a real number b.

Objective 1 Use function notation.

Video Examples

Review these examples for Objective 1:	Now Try:
1. Let $f(x) = 7x - 3$. Evaluate the function f for the following. $x = 4$ Start with the given function. Replace x with 4. $f(x) = 7x - 3$ $f(4) = 7(4) - 3$ $f(4) = 28 - 3$ $f(4) = 25$ Thus, $f(4) = 25$.	1. Let $f(x) = 8x - 7$. Evaluate the function f for the following. $x = 3$ _____
3. Let $g(x) = 5x + 6$. Find and simplify $g(n+8)$. Replace x with $n + 8$. $g(x) = 5x + 6$ $g(n+8) = 5(n+8) + 6$ $g(n+8) = 5n + 40 + 6$ $g(n+8) = 5n + 46$	3. Let $g(x) = 4x - 7$. Find and simplify $g(a-1)$. _____

Name: Date:
Instructor: Section:

4. For the function, find $f(5)$.

$f = \{(7,-27),\ (5,-25),\ (3,-23),\ (1,-21)\}$

From the ordered pair $(5, -25)$, we have $f(5) = -25$.

6. Write the equation using function notation $f(x)$. Then find $f(-5)$.

$x - 5y = 8$

Step 1 $x - 5y = 8$
$-5y = -x + 8$
$y = \frac{1}{5}x - \frac{8}{5}$

Step 2 $f(x) = \frac{1}{5}x - \frac{8}{5}$

$f(-5) = \frac{1}{5}(-5) - \frac{8}{5}$

$f(-5) = -\frac{13}{5}$

4. For the function, find $f(-6)$.

$f = \{(-2, 11),\ (-4, 17),\ (-6, 21),\ (-8, 24)\}$

_____

6. Write the equation using function notation $f(x)$. Then find $f(-3)$.

$2x + 3y = 7$

_____

Objective 1 Practice Exercises

For extra help, see Examples 1–6 on pages 614–617 of your text.

For each function f, find (a) $f(-2)$, (b) $f(0)$, and (c) $f(-x)$.

1. $f(x) = 3x - 7$

1. a. _____

b. _____

c. _____

2. $f(x) = 2x^2 + x - 5$

2. a. _____

b. _____

c. _____

3. $f(x) = 9$

3. a. _____

b. _____

c. _____

Copyright © 2016 Pearson Education, Inc.

Name: Date:
Instructor: Section:

Objective 2 Graph linear and constant functions.

Video Examples

Review this example for Objective 2:

7. Graph the function $f(x) = -2x - 3$. Give the domain and range.

    The graph of the function has slope -2 and y-intercept -3. To graph this function, plot the y-intercept $(0, -3)$ and use the definition of slope as $\frac{\text{rise}}{\text{run}}$ to find a second point on the line. Since the slope is -2, move down two units and right one unit to the point $(1, -5)$. Draw the straight line through the points to obtain the graph. The domain and range are both $(-\infty, \infty)$.

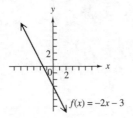

Now Try:

7. Graph the function $f(x) = \frac{1}{2}x + \frac{1}{2}$. Give the domain and range.

    _____

    _____

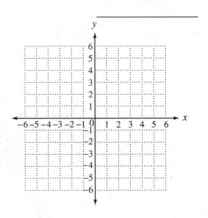

Objective 2 Practice Exercises

For extra help, see Example 7 on page 618 of your text.

Graph each function. Give the domain and range.

4. $2x - y = -2$

4. domain _____

    range _____

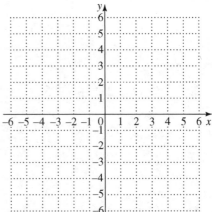

Name: Date:
Instructor: Section:

5. $y + \dfrac{1}{2}x = -2$

5. domain _____

   range _____

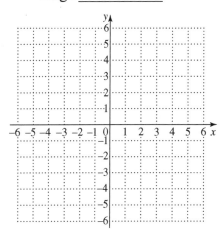

6. $y = 2$

6. domain _____

   range _____

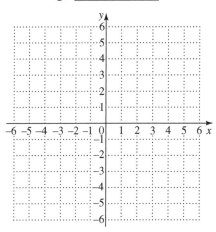

Name: Date:
Instructor: Section:

Chapter 9 RELATIONS AND FUNCTIONS

9.3 Polynomial Functions, Operations, and Composition

Learning Objectives
1. Recognize and evaluate polynomial functions.
2. Perform operations on polynomial functions.
3. Find the composition of functions.

Key Terms

Use the vocabulary terms listed below to complete each statement in exercises 1–2.

polynomial function of degree n   **composite function**

1. The function $g(f(x))$ is a _____.

2. A function defined by $f(x) = a_n x^n + a_{n-1} x^{n-1} + \cdots + a_1 x + a_0$, where $a_n \neq 0$ and n is a whole number is a _____.

Objective 1 Recognize and evaluate polynomial functions.

Video Examples

Review this example for Objective 1:

1. Let $f(x) = 6x^3 - 6x + 1$. Find $f(-3)$.

    Substitute –3 for x.

    $f(-3) = 6(-3)^3 - 6(-3) + 1$
    $= 6(-27) - 6(-3) + 1$
    $= -162 + 18 + 1$
    $= -143$

Now Try:

1. Let $p(x) = -x^4 + 3x^2 - x + 7$. Find $p(2)$.

    _____

Objective 1 Practice Exercises

For extra help, see Example 1 on page 623 of your text.

For each polynomial function, find (a) f(–2) and (b) f(3).

1. $f(x) = -x^2 - x - 5$

    1. (a) _____
       (b) _____

2. $f(x) = 2x^2 + 3x - 5$

    2. (a) _____
       (b) _____

Name:
Instructor:

Date:
Section:

3. $f(x) = 3x^4 - 5x^2$

3. (a) _____

(b) _____

Objective 2 Perform operations on polynomial functions.

Video Examples

Review these examples for Objective 2:

2. For $f(x) = 2x^2 + 4x - 5$ and
$g(x) = -x^2 + 3x - 8$, find each of the following.

   a. $(f+g)(x)$

   $(f+g)(x) = f(x) + g(x)$
   $= (2x^2 + 4x - 5) + (-x^2 + 3x - 8)$
   $= x^2 + 7x - 13$

   b. $(f-g)(x)$

   $(f-g)(x) = f(x) - g(x)$
   $= (2x^2 + 4x - 5) - (-x^2 + 3x - 8)$
   $= (2x^2 + 4x - 5) + (x^2 - 3x + 8)$
   $= 3x^2 + x + 3$

4. For $f(x) = 5x + 2$ and $g(x) = 3x^2 + 4x$, find $(fg)(x)$ and $(fg)(-2)$.

   $(fg)(x) = f(x) \cdot g(x)$
   $= (5x+2)(3x^2 + 4x)$
   $= 15x^3 + 20x^2 + 6x^2 + 8x$
   $= 15x^3 + 26x^2 + 8x$
   $(fg)(-2) = 15(-2)^3 + 26(-2)^2 + 8(-2)$
   $= -120 + 104 - 16$
   $= -32$

Now Try:

2. For $f(x) = 6x^2 - 7x + 12$ and
$g(x) = -3x^2 + x + 9$, find each of the following.

   a. $(f+g)(x)$

   _____

   b. $(f-g)(x)$

   _____

4. For $f(x) = 6x + 5$ and
$g(x) = 7x^2 + 2x$, find
$(fg)(x)$ and $(fg)(-3)$.

   _____

Copyright © 2016 Pearson Education, Inc.

Name: Date:
Instructor: Section:

5. For $f(x) = 4x^2 - 17x - 15$ and $g(x) = x - 5$, find $\left(\dfrac{f}{g}\right)(x)$ and $\left(\dfrac{f}{g}\right)(-2)$.

$\left(\dfrac{f}{g}\right)(x) = \dfrac{f(x)}{g(x)} = \dfrac{4x^2 - 17x - 15}{x - 5}$

$$\begin{array}{r} 4x+3 \\ x-5 \overline{) 4x^2 - 17x - 15} \\ \underline{4x^2 - 20x} \\ 3x - 15 \\ \underline{3x - 15} \\ 0 \end{array}$$

$\left(\dfrac{f}{g}\right)(x) = 4x + 3, \quad x \neq 5$

$\left(\dfrac{f}{g}\right)(-2) = 4(-2) + 3 = -5$

5. For $f(x) = 2x^2 - 3x - 20$ and $g(x) = x - 4$, find $\left(\dfrac{f}{g}\right)(x)$ and $\left(\dfrac{f}{g}\right)(-2)$.

_____

Objective 2 Practice Exercises

For extra help, see Examples 2–5 on pages 625–626 of your text.

For the pair of functions, find (a) $(f+g)(x)$ and (b) $(f-g)(x)$.

4. $f(x) = 2x^2 + 4x - 5, \; g(x) = -x^2 + 3x - 8$

4. a. _____

b. _____

For the pair of functions, find (a) $(fg)(x)$ and (b) $(fg)(-1)$.

5. $f(x) = 3x^2 + 2, \; g(x) = -5x$

5. a. _____

b. _____

For the pair of functions, find the quotient $\left(\dfrac{f}{g}\right)(x)$ and give any x-values that are not in the domain of the quotient function.

6. $f(x) = 4x^2 - 11x - 45, \; g(x) = x - 5$

6. _____

Name: Date:
Instructor: Section:

Objective 3   Find the composition of functions.

Video Examples

Review these examples for Objective 3:

6. Let $f(x) = 3x - 1$ and $g(x) = x^2 + 2$. Find $(f \circ g)(5)$.

$$(f \circ g)(5) = f(g(5))$$
$$= f(5^2 + 2)$$
$$= f(27)$$
$$= 3(27) - 1$$
$$= 80$$

7. Let $f(x) = 3x - 1$ and $g(x) = x^2 + 2$. Find $(f \circ g)(5)$.

$$(f \circ g)(5) = f(g(5))$$
$$= f(5^2 + 2)$$
$$= f(27)$$
$$= 3(27) - 1$$
$$= 80$$

Now Try:

6. Let $f(x) = -3x - 3$ and $g(x) = x^2 - 5$. Find $(f \circ g)(-3)$.

_____

7. Let $f(x) = -3x - 3$ and $g(x) = x^2 - 5$. Find $(f \circ g)(-3)$.

_____

Objective 3 Practice Exercises

For extra help, see Examples 6–8 on pages 627–629 of your text.

Find the following.

7. Let $f(x) = 4x - 3$ and $g(x) = 2x^2 - 1$. Find the following.
   a. $(f \circ g)(2)$
   b. $(g \circ f)(-1)$
   c. $(f \circ g)(x)$

7. a. _____

   b. _____

   c. _____

8. Let $f(x) = \dfrac{1}{x}$ and $g(x) = 3x^2 - 4x + 1$. Find the following.
   a. $(f \circ g)(2)$
   b. $(g \circ f)\left(\dfrac{1}{3}\right)$
   c. $(g \circ f)(x)$

8. a. _____

   b. _____

   c. _____

Name: Date:
Instructor: Section:

Chapter 9 RELATIONS AND FUNCTIONS

9.4 Variation

Learning Objectives
1. Write an equation expressing direct variation.
2. Find the constant of variation, and solve direct variation problems.
3. Solve inverse variation problems.
4. Solve joint variation problems.
5. Solve combined variation problems.

Key Terms

Use the vocabulary terms listed below to complete each statement in exercises 1–3.

    varies directly   **varies inversely**   **constant of variation**

1. In the equations for direct and inverse variation, k is the _____.

2. If there exists a real number k such that $y = \dfrac{k}{x}$, then y _____ as x.

3. If there exists a real number k such that $y = kx$, then y _____ as x.

Objective 1 Write an equation expressing direct variation.

For extra help, see pages 632–633 of your text.

Objective 2 Find the constant of variation, and solve direct variation problems.

Video Examples

Review these examples for Objective 2:	Now Try:

1. If 12 gallons of gasoline cost $34.68, how much does 1 gallon of gasoline cost? Write the variation equation.
 	
Let g represent the number of gallons of gasoline and let C represent the total cost of the gasoline. Then the variation equation is $C = kg$.
    $C = kg$
  $34.68 = 12k$
   $2.89 = k$
The cost per gallon is $2.89.
The variation equation is $C = 2.89g$.
	1. One week a manufacturer sold 1200 items for a total profit of $30,000. What was the profit for one item. Write the variation equation.
	
	_____
	
	_____

Name: Date:
Instructor: Section:

2. A person's weight on the moon varies directly with the person's weight on Earth. A 120-pound person would weigh about 20 pounds on the moon. How much would a 150-pound person weigh on the moon?

If m represents the person's weight on the moon and w represents the person's weight on Earth.
$$m = kw$$
$$20 = k \cdot 120 \quad \text{Let } m = 20 \text{ and } w = 120.$$
$$\frac{20}{120} = \frac{1}{6} = k \quad \text{Solve for } k; \text{ lowest terms}$$

Now, substitute $\frac{1}{6}$ for k and 150 for w in the variation equation.
$$m = kw$$
$$m = \frac{1}{6} \cdot 150 = 25$$
A 150-pound person will weigh 25 pounds on the moon.

3. The surface area of a sphere varies directly as the square of its radius. If the surface area of a sphere with a radius of 12 inches is 576π square inches, find the surface area of a sphere with a radius of 3 inches.

Step 1 A represents the surface area and r represents the radius.
$$A = kr^2$$
Step 2 Find the value of k when A is 576π and r is 12.
$$A = kr^2$$
$$576\pi = k \cdot 12^2$$
$$576\pi = 144k$$
$$\frac{576\pi}{144} = 4\pi = k$$
Step 3 Rewrite the variation equation.
$$A = 4\pi r^2$$
Step 4 Let $r = 3$ to find the surface area.
$$A = 4\pi \cdot 3^2 = 4\pi \cdot 9 = 36\pi$$
The surface area of a sphere with a radius of 3 inches is 36π square inches.

2. The pressure exerted by a certain liquid at a given point varies directly as the depth of the point beneath the surface of the liquid. The pressure at 10 feet is 50 pounds per square inch (psi). What is the pressure at 25 feet?

_____

3. The area of a circle varies directly as the square of the radius. A circle with a radius of 5 centimeters has an area of 78.5 square centimeters. Find the area if the radius changes to 7 centimeters.

_____

Name: Date:
Instructor: Section:

Objective 2 Practice Exercises

For extra help, see Examples 1–3 on pages 633–635 of your text.

Find the constant of variation, and write a direct variation equation.

1. $y = 13.75$ when $x = 55$

   1. _____

Solve each problem.

2. The circumference of a circle varies directly as the radius. A circle with a radius of 7 centimeters has a circumference of 43.96 centimeters. Find the circumference of the circle if the radius changes to 11 centimeters.

   2. _____

3. The force required to compress a spring varies directly as the change in length of the spring. If a force of 20 newtons is required to compress a spring 2 centimeters in length, how much force is required to compress a spring of length 10 centimeters?

   3. _____

Objective 3   Solve inverse variation problems.

Video Examples

Review these examples for Objective 3:

4. For a specified distance, time varies inversely with speed. If Ramona walks a certain distance on a treadmill in 40 minutes at 4.2 miles per hour, how long will it take her to walk the same distance at 3.5 miles per hour?

   Let t = time and s = speed.
   Since t varies inversely as s, there is a constant k such that $t = \dfrac{k}{s}$. Recall that 40 min = $\dfrac{40}{60}$ hr.

   $$t = \dfrac{k}{s}$$
   $$\dfrac{40}{60} = \dfrac{k}{4.2}$$
   $$2.8 = k$$

Now Try:

4. The length of a violin string varies inversely with the frequency of its vibrations. A 10-inch violin string vibrates at a frequency of 512 cycles per second. Find the frequency of an 8-inch string.

   _____

Name: Date:
Instructor: Section:

Now use $t = \dfrac{k}{s}$ to find the value of t when $s = 3.5$.

$$t = \dfrac{2.8}{3.5} = \dfrac{4}{5}$$

It takes $\dfrac{4}{5}$ hr, or 48 min to walk the same distance.

5. With constant power, the resistance used in a simple electric circuit varies inversely as the square of the current. If the resistance is 120 ohms when the current is 12 amps, find the resistance if the current is reduced to 9 amps.

   Let R represent resistance (in ohms) and I = current (in amps). Then $R = \dfrac{k}{I^2}$.

   First, we solve for the constant of variation by substituting 120 for R and 12 for I.

   $$120 = \dfrac{k}{12^2}$$

   $$k = 120 \cdot 12^2$$

   Now use the value for k and 9 for I to find R.

   $$R = \dfrac{120 \cdot 12^2}{9^2} \approx 213.3$$

   The resistance is about 213.3 ohms when the current is 9 amps.

5. If y varies inversely as x^3, and $y = 9$ when $x = 2$, find y when $x = 4$.

   _____

Objective 3 Practice Exercises

For extra help, see Examples 4–5 on page 636 of your text.

Solve each problem.

4. The illumination produced by a light source varies inversely as the square of the distance from the source. If the illumination produced 4 feet from a light source is 75 footcandles, find the illumination produced 9 feet from the same source.

4. _____

Name: Date:
Instructor: Section:

5. The weight of an object varies inversely as the square of its distance from the center of Earth. If an object 8000 miles from the center of Earth weighs 90 pounds, find its weight when it is 12,000 miles from the center of Earth.

5. _____

6. The speed of a pulley varies inversely as its diameter. One kind of pulley, with a diameter of 3 inches, turns at 150 revolutions per minute. Find the speed of a similar pulley with diameter of 5 inches.

6. _____

Objective 4   Solve joint variation problems.

Video Examples

Review this example for Objective 4:

6. For a fixed interest rate, interest varies jointly as the principal and the time in years. If $5000 invested for 4 years earns $900, how much interest will $6000 invested for 3 years earn at the same interest rate?

Let I = the interest, p = the principal, and t = the time in years. Then, $I = kpt$.
$$I = kpt$$
$$900 = k \cdot 5000 \cdot 4 \quad \text{Substitute given values.}$$
$$\frac{900}{20,000} = \frac{9}{200} = k$$

Now use $k = \frac{9}{200}$.

$$I = \frac{9}{200} \cdot 6000 \cdot 3 = 810$$

$6000 invested for three years will earn $810 in interest.

Now Try:

6. The strength of a rectangular beam varies jointly as its width and the square of its depth. If the strength of a beam 2 inches wide by 10 inches deep is 1000 pounds per square inch, what is the strength of a beam 4 inches wide and 8 inches deep?

_____

354   Copyright © 2016 Pearson Education, Inc.

Name: Date:
Instructor: Section:

Objective 4 Practice Exercises

For extra help, see Example 6 on page 637 of your text.

Solve each problem.

7. Suppose d varies jointly as f^2 and g^2, and $d = 384$ when $f = 3$ and $g = 8$. Find d when $f = 6$ and $g = 2$.

7. _____

8. The work w (in joules) done when lifting an object is jointly proportional to the product of the mass m (in kg) of the object and the height h (in meter) the object is lifted. If the work done when a 120 kg object is lifted 1.8 meters above the ground is 2116.8 joules, how much work is done when lifting a 100kg object 1.5 meters above the ground?

8. _____

9. The absolute temperature of an ideal gas varies jointly as its pressure and its volume. If the absolute temperature is 250° when the pressure is 25 pounds per square centimeter and the volume is 50 cubic centimeters, find the absolute temperature when the pressure is 50 pounds per square centimeter and the volume is 75 cubic centimeters.

9. _____

Name: Date:
Instructor: Section:

Objective 5   Solve combined variation problems.

Video Examples

Review this example for Objective 5:

7. The number of hours h that it takes w workers to assemble x machines varies directly as the number of machines and inversely as the number of workers. If four workers can assemble 12 machines in four hours, how many workers are needed to assemble 36 machines in eight hours?

The variation equation is $h = \dfrac{kx}{w}$.

To find k, let $h = 4$, $x = 12$, and $w = 4$.

$$4 = \dfrac{k \cdot 12}{4}$$

$$k = \dfrac{4 \cdot 4}{12}$$

$$k = \dfrac{4}{3}$$

Now find w when $x = 36$ and $h = 8$.

$$8 = \dfrac{\frac{4}{3} \cdot 36}{w}$$

$$8w = 48$$

$$w = 6$$

Eight workers are needed to assemble 36 machines in eight hours.

Now Try:

7. The volume of a gas varies directly as its temperature and inversely as its pressure. The volume of a gas at 85° C at a pressure of 12 kg/cm² is 300 cm³. What is the volume when the pressure is 20 kg/cm² and the temperature is 30° C?

_____

Objective 5 Practice Exercises

For extra help, see Example 7 on pages 637–638 of your text.

Solve each problem.

10. The volume of a gas varies inversely as the pressure and directly as the temperature. If a certain gas occupies a volume of 1.3 liters at 300 K and a pressure of 18 kilograms per square centimeter, find the volume at 340 K and a pressure of 24 kilograms per square centimeter.

10. _____

Name: Date:
Instructor: Section:

11. The time required to lay a sidewalk varies directly as its length and inversely as the number of people who are working on the job. If three people can lay a sidewalk 100 feet long in 15 hours, how long would it take two people to lay a sidewalk 40 feet long?

11. _____

12. When an object is moving in a circular path, the centripetal force varies directly as the square of the velocity and inversely as the radius of the circle. A stone that is whirled at the end of a string 50 centimeters long at 900 centimeters per second has a centripetal force of 3,240,000 dynes. Find the centripetal force if the stone is whirled at the end of a string 75 centimeters long at 1500 centimeters per second.

12. _____

Name: Date:
Instructor: Section:

Chapter 10 ROOTS, RADICALS, AND ROOT FUNCTIONS

10.1 Radical Expressions and Graphs

Learning Objectives
1. Find square roots.
2. Decide whether a given root is rational, irrational, or not a real number.
3. Find cube, fourth, and other roots.
4. Graph functions defined by radical expressions.
5. Find nth roots of nth powers.
6. Use a calculator to find roots.

Key Terms

Use the vocabulary terms listed below to complete each statement in exercises 1–12.

square root	principal square root	negative square root
radicand	radical	perfect square
cube root	fourth root	index (order)
radical expression	square root function	cube root function

1. 5 is the _____ of 625.

2. A _____ is a radical sign and the number or expression that appears under it.

3. The _____ of a positive number with even index n is the positive nth root of the number.

4. A _____ is the number or expression that appears inside a radical sign.

5. $f(x) = \sqrt[3]{x}$ is called the _____.

6. A number with a rational square root is called a _____.

7. The symbol $-\sqrt{}$ is used for the _____ of a number.

8. In a radical of the form $\sqrt[n]{a}$, the number n is the _____.

9. The number b is a _____ of a if $b^3 = a$.

10. The domain and range of $f(x) = \sqrt{x}$, _____, are both $[0, \infty)$.

11. The _____ of a number is a number that, when multiplied by itself, gives the original number.

12. A _____ is an algebraic expression containing a radical.

Name: Date:
Instructor: Section:

Objective 1 Find roots of numbers.

Video Examples

Review these examples for Objective 1:

2. Find each square root.

   a. $\sqrt{121}$

   $11^2 = 121$, so $\sqrt{121} = 11$.

   b. $-\sqrt{576}$

   This is the negative square root of 576.
   Since $\sqrt{576} = 24$, then $-\sqrt{576} = -24$.

   c. $\sqrt{\dfrac{16}{25}}$

   $\sqrt{\dfrac{16}{25}} = \dfrac{4}{5}$

3. Find the square of each radical expression.

   a. $\sqrt{17}$

   The square of $\sqrt{17}$ is $\left(\sqrt{17}\right)^2 = 17$.

   b. $-\sqrt{31}$

   $\left(-\sqrt{31}\right)^2 = 31$

   c. $\sqrt{w^2 + 3}$

   $\left(\sqrt{w^2 + 3}\right)^2 = w^2 + 3$

Now Try:

2. Find each square root.

   a. $\sqrt{169}$

   _____

   b. $\sqrt{1681}$

   _____

   c. $\sqrt{\dfrac{9}{49}}$

   _____

3. Find the square of each radical expression.

   a. $\sqrt{19}$

   _____

   b. $-\sqrt{37}$

   _____

   c. $\sqrt{n^2 + 5}$

   _____

Objective 1 Practice Exercises

For extra help, see Examples 1–3 on pages 650–651 of your text.

Find all square roots of each number.

1. 625

2. $\dfrac{121}{196}$

Find the square root.

3. $\sqrt{\dfrac{900}{49}}$

1. _____

2. _____

3. _____

Name: Date:
Instructor: Section:

Objective 2   Decide whether a given root is rational, irrational, or not a real number.

Video Examples

Review these examples for Objective 2:
4. Tell whether each square root is rational, irrational, or not a real number.

    a. $\sqrt{5}$

    Because 5 is not a perfect square, $\sqrt{5}$ is irrational.

    b. $\sqrt{81}$

    81 is a perfect square, 9^2, so $\sqrt{81} = 9$ is a rational number.

    c. $\sqrt{-16}$

    There is no real number whose square is −16. Therefore, $\sqrt{-16}$ is not a real number.

Now Try:
4. Tell whether each square root is rational, irrational, or not a real number.

    a. $\sqrt{11}$

    _____

    b. $\sqrt{100}$

    _____

    c. $\sqrt{-9}$

    _____

Objective 2 Practice Exercises

For extra help, see Example 4 on page 652 of your text.

Tell whether each square root is rational, irrational, *or not a real number.*

4. $\sqrt{72}$ 4. _____

5. $\sqrt{-36}$ 5. _____

6. $\sqrt{6400}$ 6. _____

Objective 3   Find cube, fourth, and other roots.

Video Examples

Review these examples for Objective 3:
5. Find each cube root.

    a. $\sqrt[3]{64}$

    Because $4^3 = 64$, $\sqrt[3]{64} = 4$.

Now Try:
5. Find each cube root.

    a. $\sqrt[3]{125}$

    _____

Name: Date:
Instructor: Section:

b. $\sqrt[3]{-64}$

$\sqrt[3]{-64} = -4$, because $(-4)^3 = -64$.

c. $\sqrt[3]{729}$

$\sqrt[3]{729} = 9$, because $9^3 = 729$.

6. Find each root.

   a. $\sqrt[4]{81}$

   $\sqrt[4]{81} = 3$, because 3 is positive and $3^4 = 81$.

   b. $-\sqrt[4]{81}$

   From part (a), $\sqrt[4]{81} = 3$, so the negative root is $-\sqrt[4]{81} = -3$.

   c. $\sqrt[4]{-81}$

   For a real number fourth root, the radicand must be nonnegative. There is no real number that equals $\sqrt[4]{-81}$.

   d. $-\sqrt[5]{1024}$

   $-\sqrt[5]{1024} = -4$

   e. $\sqrt[5]{-1024}$

   $\sqrt[5]{-1024} = -4$, because $(-4)^5 = -1024$.

b. $\sqrt[3]{-125}$

c. $\sqrt[3]{343}$

6. Find each root.

   a. $\sqrt[4]{625}$

   b. $-\sqrt[4]{625}$

   c. $\sqrt[4]{-625}$

   d. $-\sqrt[5]{3125}$

   e. $\sqrt[5]{-3125}$

Objective 3 Practice Exercises

For extra help, see Examples 5–6 on pages 653–654 of your text.

Find each root.

7. $\sqrt[3]{-64}$

8. $\sqrt[4]{256}$

9. $\sqrt[7]{-1}$

13. _____

14. _____

15. _____

362 Copyright © 2016 Pearson Education, Inc.

Name: Date:
Instructor: Section:

Objective 4 Graph functions defined by radical expressions.

Video Examples

Review these examples for Objective 4:

7. Graph each function by creating a table of values. Give the domain and the range.

   a. $f(x) = \sqrt{x-1}$

   Create a table of values.

x	$f(x) = \sqrt{x-1}$
1	$\sqrt{1-1} = 0$
5	$\sqrt{5-1} = 2$
10	$\sqrt{10-1} = 3$

   For the radicand to be nonnegative, we must have $x - 1 \geq 0$ or $x \geq 1$. Therefore, the domain is $[1, \infty)$. Function values are nonnegative, so the range is $[0, \infty)$.

   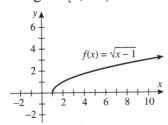

   b. $f(x) = \sqrt[3]{x} + 1$

   Create a table of values.

x	$f(x) = \sqrt[3]{x} + 1$
-8	$\sqrt[3]{-8} + 1 = -1$
-1	$\sqrt[3]{-1} + 1 = 0$
0	$\sqrt[3]{0} + 1 = 1$
1	$\sqrt[3]{1} + 1 = 2$
8	$\sqrt[3]{8} + 1 = 3$

   Both the domain and range are $(-\infty, \infty)$.

   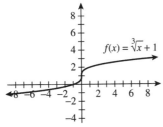

Now Try:

7. Graph each function by creating a table of values. Give the domain and the range.

   a. $f(x) = \sqrt{x} - 1$

   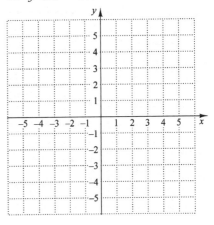

   domain: _____

   range: _____

   b. $f(x) = \sqrt[3]{x+1}$

   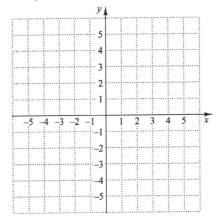

   domain: _____

   range: _____

Copyright © 2016 Pearson Education, Inc.

Name:
Instructor:
Date:
Section:

Objective 4 Practice Exercises

For extra help, see Example 7 on page 655 of your text.

Graph each function and give its domain and its range.

10. $f(x) = \sqrt{x} + 2$

10.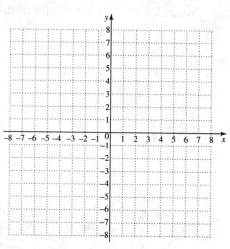

domain: _____

range: _____

11. $f(x) = \sqrt[3]{x} - 2$

11.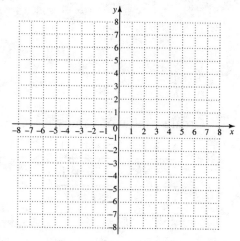

domain: _____

range: _____

364 Copyright © 2016 Pearson Education, Inc.

Name: Date:
Instructor: Section:

12. $f(x) = \sqrt[3]{x} + 2$

12.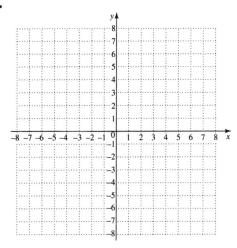

domain: _____

range: _____

Objective 5 Find *n*th roots of *n*th powers.

Video Examples

Review these examples for Objective 5:

8. Find each square root. In parts (c) and (d), *m* is a real number.

   a. $\sqrt{33^2}$

   $\sqrt{33^2} = |33| = 33$

   b. $\sqrt{(-33)^2}$

   $\sqrt{(-33)^2} = |-33| = 33$

   c. $\sqrt{m^2}$

   $\sqrt{m^2} = |m|$

   d. $\sqrt{(-m)^2}$

   $\sqrt{(-m)^2} = |-m| = |m|$

Now Try:

8. Find each square root. In parts (c) and (d), *n* is a real number.

   a. $\sqrt{73^2}$

   _____

   b. $\sqrt{(-37)^2}$

   _____

   c. $\sqrt{n^2}$

   _____

   d. $\sqrt{(-n)^2}$

   _____

Name: Date:
Instructor: Section:

9. Simplify each root.

   a. $\sqrt[4]{(-5)^4}$

   n is even. Use absolute value.
   $\sqrt[4]{(-5)^4} = |-5| = 5$

   b. $\sqrt[5]{(-3)^5}$

   n is odd.
   $\sqrt[5]{(-3)^5} = -3$

   c. $-\sqrt[6]{(-8)^6}$

   n is even. Use absolute value.
   $-\sqrt[6]{(-8)^6} = -|-8| = -8$

   d. $-\sqrt{r^{12}}$

   $-\sqrt{r^{12}} = -|r^6| = -r^6$

   No absolute value bars are needed here since r^6 is nonnegative for any real number value of r.

   e. $\sqrt[5]{s^{20}}$

   $\sqrt[5]{s^{20}} = s^4$, because $s^{20} = (s^4)^5$.

   f. $\sqrt[4]{x^{20}}$

   $\sqrt[4]{x^{20}} = |x^5|$

9. Simplify each root.

   a. $\sqrt[8]{(-4)^8}$

   _____

   b. $\sqrt[7]{(-5)^7}$

   _____

   c. $-\sqrt[4]{(-2)^4}$

   _____

   d. $-\sqrt{x^8}$

   _____

   e. $\sqrt[3]{w^{30}}$

   _____

   f. $\sqrt[6]{x^{30}}$

   _____

Objective 5 Practice Exercises

For extra help, see Examples 8–9 on page 656 of your text.

Simplify each root.

13. $\sqrt{(-9)^2}$

10. _____

14. $-\sqrt[5]{x^5}$

11. _____

15. $-\sqrt[4]{x^{16}}$

12. _____

Name: Date:
Instructor: Section:

Objective 6   Use a calculator to find roots.

Video Examples

Review these examples for Objective 6:

10. Use a calculator to approximate each radical to three decimal places.

   a. $\sqrt{12}$

   Use the square root key on a calculator.
   $\sqrt{12} \approx 3.464$

   b. $-\sqrt{596}$

   $-\sqrt{596} \approx -24.413$

   c. $\sqrt[3]{61}$

   $\sqrt[3]{61} \approx 3.936$

   d. $\sqrt[4]{6902}$

   $\sqrt[4]{6902} \approx 9.115$

Now Try:

10. Use a calculator to approximate each radical to three decimal places.

   a. $\sqrt{14}$

   _____

   b. $-\sqrt{678}$

   _____

   c. $\sqrt[3]{431}$

   _____

   d. $\sqrt[4]{8142}$

   _____

Objective 6 Practice Exercises

For extra help, see Examples 6–7 on page 657 of your text.

Use a calculator to find a decimal approximation for each radical. Give the answer to the nearest thousandth.

16. $\sqrt[3]{701}$ 16. _____

17. $-\sqrt{990}$ 17. _____

18.   The time t in seconds for one complete swing of a simple pendulum, where L is the length of the pendulum in feet is $t = 2\pi\sqrt{\dfrac{L}{32}}$. Find the time of a complete swing of a 4-ft pendulum to the nearest tenth of a second.   18. _____

Name: Date:
Instructor: Section:

Chapter 10 ROOTS, RADICALS, AND ROOT FUNCTIONS

10.2 Rational Exponents

Learning Objectives
1   Use exponential notation for *n*th roots.
2   Define and use expressions of the form $a^{m/n}$.
3   Convert between radicals and rational exponents.
4   Use the rules for exponents with rational exponents.

Key Terms

Use the vocabulary terms listed below to complete each statement in exercises 1–3.

    product rule for exponents   quotient rule for exponents

    power rule for exponents

1. $(x^2 y^3)^4 = x^8 y^{12}$ is an example of the _____.

2. $w^5 w^3 = w^8$ is an example of the _____.

3. $\dfrac{z^6}{z^4} = z^2$ is an example of the _____.

Objective 1   Use exponential notation for *n*th roots.

Video Examples

Review these examples for Objective 1:	Now Try:
1. Evaluate each exponential.	1. Evaluate each exponential.
a. $25^{1/2}$	a. $121^{1/2}$
$25^{1/2} = \sqrt{25} = 5$	_____
b. $(-8)^{1/3}$	b. $(-243)^{1/5}$
$(-8)^{1/3} = \sqrt[3]{-8} = -2$	_____
c. $(-81)^{1/4}$	c. $(-1024)^{1/10}$
$(-81)^{1/4} = \sqrt[4]{-81}$ is not a real number.	_____

Name: Date:
Instructor: Section:

Objective 1 Practice Exercises

For extra help, see Example 1 on page 662 of your text.

Evaluate each exponential.

1. $-256^{1/4}$

   1. _____

2. $16^{1/2}$

   2. _____

3. $(-3375)^{1/3}$

   3. _____

Objective 2 Define and use expressions of the form $a^{m/n}$.

Video Examples

Review these examples for Objective 2:

2. Evaluate each polynomial.

   a. $100^{3/2}$

   $100^{3/2} = (100^{1/2})^3 = 10^3 = 1000$

   b. $-729^{5/6}$

   $-729^{5/6} = -(729^{1/6})^5 = -(3)^5 = -243$

3. Evaluate the exponential.

   $625^{-3/4}$

   $625^{-3/4} = \dfrac{1}{625^{3/4}} = \dfrac{1}{(625^{1/4})^3} = \dfrac{1}{(\sqrt[4]{625})^3}$

   $= \dfrac{1}{5^3} = \dfrac{1}{125}$

Now Try:

2. Evaluate each polynomial.

   a. $27^{2/3}$

   _____

   b. $-36^{3/2}$

   _____

3. Evaluate the exponential.

   $32^{-2/5}$

   _____

Objective 2 Practice Exercises

For extra help, see Examples 2–3 on pages 663–664 of your text.

Evaluate each exponential.

4. $-81^{5/4}$

   4. _____

Name: Date:
Instructor: Section:

5. $36^{5/2}$ 5. _____

6. $\left(\dfrac{125}{27}\right)^{-2/3}$ 6. _____

Objective 3 Convert between radicals and rational exponents.

Video Examples

Review these examples for Objective 3:

4. Write each radical as an exponential and simplify. Assume that all variables represent positive real numbers. Use the definition that takes the root first.

   a. $17^{1/2}$

   $17^{1/2} = \sqrt{17}$

   b. $2x^{2/5} - (4x)^{5/6}$

   $2x^{2/5} - (4x)^{5/6} = 2\left(\sqrt[5]{x}\right)^2 - \left(\sqrt[6]{4x}\right)^5$

   c. $\sqrt[3]{3^6}$

   $\sqrt[3]{3^6} = 3^{6/3} = 3^2 = 9$

Now Try:

4. Write each radical as an exponential and simplify. Assume that all variables represent positive real numbers. Use the definition that takes the root first.

   a. $23^{1/3}$

   _____

   b. $(2x)^{4/3} - 3x^{2/5}$

   _____

   c. $\sqrt{10^4}$

   _____

Objective 3 Practice Exercises

For extra help, see Example 4 on page 665 of your text.

Write with radicals. Assume that all variables represent positive real numbers.

7. $4y^{2/5} + (5x)^{1/5}$ 7. _____

8. $(2x^4 - 3y^2)^{-4/3}$ 8. _____

Name: Date:
Instructor: Section:

Simplify the radical by rewriting it with a rational exponent. Write answer in radical form if necessary. Assume that variables represent positive real numbers.

9. $\sqrt[8]{a^2}$

9. _____

Objective 4   Use the rules for exponents with rational exponents.

Video Examples

Review these examples for Objective 4:

5. Write with only positive exponents. Assume that all variables represent positive real numbers.

   a. $13^{4/5} \cdot 13^{1/2}$

   $13^{4/5} \cdot 13^{1/2} = 13^{4/5+1/2} = 13^{13/10}$

   b. $\left(\dfrac{c^6 x^3}{c^{-2} x^{1/2}}\right)^{-3/4}$

   $\left(\dfrac{c^6 x^{1/2}}{c^{-2} x^3}\right)^{-3/4} = \left(c^{6-(-2)} x^{1/2-3}\right)^{-3/4}$

   $= \left(c^8 x^{-5/2}\right)^{-3/4}$

   $= \left(c^8\right)^{-3/4} \left(x^{-5/2}\right)^{-3/4}$

   $= c^{-6} x^{15/8}$

   $= \dfrac{x^{15/8}}{c^6}$

Now Try:

5. Write with only positive exponents. Assume that all variables represent positive real numbers.

   a. $5^{3/4} \cdot 5^{7/4}$

   _____

   b. $\left(\dfrac{x^{-1} y^{2/3}}{x^{1/3} y^{1/2}}\right)^{-3/2}$

   _____

Name: Date:
Instructor: Section:

6. Write all radicals as exponentials, and then apply the rules for rational exponents. Leave answers in exponential form. Assume that all variables represent positive real numbers.

   a. $\sqrt[4]{x^3} \cdot \sqrt[5]{x}$

   $\sqrt[4]{x^3} \cdot \sqrt[5]{x} = x^{3/4} \cdot x^{1/5}$
   $= x^{3/4+1/5}$
   $= x^{15/20+4/20}$
   $= x^{19/20}$

   b. $\sqrt{\sqrt[4]{y^3}}$

   $\sqrt{\sqrt[4]{y^3}} = \sqrt{y^{3/4}} = \left(y^{3/4}\right)^{1/2} = y^{3/8}$

6. Write all radicals as exponentials, and then apply the rules for rational exponents. Leave answers in exponential form. Assume that all variables represent positive real numbers.

   a. $\sqrt[6]{x^3} \cdot \sqrt[3]{x^2}$

   _____

   b. $\sqrt[3]{\sqrt[4]{x^3}}$

   _____

Objective 4 Practice Exercises

For extra help, see Examples 5–6 on pages 666–667 of your text.

Use the rules of exponents to simplify each expression. Write all answers with positive exponents. Assume that variables represent positive real numbers.

10. $y^{7/3} \cdot y^{-4/3}$

10. _____

11. $\dfrac{a^{2/3} \cdot a^{-1/3}}{\left(a^{-1/6}\right)^3}$

11. _____

12. $\dfrac{\left(x^{-3} y^2\right)^{2/3}}{\left(x^2 y^{-5}\right)^{2/5}}$

12. _____

Name: Date:
Instructor: Section:

Chapter 10 ROOTS, RADICALS, AND ROOT FUNCTIONS

10.3 Simplifying Radicals, the Distance Formula, and Circles

Learning Objectives
1. Use the product rule for radicals.
2. Use the quotient rule for radicals.
3. Simplify radicals.
4. Simplify products and quotients of radicals with different indexes.
5. Use the Pythagorean theorem.
6. Use the distance formula.
7. Find an equation of a circle given its center and radius.

Key Terms

Use the vocabulary terms listed below to complete each statement in exercises 1–6.

index   radicand   hypotenuse   legs

radius   circle   center

1. In a right triangle, the side opposite the right angle is called the
   _____.

2. In the expression $\sqrt[4]{x^2}$, the "4" is the _____ and x^2 is the _____.

3. In a right triangle, the sides that form the right angle are called the
   _____.

4. A(n) _____ is the set of all points in a plane that lie a fixed distance from a fixed point.

5. A fixed point such that every point on a circle is a fixed distance from it is the
   _____.

6. The distance from the center of a circle to a point on the circle is called the
   _____.

Objective 1   Use the product rule for radicals.

Video Examples

Review these examples for Objective 1:
1. Multiply. Assume that all variables represent positive real numbers.

   a. $\sqrt{13} \cdot \sqrt{5}$

   $\sqrt{13} \cdot \sqrt{5} = \sqrt{13 \cdot 5} = \sqrt{65}$

Now Try:
1. Multiply. Assume that all variables represent positive real numbers.

   a. $\sqrt{2} \cdot \sqrt{7}$

   _____

Name: Date:
Instructor: Section:

b. $\sqrt{5x} \cdot \sqrt{2yz}$

$\sqrt{5x} \cdot \sqrt{2yz} = \sqrt{10xyz}$

2. Multiply. Assume that all variables represent positive real numbers.

   a. $\sqrt[4]{2} \cdot \sqrt[4]{2x}$

   $\sqrt[4]{2} \cdot \sqrt[4]{2x} = \sqrt[4]{2 \cdot 2x} = \sqrt[4]{4x}$

   b. $\sqrt[3]{8x} \cdot \sqrt[3]{2y^2}$

   $\sqrt[3]{8x} \cdot \sqrt[3]{2y^2} = \sqrt[3]{8x \cdot 2y^2} = \sqrt[3]{16xy^2}$

   c. $\sqrt[5]{6r^2} \cdot \sqrt[5]{4r^2}$

   $\sqrt[5]{6r^2} \cdot \sqrt[5]{4r^2} = \sqrt[5]{6r^2 \cdot 4r^2} = \sqrt[5]{24r^4}$

   d. $\sqrt[5]{2} \cdot \sqrt[4]{6}$

   $\sqrt[5]{2} \cdot \sqrt[4]{6}$ cannot be simplified using the product rule for radicals, because the indexes (5 and 4) are different.

b. $\sqrt{3} \cdot \sqrt{11mn}$

_____

2. Multiply. Assume that all variables represent positive real numbers.

   a. $\sqrt[3]{3} \cdot \sqrt[3]{7}$

   _____

   b. $\sqrt[3]{7x} \cdot \sqrt[3]{5y}$

   _____

   c. $\sqrt[5]{4w} \cdot \sqrt[5]{2w^3}$

   _____

   d. $\sqrt{3} \cdot \sqrt[3]{64}$

   _____

Objective 1 Practice Exercises

For extra help, see Examples 1–2 on page 671 of your text.

Multiply. Assume that variables represent positive real numbers.

1. $\sqrt{7x} \cdot \sqrt{6t}$ 1. _____

2. $\sqrt[5]{6r^2t^3} \cdot \sqrt[5]{4r^2t}$ 2. _____

3. $\sqrt{3} \cdot \sqrt[3]{7}$ 3. _____

Name: Date:
Instructor: Section:

Objective 2 Use the quotient rule for radicals.

Video Examples

Review these examples for Objective 2:
3. Simplify. Assume that all variables represent positive real numbers.

   a. $\sqrt{\dfrac{64}{9}}$

   $\sqrt{\dfrac{64}{9}} = \dfrac{\sqrt{64}}{\sqrt{9}} = \dfrac{8}{3}$

   b. $\sqrt{\dfrac{5}{16}}$

   $\sqrt{\dfrac{5}{16}} = \dfrac{\sqrt{5}}{\sqrt{16}} = \dfrac{\sqrt{5}}{4}$

   c. $\sqrt[3]{-\dfrac{27}{8}}$

   $\sqrt[3]{-\dfrac{27}{8}} = \dfrac{\sqrt[3]{-27}}{\sqrt[3]{8}} = \dfrac{-3}{2} = -\dfrac{3}{2}$

   d. $\sqrt[5]{-\dfrac{a^3}{243}}$

   $\sqrt[5]{-\dfrac{a^3}{243}} = \dfrac{\sqrt[5]{a^3}}{\sqrt[5]{-243}} = \dfrac{\sqrt[5]{a^3}}{-3} = -\dfrac{\sqrt[5]{a^3}}{3}$

   e. $\sqrt{\dfrac{z^4}{36}}$

   $\sqrt{\dfrac{z^4}{36}} = \dfrac{\sqrt{z^4}}{\sqrt{36}} = \dfrac{z^2}{6}$

Now Try:
3. Simplify. Assume that all variables represent positive real numbers.

   a. $\sqrt{\dfrac{36}{49}}$

   _____

   b. $\sqrt{\dfrac{13}{81}}$

   _____

   c. $\sqrt[3]{-\dfrac{343}{125}}$

   _____

   d. $\sqrt[3]{-\dfrac{a^6}{125}}$

   _____

   e. $\sqrt[4]{\dfrac{m}{81}}$

   _____

Objective 2 Practice Exercises

For extra help, see Example 3 on pages 671–672 of your text.

Simplify each radical. Assume that variables represent positive real numbers.

4. $\sqrt[3]{\dfrac{27}{8}}$

4. _____

Copyright © 2016 Pearson Education, Inc.

Name: Date:
Instructor: Section:

5. $\sqrt[5]{\dfrac{7x}{32}}$

5. _____

6. $\sqrt[3]{-\dfrac{x^9}{216}}$

6. _____

Objective 3 Simplify radicals.

Video Examples

Review these examples for Objective 3:
4. Simplify.

   a. $\sqrt{90}$

$$\sqrt{90} = \sqrt{9 \cdot 10}$$
$$= \sqrt{9} \cdot \sqrt{10}$$
$$= 3\sqrt{10}$$

   b. $\sqrt{288}$

$$\sqrt{288} = \sqrt{144 \cdot 2}$$
$$= \sqrt{144} \cdot \sqrt{2}$$
$$= 12\sqrt{2}$$

   c. $\sqrt{35}$

No perfect square (other than 1) divides into 35, so $\sqrt{35}$ cannot be simplified further.

   d. $\sqrt[3]{81}$

$$\sqrt[3]{81} = \sqrt[3]{27 \cdot 3} = \sqrt[3]{27} \cdot \sqrt[3]{3} = 3\sqrt[3]{3}$$

   e. $-\sqrt[4]{3125}$

$$-\sqrt[4]{3125} = -\sqrt[4]{5^5} = \sqrt[4]{5^4 \cdot 5}$$
$$= -\sqrt[4]{5^4} \cdot \sqrt[4]{5}$$
$$= -5\sqrt[4]{5}$$

Now Try:
4. Simplify.

   a. $\sqrt{84}$

_____

   b. $\sqrt{162}$

_____

   c. $\sqrt{95}$

_____

   d. $\sqrt[3]{256}$

_____

   e. $-\sqrt[5]{512}$

_____

Name: Date:
Instructor: Section:

5. Simplify. Assume that all variables represent positive real numbers.

   a. $\sqrt{81x^3}$

   $\sqrt{81x^3} = \sqrt{9^2 \cdot x^2 \cdot x} = 9x\sqrt{x}$

   b. $\sqrt{56x^7y^6}$

   $\sqrt{56x^7y^6} = \sqrt{4 \cdot 14 \cdot (x^3)^2 \cdot x \cdot (y^3)^2}$
   $= 2x^3y^3\sqrt{14x}$

   c. $\sqrt[3]{-270b^4c^8}$

   $\sqrt[3]{-270b^4c^8} = \sqrt[3]{(-27b^3c^6)(10bc^2)}$
   $= \sqrt[3]{-27b^3c^6} \cdot \sqrt[3]{10bc^2}$
   $= -3bc^2\sqrt[3]{10bc^2}$

   d. $-\sqrt[6]{448a^7b^7}$

   $-\sqrt[6]{448a^7b^7} = -\sqrt[6]{(64a^6b^6)(7ab)}$
   $= -\sqrt[6]{64a^6b^6} \cdot \sqrt[6]{7ab}$
   $= -2ab\sqrt[6]{7ab}$

6. Simplify. Assume that all variables represent positive real numbers.

   a. $\sqrt[24]{5^4}$

   $\sqrt[24]{5^4} = (5^4)^{1/24} = 5^{4/24} = 5^{1/6} = \sqrt[6]{5}$

   b. $\sqrt[12]{x^8}$

   $\sqrt[12]{x^8} = (x^8)^{1/12} = x^{8/12} = x^{2/3} = \sqrt[3]{x^2}$

5. Simplify. Assume that all variables represent positive real numbers.

   a. $\sqrt{100y^3}$

   _____

   b. $\sqrt{48m^5r^9}$

   _____

   c. $\sqrt[3]{-32n^7t^5}$

   _____

   d. $-\sqrt[4]{405x^3y^9}$

   _____

6. Simplify. Assume that all variables represent positive real numbers.

   a. $\sqrt[12]{11^9}$

   _____

   b. $\sqrt[30]{z^{24}}$

   _____

Objective 3 Practice Exercises

For extra help, see Examples 4–6 on pages 672–674 of your text.

Simplify each radical. Assume that variables represent positive real numbers.

7. $\sqrt[42]{x^{28}}$

7.

Name: Date:
Instructor: Section:

8. $\sqrt{8x^3y^6z^{11}}$

8. _____

9. $\sqrt[3]{1250a^5b^7}$

9. _____

Objective 4 Simplify products and quotients of radicals with different indexes.

Video Examples

Review this example for Objective 4:

7. Simplify $\sqrt{3} \cdot \sqrt[5]{6}$.

Because the different indexes, 2 and 5, have least common multiple index of 10, we use rational exponents to write each radical as a tenth root.

$\sqrt{3} = 3^{1/2} = 3^{5/10} = \sqrt[10]{3^5} = \sqrt[10]{243}$

$\sqrt[5]{6} = 6^{1/5} = 6^{2/10} = \sqrt[10]{6^2} = \sqrt[10]{36}$

$\sqrt{3} \cdot \sqrt[5]{6} = \sqrt[10]{243} \cdot \sqrt[10]{36}$

$\phantom{\sqrt{3} \cdot \sqrt[5]{6}} = \sqrt[10]{243 \cdot 36}$

$\phantom{\sqrt{3} \cdot \sqrt[5]{6}} = \sqrt[10]{8748}$

Now Try:

7. Simplify $\sqrt[3]{3} \cdot \sqrt[6]{7}$.

_____

Objective 4 Practice Exercises

For extra help, see Example 7 on page 674 of your text.

Simplify each radical. Assume that variables represent positive real numbers.

10. $\sqrt{r} \cdot \sqrt[3]{r}$

10. _____

11. $\sqrt[4]{2} \cdot \sqrt[8]{7}$

11. _____

12. $\sqrt{3} \cdot \sqrt[5]{64}$

12. _____

Name: Date:
Instructor: Section:

Objective 5   Use the Pythagorean theorem.

Video Examples

Review this example for Objective 5:	Now Try:
8. Use the Pythagorean theorem to find the length of the unknown side of the triangle. $$a^2 + b^2 = c^2$$ $$12^2 + b^2 = 25^2$$ $$144 + b^2 = 625$$ $$b^2 = 481$$ $$b = \sqrt{481}$$ The length of the side is $\sqrt{481}$.	8. Use the Pythagorean theorem to find the length of the unknown side of the triangle. _____

Objective 5 Practice Exercises

For extra help, see Example 8 on page 675 of your text.

Find the unknown length in each right triangle. Simplify the answer if necessary.

13.

13. _____

14.

14. _____

15.

15. _____

Name: Date:
Instructor: Section:

Objective 6   Use the distance formula.

Video Examples

Review this example for Objective 6:

9. Find the distance between the points $(2,-2)$ and $(-6, 1)$.

    Use the distance formula. Let $(x_1, y_1) = (2, -2)$ and $(x_2, y_2) = (-6, 1)$.

    $$d = \sqrt{(x_2 - x_1)^2 + (y_2 - y_1)^2}$$
    $$= \sqrt{(-6-2)^2 + [1-(-2)]^2}$$
    $$= \sqrt{(-8)^2 + 3^2}$$
    $$= \sqrt{64 + 9}$$
    $$= \sqrt{73}$$

Now Try:

9. Find the distance between the points $(-1,-2)$ and $(-4, 3)$.

    _____

Objective 6 Practice Exercises

For extra help, see Example 9 on page 677 of your text.

Find the distance between each pair of points.

16. $(3, 4)$ and $(-1, -2)$ 16. _____

17. $(-2, -3)$ and $(-5, 1)$ 17. _____

18. $(4, 2)$ and $(3, -1)$ 18. _____

Name:
Instructor:
Date:
Section:

Objective 7   Find an equation of a circle given its center and radius.

Video Examples

Review these examples for Objective 7:

10. Write an equation of the circle with radius 2 and center at (0, 0) and graph.

If the point (x, y) is on the circle, then the distance from (x, y) to the center $(0, 0)$ is 2.

$$\sqrt{(x_2-x_1)^2+(y_2-y_1)^2}=d$$
$$\sqrt{(x-0)^2+(y-0)^2}=2$$
$$\left(\sqrt{x^2+y^2}\right)^2=2^2$$
$$x^2+y^2=4$$

The equation of this circle is $x^2+y^2=4$.

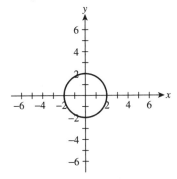

11. Find an equation of the circle with center $(-3, 2)$ and radius 3 and graph it.

$$\sqrt{(x_2-x_1)^2+(y_2-y_1)^2}=d$$
$$\sqrt{(x-(-3))^2+(y-2)^2}=3$$
$$\left(\sqrt{(x+3)^2+(y-2)^2}\right)^2=3^2$$
$$(x+3)^2+(y-2)^2=9$$

To graph the circle, plot the center $(-3, 2)$, then move three units right, left, up, and down from the center, plotting the points $(0, 2)$, $(-3, 5)$, $(-6, 2)$, and $(-3, -1)$. Draw a smooth curve through the points.

Now Try:

10. Find an equation of the circle with radius 5 and center $(0, 0)$ and graph it.

_____

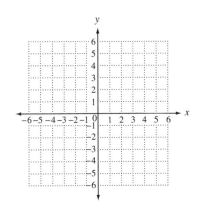

11. Find an equation of the circle with center $(-5, 4)$ and radius 4 and graph it.

_____

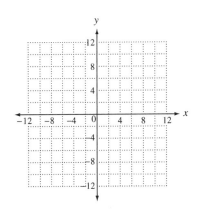

Name: Date:
Instructor: Section:

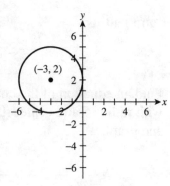

12. Write an equation of a circle with center (−2,−4) and radius $2\sqrt{5}$.

   Use the center-radius form.
   $$(x-h)^2 + (y-k)^2 = r^2$$
   $$[x-(-2)]^2 + [y-(-4)]^2 = (2\sqrt{5})^2$$
   $$(x+2)^2 + (y+4)^2 = 20$$

12. Write an equation of a circle with center (3,−1) and radius $\sqrt{6}$.

   _____

Objective 7 Practice Exercises

For extra help, see Examples 10–12 on pages 677–678 of your text.

Find the equation of a circle satisfying the given conditions.

19. center: (3,−4); radius: 5

19. _____

20. center: (−2,−2); radius: 3

20. _____

21. center: (0, 3); radius: $\sqrt{2}$

21. _____

Name: Date:
Instructor: Section:

Chapter 10 ROOTS, RADICALS, AND ROOT FUNCTIONS

10.4 Adding and Subtracting Radical Expressions

Learning Objectives
1   Simplify radical expressions involving addition and subtraction.

Key Terms

Use the vocabulary terms listed below to complete each statement in exercises 1–2.

like radicals **unlike radicals**

1. The expressions $2\sqrt{2}$ and $6\sqrt[3]{2}$ are _____.

2. The expressions $2\sqrt{2}$ and $7\sqrt{2}$ are _____.

Objective 1   Simplify radical expressions involving addition and subtraction.

Video Examples

Review these examples for Objective 1:
1. Add or subtract to simplify each radical expression.

   a. $3\sqrt{13} + 5\sqrt{52}$

   $3\sqrt{13} + 5\sqrt{52} = 3\sqrt{13} + 5\sqrt{4}\sqrt{13}$
   $= 3\sqrt{13} + 5 \cdot 2\sqrt{13}$
   $= 3\sqrt{13} + 10\sqrt{13}$
   $= (3+10)\sqrt{13}$
   $= 13\sqrt{13}$

   b. $\sqrt{48x} - \sqrt{12x}$, $x \geq 0$

   $\sqrt{48x} - \sqrt{12x} = \sqrt{16} \cdot \sqrt{3x} - \sqrt{4} \cdot \sqrt{3x}$
   $= 4\sqrt{3x} - 2\sqrt{3x}$
   $= (4-2)\sqrt{3x}$
   $= 2\sqrt{3x}$

   c. $7\sqrt{3} - 6\sqrt{21}$

   The radicands differ and are already simplified, so this expression cannot be simplified further.

Now Try:
1. Add or subtract to simplify each radical expression.

   a. $3\sqrt{54} - 5\sqrt{24}$

   _____

   b. $3\sqrt{18z} + 2\sqrt{8z}$, $z \geq 0$

   _____

   c. $3\sqrt{7} + 2\sqrt{6}$

   _____

Name: Date:
Instructor: Section:

2. Simplify. Assume that all variables represent positive real numbers.

   a. $7\sqrt[4]{32} - 9\sqrt[4]{2}$

   $\begin{aligned} 7\sqrt[4]{32} - 9\sqrt[4]{2} &= 7\sqrt[4]{16} \cdot \sqrt[4]{2} - 9\sqrt[4]{2} \\ &= 7 \cdot 2 \cdot \sqrt[4]{2} - 9\sqrt[4]{2} \\ &= 14\sqrt[4]{2} - 9\sqrt[4]{2} \\ &= (14 - 9)\sqrt[4]{2} \\ &= 5\sqrt[4]{2} \end{aligned}$

   b. $6\sqrt[3]{27x^5 r} + 2x\sqrt[3]{x^2 r}$

   $\begin{aligned} &6\sqrt[3]{27x^5 r} + 2x\sqrt[3]{x^2 r} \\ &= 6 \cdot \sqrt[3]{27x^3} \cdot \sqrt[3]{x^2 r} + 2x\sqrt[3]{x^2 r} \\ &= 18x\sqrt[3]{x^2 r} + 2x\sqrt[3]{x^2 r} \\ &= (18x + 2x)\sqrt[3]{x^2 r} \\ &= 20x\sqrt[3]{x^2 r} \end{aligned}$

   c. $3\sqrt{40x^5} + 5\sqrt[3]{48x^5}$

   $\begin{aligned} &3\sqrt{40x^5} + 5\sqrt[3]{48x^5} \\ &= 3 \cdot \sqrt{4x^4 \cdot 10x} + 5\sqrt[3]{8x^3 \cdot 6x^2} \\ &= 3 \cdot \sqrt{4x^4} \cdot \sqrt{10x} + 5\sqrt[3]{8x^3} \cdot \sqrt[3]{6x^2} \\ &= 3 \cdot 2x^2 \cdot \sqrt{10x} + 5 \cdot 2x \cdot \sqrt[3]{6x^2} \\ &= 6x^2\sqrt{10x} + 10x\sqrt[3]{6x^2} \end{aligned}$

2. Simplify. Assume that all variables represent positive real numbers.

   a. $7\sqrt[3]{54} - 6\sqrt[3]{128}$

   _____

   b. $\sqrt[4]{32y^2 z^5} + 3z\sqrt[4]{2y^2 z}$

   _____

   c. $2\sqrt[3]{54x^7} + 2\sqrt{27x^7}$

   _____

Name: Date:
Instructor: Section:

3. Simplify. Assume that all variables represent positive real numbers.

    a. $\dfrac{\sqrt{32}}{3}+\dfrac{\sqrt{8}}{\sqrt{18}}$

$$\dfrac{\sqrt{32}}{3}+\dfrac{\sqrt{8}}{\sqrt{18}}=\dfrac{\sqrt{16\cdot 2}}{3}+\dfrac{\sqrt{4\cdot 2}}{\sqrt{9\cdot 2}}$$

$$=\dfrac{4\sqrt{2}}{3}+\dfrac{2\sqrt{2}}{3\sqrt{2}}$$

$$=\dfrac{4\sqrt{2}}{3}+\dfrac{2}{3}$$

$$=\dfrac{4\sqrt{2}+2}{3}$$

    b. $\sqrt[3]{\dfrac{81}{y^6}}+5\sqrt[3]{\dfrac{27}{y^3}}$

$$\sqrt[3]{\dfrac{81}{y^6}}+5\sqrt[3]{\dfrac{27}{y^3}}=\dfrac{\sqrt[3]{81}}{\sqrt[3]{y^6}}+5\dfrac{\sqrt[3]{27}}{\sqrt[3]{y^3}}$$

$$=\dfrac{3\sqrt[3]{3}}{y^2}+5\left(\dfrac{3}{y}\right)$$

$$=\dfrac{3\sqrt[3]{3}}{y^2}+\dfrac{15}{y}$$

$$=\dfrac{3\sqrt[3]{3}+15y}{y^2}$$

3. Simplify. Assume that all variables represent positive real numbers.

    a. $\sqrt{\dfrac{10}{18}}+\dfrac{\sqrt{15}}{\sqrt{27}}$

_____

    b. $\sqrt[3]{\dfrac{216}{w^6}}+\sqrt{\dfrac{121}{w^4}}$

_____

Objective 1 Practice Exercises

For extra help, see Examples 1–3 on pages 683–685 of your text.

Add or subtract. Assume that all variables represent positive real numbers.

1. $\sqrt{100x}-\sqrt{9x}+\sqrt{25x}$ 1. _____

2. $2\sqrt[3]{16r}+\sqrt[3]{54r}-\sqrt[3]{16r}$ 2. _____

3. $\sqrt[3]{\dfrac{y^7}{125}}+y^2\sqrt[3]{\dfrac{y}{27}}$ 3. _____

Copyright © 2016 Pearson Education, Inc.

Name: Date:
Instructor: Section:

Chapter 10 ROOTS, RADICALS, AND ROOT FUNCTIONS

10.5 Multiplying and Dividing Radical Expressions

Learning Objectives
1 Multiply radical expressions.
2 Rationalize denominators with one radical term.
3 Rationalize denominators with binomials involving radicals.
4 Write radical quotients in lowest terms.

Key Terms

Use the vocabulary terms listed below to complete each statement in exercises 1–2.

rationalizing the denominator   **conjugate**

1. The _____ of $a + b$ is $a - b$.

2. The process of removing radicals from the denominator so that the denominator contains only rational quantities is called _____.

Objective 1   Multiply radical expressions.

Video Examples

Review these examples for Objective 1:
2. Multiply, using the FOIL method.

   a. $(\sqrt{6}+\sqrt{5})(\sqrt{6}-\sqrt{5})$

   This is the difference of squares.

   $(\sqrt{6}+\sqrt{5})(\sqrt{6}-\sqrt{5}) = (\sqrt{6})^2 - (\sqrt{5})^2$
   $= 6 - 5$
   $= 1$

   b. $(5-\sqrt{3})(\sqrt{2}+\sqrt{5})$

   $(5-\sqrt{3})(\sqrt{2}+\sqrt{5})$
   $= 5\cdot\sqrt{2} + 5\cdot\sqrt{5} - \sqrt{3}\cdot\sqrt{2} - \sqrt{3}\cdot\sqrt{5}$
   $= 5\sqrt{2} + 5\sqrt{5} - \sqrt{6} - \sqrt{15}$

Now Try:
2. Multiply, using the FOIL method.

   a. $(\sqrt{14}-\sqrt{2})(\sqrt{14}+\sqrt{2})$

   _____

   b. $(3-\sqrt{2})(2+\sqrt{7})$

   _____

Name: Date:
Instructor: Section:

c. $\left(\sqrt{11}-6\right)^2$

$$\begin{aligned}\left(\sqrt{11}-6\right)^2 &= \left(\sqrt{11}-6\right)\left(\sqrt{11}-6\right)\\ &= \sqrt{11}\cdot\sqrt{11}-6\sqrt{11}-6\sqrt{11}+6\cdot 6\\ &= 11-12\sqrt{11}+36\\ &= 47-12\sqrt{11}\end{aligned}$$

c. $\left(3-\sqrt{2}\right)^2$

_____

Objective 1 Practice Exercises

For extra help, see Examples 1–2 on pages 688–689 of your text.

Multiply each product, then simplify. Assume that variables represent positive real numbers.

1. $\left(\sqrt{5}+\sqrt{6}\right)\left(\sqrt{2}-4\right)$

1. _____

2. $\left(\sqrt{2}-\sqrt{12}\right)^2$

2. _____

3. $\left(2+\sqrt[3]{5}\right)\left(2-\sqrt[3]{5}\right)$

3. _____

Name: Date:
Instructor: Section:

Objective 2 Rationalize denominators with one radical term.

Video Examples

Review these examples for Objective 2:

3. Rationalize the denominator.

$$\frac{4}{\sqrt{3}}$$

$$\frac{4}{\sqrt{3}} = \frac{4 \cdot \sqrt{3}}{\sqrt{3} \cdot \sqrt{3}} = \frac{4\sqrt{3}}{3}$$

4. Simplify the radical.

$$-\sqrt{\frac{27}{98}}$$

$$-\sqrt{\frac{27}{98}} = -\frac{\sqrt{27}}{\sqrt{98}}$$

$$= -\frac{\sqrt{9 \cdot 3}}{\sqrt{49 \cdot 2}}$$

$$= -\frac{3\sqrt{3}}{7\sqrt{2}}$$

$$= -\frac{3\sqrt{3} \cdot \sqrt{2}}{7\sqrt{2} \cdot \sqrt{2}}$$

$$= -\frac{3\sqrt{6}}{7 \cdot 2}$$

$$= -\frac{3\sqrt{6}}{14}$$

5. Simplify.

$$\sqrt[3]{\frac{16}{9}}$$

$$\sqrt[3]{\frac{16}{9}} = \frac{\sqrt[3]{8 \cdot 2}}{\sqrt[3]{9}} = \frac{2\sqrt[3]{2}}{\sqrt[3]{9}}$$

$$= \frac{2\sqrt[3]{2} \cdot \sqrt[3]{3}}{\sqrt[3]{9} \cdot \sqrt[3]{3}}$$

$$= \frac{2\sqrt[3]{6}}{\sqrt[3]{27}}$$

$$= \frac{2\sqrt[3]{6}}{3}$$

Now Try:

3. Rationalize the denominator.

$$\frac{2}{\sqrt{15}}$$

4. Simplify the radical.

$$-\sqrt{\frac{45}{32}}$$

5. Simplify.

$$\sqrt[3]{\frac{8}{100}}$$

Name: Date:
Instructor: Section:

Objective 2 Practice Exercises

For extra help, see Examples 3–5 on pages 690–692 of your text.

Simplify. Assume that variables represent positive real numbers.

4. $\sqrt{\dfrac{5a^2b^3}{6}}$ 4. _____

5. $\sqrt{\dfrac{7y^2}{12b}}$ 5. _____

6. $\sqrt[3]{\dfrac{5}{49x}}$ 6. _____

Name: Date:
Instructor: Section:

Objective 3 Rationalize denominators with binomials involving radicals.

Video Examples

Review this example for Objective 3:
6. Rationalize the denominator.

$$\frac{5}{5+\sqrt{2}}$$

$$\frac{5}{5+\sqrt{2}} = \frac{5(5-\sqrt{2})}{(5+\sqrt{2})(5-\sqrt{2})}$$

$$= \frac{5(5-\sqrt{2})}{25-2}$$

$$= \frac{5(5-\sqrt{2})}{23}$$

Now Try:
6. Rationalize the denominator.

$$\frac{2}{\sqrt{3}-2}$$

_____

Objective 3 Practice Exercises

For extra help, see Example 6 on page 693 of your text.

Rationalize each denominator. Write quotients in lowest terms. Assume that variables represent positive real numbers.

7. $\dfrac{4}{\sqrt{3}+2}$

7. _____

8. $\dfrac{5}{\sqrt{3}-\sqrt{10}}$

8. _____

9. $\dfrac{\sqrt{6}+2}{\sqrt{2}-4}$

9. _____

Name: Date:
Instructor: Section:

Objective 4 Write radical quotients in lowest terms.

Video Examples

Review this example for Objective 4:
7. Write the quotient in lowest terms.

$$\frac{72\sqrt{2} - 16\sqrt{7}}{24}$$

$$\frac{72\sqrt{2} - 16\sqrt{7}}{24} = \frac{8(9\sqrt{2} - 2\sqrt{7})}{24}$$

$$= \frac{9\sqrt{2} - 2\sqrt{7}}{3}$$

Now Try:
7. Write the quotient in lowest terms.

$$\frac{9 + 6\sqrt{15}}{12}$$

_____

Objective 4 Practice Exercises

For extra help, see Example 7 on page 694 of your text.

Write each quotient in lowest terms. Assume that variables represent positive real numbers.

10. $\dfrac{7 - \sqrt{98}}{14}$

10. _____

11. $\dfrac{16 - 12\sqrt{72}}{24}$

11. _____

12. $\dfrac{2x - \sqrt{8x^2}}{4x}$

12. _____

Name: Date:
Instructor: Section:

Chapter 10 ROOTS, RADICALS, AND ROOT FUNCTIONS

10.6 Solving Equations with Radicals

Learning Objectives
1. Solve radical equations using the power rule.
2. Solve radical equations with indexes greater than 2.
3. Use the power rule to solve a formula for a specified variable.

Key Terms

Use the vocabulary terms listed below to complete each statement in exercises 1–2.

radical equation   **extraneous solution**

1. A(n) _____ is a potential solution to an equation that does not satisfy the equation.

2. An equation with a variable in the radicand is a(n) _____.

Objective 1 Solve radical equations using the power rule.

Video Examples

Review these examples for Objective 1:

1. Solve $\sqrt{3w+4} = 7$.

   $$\left(\sqrt{3w+4}\right)^2 = 7^2$$
   $$3w+4 = 49$$
   $$3w = 45$$
   $$w = 15$$

   Check $\sqrt{3w+4} = 7$
   $$\sqrt{3(15)+4} \stackrel{?}{=} 7$$
   $$\sqrt{49} \stackrel{?}{=} 7$$
   $$7 = 7 \quad \text{True}$$

   Since 15 satisfies the original equation, the solution set is {15}.

2. Solve $\sqrt{12p+1} + 7 = 0$.

   Step 1 $\sqrt{12p+1} = -7$

   Step 2 $\left(\sqrt{12p+1}\right)^2 = (-7)^2$

Now Try:

1. Solve $\sqrt{7x-6} = 8$.

   _____

2. Solve $\sqrt{4x-19} + 5 = 0$.

   _____

Name:
Instructor:

Date:
Section:

Step 3 $\quad 12p+1=49$
$\quad\quad\quad\quad 12p=48$
$\quad\quad\quad\quad\quad p=4$

Step 4 Check $\sqrt{12p+1}+7=0$
$\sqrt{12(4)+1}+7\stackrel{?}{=}0$
$\sqrt{49}+7\stackrel{?}{=}0$
$14=0\quad$ False

The false result shows that the proposed solution 4 is not a solution of the original equation. It is extraneous. The solution set is $\varnothing$.

3. Solve $\sqrt{x+3}=x-3$.

   Step 1 The radical is isolated on the left side of the equation.

   Step 2 Square each side.
   $$\left(\sqrt{x+3}\right)^2=(x-3)^2$$
   $$x+3=x^2-6x+9$$

   Step 3 Write the equation in standard form and solve.
   $$0=x^2-7x+6$$
   $$0=(x-1)(x-6)$$
   $$x-1=0\quad\text{or}\quad x-6=0$$
   $$x=1\quad\text{or}\quad x=6$$

   Step 4 Check each proposed solution in the original equation.

   $\sqrt{x+3}=x-3\quad\quad\quad \sqrt{x+3}=x-3$
   $\sqrt{1+3}\stackrel{?}{=}1-3\quad\quad\quad \sqrt{6+3}\stackrel{?}{=}6-3$
   $\sqrt{4}\stackrel{?}{=}-2\quad\quad\quad\quad\quad \sqrt{9}\stackrel{?}{=}3$
   $2=-2\quad$ False $\quad\quad 3=3\quad$ True

   The solution set is $\{6\}$. The other proposed solution, 1, is extraneous.

3. Solve $\sqrt{x+11}=x-1$.

   _____

Name: Date:
Instructor: Section:

5. Solve $\sqrt{3x} - 4 = \sqrt{x-2}$.

$$(\sqrt{3x} - 4)^2 = (\sqrt{x-2})^2$$
$$3x - 8\sqrt{3x} + 16 = x - 2$$
$$-8\sqrt{3x} = -2x - 18$$
$$(-8\sqrt{3x})^2 = (-2x - 18)^2$$
$$192x = 4x^2 + 72x + 324$$
$$0 = 4x^2 - 120x + 324$$
$$0 = 4(x^2 - 30x + 81)$$
$$0 = 4(x-3)(x-27)$$
$$x - 3 = 0 \quad \text{or} \quad x - 27 = 0$$
$$x = 3 \quad \text{or} \quad x = 27$$

Check

$\sqrt{3x} - 4 = \sqrt{x-2}$ $\sqrt{3x} - 4 = \sqrt{x-2}$
$\sqrt{3(3)} - 4 \stackrel{?}{=} \sqrt{3-2}$ $\sqrt{3(27)} - 4 \stackrel{?}{=} \sqrt{27-2}$
$3 - 4 \stackrel{?}{=} \sqrt{1}$ $9 - 4 \stackrel{?}{=} \sqrt{25}$
$-1 = 1$ False $5 = 5$ True

The proposed solution, 27, is valid, but 3 is extraneous and must be rejected. The solution set is {27}.

5. Solve $\sqrt{3x+4} = \sqrt{9x} - 2$.

5. _____

Objective 1 Practice Exercises

For extra help, see Examples 1–5 on pages 699–702 of your text.

Solve each equation.

1. $\sqrt{4x - 19} = 5$

1. _____

2. $\sqrt{12p + 1} + 7 = 0$

2. _____

Name: Date:
Instructor: Section:

3. $\sqrt{k+10}+\sqrt{2k+19}=2$

3. _____

Objective 2 Solve radical equations with indexes greater than 2.

Video Examples

Review this example for Objective 2:

6. Solve $\sqrt[3]{5r-6}=\sqrt[3]{3r+4}$.

$$\left(\sqrt[3]{5r-6}\right)^3=\left(\sqrt[3]{3r+4}\right)^3$$
$$5r-6=3r+4$$
$$2r=10$$
$$r=5$$

Check $\sqrt[3]{5r-6}=\sqrt[3]{3r+4}$
$$\sqrt[3]{5(5)-6}\stackrel{?}{=}\sqrt[3]{3(5)+4}$$
$$\sqrt[3]{19}=\sqrt[3]{19} \quad \text{True}$$

The solution set is $\{5\}$.

Now Try:

6. Solve $\sqrt[4]{8x+5}=\sqrt[4]{7x+7}$.

_____

Objective 2 Practice Exercises

For extra help, see Example 6 on pages 702–703 of your text.

Solve each equation.

4. $\sqrt[3]{2a-63}+5=0$

4. _____

5. $\sqrt[5]{5a+1}-\sqrt[5]{2a-11}=0$

5. _____

6. $\sqrt[4]{8x+5}=\sqrt[4]{7x+7}$

6. _____

Name: Date:
Instructor: Section:

Objective 3   Use the power rule to solve a formula for a specified variable.

Video Examples

Review this example for Objective 3:

7. Solve the formula $d = \sqrt{\dfrac{H}{1.6n}}$ for n.

$$d^2 = \left(\sqrt{\dfrac{H}{1.6n}}\right)^2$$

$$d^2 = \dfrac{H}{1.6n}$$

$$1.6d^2 n = H$$

$$n = \dfrac{H}{1.6d^2}$$

Now Try:

7. Solve the formula $r = \sqrt{\dfrac{3v}{\pi h}}$ for h.

_____

Objective 3 Practice Exercises

For extra help, see Example 7 on page 703 of your text.

Solve each equation for the indicated variable.

7. $Z = \sqrt{\dfrac{L}{C}}$, for L 7. _____

8. $f = \dfrac{1}{2\pi\sqrt{LC}}$, for C 8. _____

9. $N = \dfrac{1}{2\pi}\sqrt{\dfrac{a}{r}}$, for r 9. _____

Name: Date:
Instructor: Section:

Chapter 10 ROOTS, RADICALS, AND ROOT FUNCTIONS

10.7 Complex Numbers

Learning Objectives
1. Simplify numbers of the form $\sqrt{-b}$, where $b > 0$.
2. Recognize subsets of the complex numbers.
3. Add and subtract complex numbers.
4. Multiply complex numbers.
5. Divide complex numbers.
6. Simplify powers of i.

Key Terms

Use the vocabulary terms listed below to complete each statement in exercises 1–7.

complex number   **real part**   **imaginary part**

pure imaginary number   **standard form (of a complex number)**

nonreal complex number   **complex conjugate**

1. A _____ is a number that can be written in the form $a + bi$, where a and b are real numbers.

2. The _____ of $a + bi$ is $a - bi$.

3. The _____ of $a + bi$ is bi.

4. The _____ of $a + bi$ is a.

5. A complex number is in _____ if it is written in the form $a + bi$.

6. A complex number $a + bi$ with $a = 0$ and $b \neq 0$ is called a _____.

7. A complex number $a + bi$ with $b \neq 0$ is called a _____.

Objective 1   Simplify numbers of the form $\sqrt{-b}$, where $b > 0$.

Video Examples

Review these examples for Objective 1:
1. Write the number as a product of a real number and i.

    $\sqrt{-36}$

    $\sqrt{-36} = i\sqrt{36} = 6i$

Now Try:
1. Write the number as a product of a real number and i.

    $\sqrt{-16}$

    _____

Name: Date:
Instructor: Section:

2. Multiply.

$\sqrt{-6} \cdot \sqrt{-7}$

$\sqrt{-6} \cdot \sqrt{-7} = i\sqrt{6} \cdot i\sqrt{7}$
$= i^2 \sqrt{6 \cdot 7}$
$= (-1)\sqrt{42}$
$= -\sqrt{42}$

3. Divide.

$\dfrac{\sqrt{-125}}{\sqrt{-5}}$

$\dfrac{\sqrt{-125}}{\sqrt{-5}} = \dfrac{i\sqrt{125}}{i\sqrt{5}}$
$= \sqrt{\dfrac{125}{5}}$
$= \sqrt{25}$
$= 5$

2. Multiply.

$\sqrt{-5} \cdot \sqrt{-6}$

3. Divide.

$\dfrac{\sqrt{-200}}{\sqrt{-8}}$

Objective 1 Practice Exercises

For extra help, see Examples 1–3 on pages 7066–707 of your text.

Write the number as a product of a real number and i. Simplify all radical expressions.

1. $-\sqrt{-162}$

1. _____

Multiply or divide as indicated

2. $\sqrt{-5} \cdot \sqrt{-3} \cdot \sqrt{-7}$

2. _____

3. $\dfrac{\sqrt{-42} \cdot \sqrt{-6}}{\sqrt{-7}}$

3. _____

Name: Date:
Instructor: Section:

Objective 2 Recognize subsets of the complex numbers.

For extra help, see page 707 of your text.

Classify each of the following complex numbers as real *or* imaginary.

4. $\sqrt{5}$ 4. _____

5. $\sqrt{3} - i\sqrt{5}$ 5. _____

6. $i\sqrt{7}$ 6. _____

Objective 3 Add and subtract complex numbers.

Video Examples

Review these examples for Objective 3: **Now Try:**
4. Add. 4. Add.

$(2+9i)+(10-3i)$ $(4-7i)+(6-2i)$

$(2+9i)+(10-3i) = (2+10)+(9-3)i$
$\qquad = 12+6i$ _____

5. Subtract. 5. Subtract.

$(7-9i)-(-5-6i)$ $(12+2i)-(-12-2i)$

$(7-9i)-(-5-6i) = [7-(-5)]+[-9-(-6)]i$
$\qquad = (7+5)+(-9+6)i$
$\qquad = 12-3i$ _____

Objective 3 Practice Exercises

For extra help, see Examples 4–5 on page 708 of your text.

Add or subtract as indicated. Write answers in standard form.

7. $(-7-2i)-(-3-3i)$ 7. _____

8. $4i-(9+5i)+(2+3i)$ 8. _____

9. $(7-9i)-(5-6i)$ 9. _____

Name: Date:
Instructor: Section:

Objective 4   Multiply complex numbers.

Video Examples

Review this example for Objective 4:

6. Multiply.

   $6i(2-7i)$

   $6i(2-7i) = 6i(2) + 6i(-7i)$
   $= 12i - 42i^2$
   $= 12i - 42(-1)$
   $= 42 + 12i$

Now Try:

6. Multiply.

   $2i(4+7i)$

   _____

Objective 4 Practice Exercises

For extra help, see Example 6 on pages 708–709 of your text.

Multiply.

10. $(2-5i)(2+5i)$

10. _____

11. $(1+3i)^2$

11. _____

12. $(12+2i)(-1+i)$

12. _____

Name: Date:
Instructor: Section:

Objective 5 Divide complex numbers.

Video Examples

Review this example for Objective 5:
7. Find the quotient.

$$\frac{6-i}{2-3i}$$

Multiply the numerator and denominator by $2+3i$, the conjugate of the denominator.

$$\frac{6-i}{2-3i} = \frac{(6-i)(2+3i)}{(2-3i)(2+3i)}$$

$$= \frac{12+18i-2i-3i^2}{2^2+3^2}$$

$$= \frac{12+16i-3(-1)}{4+9}$$

$$= \frac{15+16i}{13}, \text{ or } \frac{15}{13}+\frac{16}{13}i$$

Now Try:
7. Find the quotient.

$$\frac{4+i}{5-2i}$$

Objective 5 Practice Exercises

For extra help, see Example 7 on pages 709–710 of your text.

Write each quotient in the form a + bi.

13. $\dfrac{3-2i}{2+i}$

13. _____

14. $\dfrac{5+2i}{9-4i}$

14. _____

15. $\dfrac{6-i}{2-3i}$

15. _____

Name: Date:
Instructor: Section:

Objective 6   Simplify powers of *i*.

Video Examples

Review this example for Objective 6:
8. Find the power of *i*.

i^{100}

$i^{100} = (i^4)^{25} = 1^{25} = 1$

Now Try:
8. Find the power of *i*.

i^{48}

_____

Objective 6 Practice Exercises

For extra help, see Example 8 on page 710 of your text.

Find each power of i.

16. i^{14}

16. _____

17. i^{113}

17. _____

18. i^{-21}

18. _____

Name: Date:
Instructor: Section:

Chapter 11 QUADRATIC EQUATIONS, INEQUALITIES, AND FUNCTIONS

11.1 Solving Quadratic Equations by the Square Root Property

Learning Objectives
1. Review the zero-factor property.
2. Solve equations of the form $x^2 = k$, where $k > 0$.
3. Solve quadratic equations of the form $(ax+b)^2 = k$, where $k > 0$.
4. Solve quadratic equations with solutions that are not real numbers.

Key Terms

Use the vocabulary terms listed below to complete each statement in exercises 1–2.

quadratic equation   **zero-factor property**

1. An equation that can be written in the form $ax^2 + bx + c = 0$ is a _____.

2. The _____ states that if a product equals 0, then at least one of the factors of the product also equals zero.

Objective 1 Review the zero-factor property.

Video Examples

Review these examples for Objective 1:	Now Try:
1. Solve each equation by the zero-factor property. a. $x^2 + 5x + 4 = 0$ $x^2 + 5x + 4 = 0$ $(x+4)(x+1) = 0$ $x + 4 = 0$ or $x + 1 = 0$ $x = -4$ or $x = -1$ The solution set is $\{-4, -1\}$. b. $x^2 = 64$ $x^2 = 64$ $x^2 - 64 = 0$ $(x+8)(x-8) = 0$ $x + 8 = 0$ or $x - 8 = 0$ $x = -8$ or $x = 8$ The solution set is $\{-8, 8\}$.	1. Solve each equation by the zero-factor property. a. $x^2 + 8x + 7 = 0$ _____ b. $x^2 = 100$ _____

Copyright © 2016 Pearson Education, Inc. 403

Name: Date:
Instructor: Section:

Objective 1 Practice Exercises

For extra help, see Example 1 on page 724 of your text.

Solve each equation by using the zero-factor property.

1. $x^2 + 6x + 8 = 0$ 1. _____

2. $x^2 = 121$ 2. _____

3. $x^2 + 2x - 35 = 0$ 3. _____

Objective 2 Solve equations of the form $x^2 = k$, where $k > 0$.

Video Examples

Review these examples for Objective 2:
2. Solve each equation. Write radicals in simplified form.

   a. $x^2 = 36$

   By the square root property, if $x^2 = 36$, then
   $x = \sqrt{36} = 6$ or $x = -\sqrt{36} = -6$
   The solution set is $\{-6, 6\}$, or $\{\pm 6\}$.

   b. $x^2 = 17$

   The solutions are $x = \sqrt{17}$ or $x = -\sqrt{17}$
   The solution set is $\{-\sqrt{17}, \sqrt{17}\}$, or $\{\pm\sqrt{17}\}$.

   c. $5p^2 - 100 = 0$

   $5p^2 - 100 = 0$
   $5p^2 = 100$
   $p^2 = 20$
   $p = \sqrt{20}$ or $p = -\sqrt{20}$
   $p = 2\sqrt{5}$ or $p = -2\sqrt{5}$

Now Try:
2. Solve each equation. Write radicals in simplified form.

   a. $x^2 = 81$

   _____

   b. $x^2 = 23$

   _____

   c. $3x^2 - 54 = 0$

   _____

Name: Date:
Instructor: Section:

Check

$$5p^2 - 100 = 0 \qquad\qquad 5p^2 - 100 = 0$$
$$5(2\sqrt{5})^2 - 100 \stackrel{?}{=} 0 \qquad 5(-2\sqrt{5})^2 - 100 \stackrel{?}{=} 0$$
$$5(20) - 100 \stackrel{?}{=} 0 \qquad\quad 5(20) - 100 \stackrel{?}{=} 0$$
$$0 = 0 \ \text{True} \qquad\qquad\quad \text{True} \ 0 = 0$$

The solution set is $\{2\sqrt{5}, -2\sqrt{5}\}$, or $\{\pm 2\sqrt{5}\}$.

d. $3x^2 + 11 = 35$

$$3x^2 + 11 = 35$$
$$3x^2 = 24$$
$$x^2 = 8$$
$$x = \sqrt{8} \quad \text{or} \quad x = -\sqrt{8}$$
$$x = 2\sqrt{2} \quad \text{or} \quad x = -2\sqrt{2}$$

The solution set is $\{2\sqrt{2}, -2\sqrt{2}\}$, or $\{\pm 2\sqrt{2}\}$.

d. $4y^2 + 3 = 51$

_____

3. Use Galileo's formula to determine how long it will take a penny dropped from the 86th floor Observatory deck of the Empire State Building to reach the ground. The deck is 1050 feet above the ground. Round your answer to the nearest tenth.

Galileo's formula is $d = 16t^2$, where d is the distance in feet that an object falls, and t is the time in seconds.

$$d = 16t^2$$
$$1050 = 16t^2$$
$$65.625 = t^2$$
$$t = \sqrt{65.625} \ \text{ or } \ t = -\sqrt{65.625}$$

Time cannot be negative, so we discard $t = -\sqrt{65.625}$. Using a calculator, $\sqrt{65.625} \approx 8.1$, so $t \approx 8.1$. The penny would fall to the ground in about 8.1 seconds.

3. A child dropped a ball from a hotel balcony that is 113 ft above the ground. Use Galileo's formula to determine how long it takes for the ball to reach the ground. Round your answer to the nearest tenth.

_____

Name: Date:
Instructor: Section:

Objective 2 Practice Exercises

For extra help, see Examples 2–3 on pages 725–726 of your text.

Solve each equation by using the square root property. Express all radicals in simplest form.

4. $r^2 = 900$

4. _____

5. $s^2 - 98 = 0$

5. _____

6. $p^2 = -144$

6. _____

Objective 3 Solve equations of the form $(ax+b)^2 = k$, where $k > 0$.

Video Examples

Review these examples for Objective 3:

5. Solve $(5r-3)^2 = 12$.

$5r - 3 = \sqrt{12}$ or $5r - 3 = -\sqrt{12}$
$5r - 3 = 2\sqrt{3}$ or $5r - 3 = -2\sqrt{3}$
$5r = 3 + 2\sqrt{3}$ or $5r = 3 - 2\sqrt{3}$
$r = \dfrac{3 + 2\sqrt{3}}{5}$ or $r = \dfrac{3 - 2\sqrt{3}}{5}$

Check
$$(5r-3)^2 = 12$$
$$\left[5 \cdot \dfrac{3+2\sqrt{3}}{5} - 3\right]^2 \stackrel{?}{=} 12$$
$$(3 + 2\sqrt{3} - 3)^2 \stackrel{?}{=} 12$$
$$(2\sqrt{3})^2 \stackrel{?}{=} 12$$
$$12 = 12 \quad \text{True}$$

The check of the other solution is similar. The solution set is

The solution set is $\left\{\dfrac{3+2\sqrt{3}}{5}, \dfrac{3-2\sqrt{3}}{5}\right\}$.

Now Try:

5. Solve $(7r-3)^2 = 32$.

_____

Name: Date:
Instructor: Section:

Objective 3 Practice Exercises

For extra help, see Examples 3–5 on pages 726–727 of your text.

Solve each equation by using the square root property. Express all radicals in simplest form.

7. $(y+2)^2 = 16$ 7. _____

8. $(7p-4)^2 = 289$ 8. _____

9. $(10m-5)^2 - 9 = 0$ 9. _____

Objective 4 Solve quadratic equations with solutions that are not real numbers.

Video Examples

Review these examples for Objective 4:
6. Solve each equation.

  a. $x^2 = -48$

  $$x^2 = -48$$
  $x = \sqrt{-48}$ or $x = -\sqrt{-48}$
  $x = 4i\sqrt{3}$ or $x = -4i\sqrt{3}$
  The solution set is $\{-4i\sqrt{3},\ 4i\sqrt{3}\}$.

  b. $(x-3)^2 = -25$

  $$(x-3)^2 = -25$$
  $x-3 = \sqrt{-25}$ or $x-3 = -\sqrt{-25}$
  $x-3 = 5i$ or $x-3 = -5i$
  $x = 3+5i$ or $x = 3-5i$
  The solution set is $\{3-5i,\ 3+5i\}$.

Now Try:
6. Solve each equation.

  a. $y^2 = -32$

  _____

  b. $(x+2)^2 = -49$

  _____

Copyright © 2016 Pearson Education, Inc.

Name: Date:
Instructor: Section:

Objective 4 Practice Exercises

For extra help, see Example 6 on pages 727–728 of your text.

Find the complex solutions of each equation.

10. $(10m-5)^2 + 9 = 0$ 10. _____

11. $(m+1)^2 = -36$ 11. _____

12. $(x-1)^2 + 2 = 0$ 12. _____

Name: Date:
Instructor: Section:

Chapter 11 QUADRATIC EQUATIONS, INEQUALITIES, AND FUNCTIONS

11.2 Solving Quadratic Equations by Completing the Square

Learning Objectives
1. Solve quadratic equations by completing the square when the coefficient of the second-degree term is 1.
2. Solve quadratic equations by completing the square when the coefficient of the second-degree term is not 1.
3. Simplify the terms of an equation before solving.

Key Terms

Use the vocabulary terms listed below to complete each statement in exercises 1–3.

completing the square   perfect square trinomial   square root property

1. A _____ can be written in the form $x^2 + 2kx + k^2$ or $x^2 - 2kx + k^2$

2. The _____ says that, if k is positive and $a^2 = k$, then $a = \pm\sqrt{k}$.

3. Use the process called _____ in order to rewrite an equation so it can be solved using the square root property.

Video Examples

Objective 1 Solve quadratic equations by completing the square when the coefficient of the second-degree term is 1.

Review this example for Objective 1:	**Now Try:**
3. Solve $x^2 + 5x + 2 = 0$.	3. Solve $x^2 - 11x + 8 = 0$.
Since the coefficient of the second-degree term is 1, begin with Step 2.	_____
Step 2 $x^2 + 5x = -2$	
Step 3 Take half the coefficient of the first-degree term and square the result.	
$\left[\frac{1}{2}(5)\right]^2 = \left(\frac{5}{2}\right)^2 = \frac{25}{4}$	
$x^2 + 5x + \frac{25}{4} = -2 + \frac{25}{4}$	
$\left(x + \frac{5}{2}\right)^2 = \frac{17}{4}$	

Copyright © 2016 Pearson Education, Inc.

Name: Date:
Instructor: Section:

Step 4

$$x+\frac{5}{2}=\sqrt{\frac{17}{4}} \quad \text{or} \quad x+\frac{5}{2}=-\sqrt{\frac{17}{4}}$$

$$x+\frac{5}{2}=\frac{\sqrt{17}}{2} \quad \text{or} \quad x+\frac{5}{2}=-\frac{\sqrt{17}}{2}$$

$$x=-\frac{5}{2}+\frac{\sqrt{17}}{2} \quad \text{or} \quad x=-\frac{5}{2}-\frac{\sqrt{17}}{2}$$

A check shows the solution set is
$$\left\{-\frac{5}{2}-\frac{\sqrt{17}}{2},\ -\frac{5}{2}+\frac{\sqrt{17}}{2}\right\}.$$

Objective 1 Practice Exercises

For extra help, see Examples 1–3 on pages 730–732 of your text.

Solve each equation by completing the square.

1. $r^2 + 8r = -4$

    1. _____

2. $x^2 - 4x = 2$

    2. _____

3. $x^2 + 2x = 63$

    3. _____

Objective 2 Solve quadratic equations by completing the square when the coefficient of the second-degree term is not 1.

Video Examples

Review these examples for Objective 2: | **Now Try:**
4. Solve $9x^2 + 18x = 7$. | 4. Solve $16x^2 - 64x = 55$.

Step 1 Before completing the square, the coefficient of x^2 must be 1, not 9. Divide each side by 9.

$$x^2 + 2x = \frac{7}{9}$$

Step 2 The equation is already in the correct form, with the variable terms on one side and the constant on the other.

Name: Date:
Instructor: Section:

Step 3 Complete the square. Take half the coefficient of *x*, and square it.

$$\frac{1}{2}(2) = 1 \quad \text{and} \quad 1^2 = 1$$

$$x^2 + 2x + 1 = \frac{7}{9} + 1$$

$$(x+1)^2 = \frac{16}{9}$$

Step 4 Solve the equation by using the square root property.

$$x+1 = \sqrt{\frac{16}{9}} \quad \text{or} \quad x+1 = -\sqrt{\frac{16}{9}}$$

$$x+1 = \frac{4}{3} \qquad\qquad x+1 = -\frac{4}{3}$$

$$x = \frac{1}{3} \quad \text{or} \qquad x = -\frac{7}{3}$$

The two solutions $-\frac{7}{3}$ and $\frac{1}{3}$ check, so the solution set is $\left\{-\frac{7}{3}, \frac{1}{3}\right\}$.

5. Solve $3x^2 - 6x - 2 = 0$.

$$x^2 - 2x - \frac{2}{3} = 0 \qquad \text{Step 1}$$

$$x^2 - 2x = \frac{2}{3} \qquad \text{Step 2}$$

$$\left[\frac{1}{2}(-2)\right]^2 = (-1)^2 = 1 \qquad \text{Step 3}$$

$$x^2 - 2x + 1 = \frac{2}{3} + 1$$

$$(x-1)^2 = \frac{5}{3}$$

$$x - 1 = \sqrt{\frac{5}{3}} \quad \text{or} \quad x - 1 = -\sqrt{\frac{5}{3}} \qquad \text{Step 4}$$

$$x = 1 + \sqrt{\frac{5}{3}} \quad \text{or} \quad x = 1 - \sqrt{\frac{5}{3}}$$

$$x = 1 + \frac{\sqrt{15}}{3} \quad \text{or} \quad x = 1 - \frac{\sqrt{15}}{3}$$

$$x = \frac{3 + \sqrt{15}}{3} \quad \text{or} \quad x = \frac{3 - \sqrt{15}}{3}$$

The solution set is $\left\{\frac{3-\sqrt{15}}{3}, \frac{3+\sqrt{15}}{3}\right\}$.

5. Solve $2p^2 + 6p - 1 = 0$.

_____

Name: Date:
Instructor: Section:

Objective 2 Practice Exercises

For extra help, see Examples 4–6 on pages 734–735 of your text.

Solve each equation by completing the square.

4. $6x^2 - x = 15$

4. _____

5. $6q^2 + 4q = 1$

5. _____

6. $3t^2 + t - 2 = 0$

6. _____

Objective 3   Simplify the terms of an equation before solving.

Video Examples

Review this example for Objective 3:

7. Solve $x(x+5) = 7$.

$$x(x+5) = 7$$
$$x^2 + 5x = 7$$
$$x^2 + 5x + \frac{25}{4} = 7 + \frac{25}{4}$$
$$\left(x + \frac{5}{2}\right)^2 = \frac{53}{4}$$
$$x + \frac{5}{2} = \sqrt{\frac{53}{4}} \quad \text{or} \quad x + \frac{5}{2} = -\sqrt{\frac{53}{4}}$$
$$x = -\frac{5}{2} + \frac{\sqrt{53}}{2} \quad \text{or} \quad x = -\frac{5}{2} - \frac{\sqrt{53}}{2}$$
$$x = \frac{-5 + \sqrt{53}}{2} \quad \text{or} \quad x = \frac{-5 - \sqrt{53}}{2}$$

A check confirms the solution set is $\left\{\frac{-5 \pm \sqrt{53}}{2}\right\}$.

Now Try:

7. Complete the square to solve $x(x+3) = 6$.

_____

Name: Date:
Instructor: Section:

Objective 3 Practice Exercises

For extra help, see Examples 7–8 on page 735 of your text.

Simplify each of the following equations and then solve by completing the square.

7. $6y^2 + 3y = 4y^2 + y + 6$ 7. _____

8. $(b-1)(b+7) = 9$ 8. _____

9. $(s+3)(s+1) = 1$ 9. _____

Name: Date:
Instructor: Section:

Chapter 11 QUADRATIC EQUATIONS, INEQUALITIES, AND FUNCTIONS

11.3 Solving Quadratic Equations by the Quadratic Formula

Learning Objectives
1. Derive the quadratic formula.
2. Solve quadratic equations by using the quadratic formula.
3. Use the discriminant to determine the number and type of solutions.

Key Terms

Use the vocabulary terms listed below to complete each statement in exercises 1–2.

quadratic formula **discriminant**

1. The expression under the radical in the quadratic formula is called the _____.

2. The formula $x = \dfrac{-b \pm \sqrt{b^2 - 4ac}}{2a}$ is called the _____.

Objective 1 Derive the quadratic formula.

For extra help, see pages 738–739 of your text.

Objective 2 Solve quadratic equations by using the quadratic formula.

Video Examples

Review these examples for Objective 2:

1. Solve $5x^2 - 13x - 6 = 0$.

   Use the quadratic formula with $a = 5$, $b = -13$, and $c = -6$.

   $x = \dfrac{-b \pm \sqrt{b^2 - 4ac}}{2a}$

   $x = \dfrac{-(-13) \pm \sqrt{(-13)^2 - 4(5)(-6)}}{2(5)}$

   $x = \dfrac{13 \pm \sqrt{169 + 120}}{10}$

   $x = \dfrac{13 \pm \sqrt{289}}{10}$

   $x = \dfrac{13 \pm 17}{10}$

   There are two solutions.

Now Try:

1. Solve $6x^2 - 17x + 12 = 0$.

   _____

Name: Date:
Instructor: Section:

$x = \dfrac{13+17}{10} = 3$ or $x = \dfrac{13-17}{10} = \dfrac{-4}{10} = -\dfrac{2}{5}$

The solution set is $\left\{-\dfrac{2}{5}, 3\right\}$.

2. Solve $4x^2 = -4x + 1$.

First write the equation in standard form as $4x^2 + 4x - 1 = 0$.
$a = 4$, $b = 4$, $c = -1$

$x = \dfrac{-b \pm \sqrt{b^2 - 4ac}}{2a}$

$x = \dfrac{-4 \pm \sqrt{(4)^2 - 4(4)(-1)}}{2(4)}$

$x = \dfrac{-4 \pm \sqrt{16 + 16}}{8} = \dfrac{-4 \pm \sqrt{32}}{8}$

$x = \dfrac{-4 \pm 4\sqrt{2}}{8}$

$x = \dfrac{4(-1 \pm \sqrt{2})}{4(2)}$

$x = \dfrac{-1 \pm \sqrt{2}}{2}$

The solution set is $\left\{\dfrac{-1-\sqrt{2}}{2}, \dfrac{-1+\sqrt{2}}{2}\right\}$.

2. Solve $2x^2 = 2x + 3$.

_____

3. Solve $(5x-2)(x+2) = -9$.

$(5x-2)(x+2) = -9$
$5x^2 + 8x - 4 = -9$
$5x^2 + 8x + 5 = 0$

From the standard form, we identify $a = 5$, $b = 8$, and $c = 5$.

$x = \dfrac{-b \pm \sqrt{b^2 - 4ac}}{2a}$

$x = \dfrac{-8 \pm \sqrt{(8)^2 - 4(5)(5)}}{2(5)}$

$x = \dfrac{-8 \pm \sqrt{-36}}{10} = \dfrac{-8 \pm 6i}{10}$

$x = \dfrac{2(-4 \pm 3i)}{2(5)}$

$x = \dfrac{-4 \pm 3i}{5} = -\dfrac{4}{5} \pm \dfrac{3}{5}i$

The solution set is $\left\{-\dfrac{4}{5} - \dfrac{3}{5}i, -\dfrac{4}{5} + \dfrac{3}{5}i\right\}$.

3. Solve $(2x-6)(x+1) = -16$.

_____

Name: Date:
Instructor: Section:

Objective 2 Practice Exercises

For extra help, see Examples 1–3 on pages 739–741 of your text.

Use the quadratic formula to solve each equation. (All solutions for these equations are real numbers.)

1. $(z+2)^2 = 2(5z-2)$
   1. _____

2. $5k^2 + 4k - 2 = 0$
   2. _____

3. $34 - 10x = -x^2$
   3. _____

Objective 3 Use the discriminant to determine the number and type of solutions.

Video Examples

Review these examples for Objective 3:

4. Find the discriminant. Use it to predict the number and type of solutions for each equation. Then tell whether the equation can be solved by factoring or whether the quadratic formula should be used.

   a. $3x^2 + x - 2 = 0$

   First identify the values of a, b, and c.
   $a = 3$, $b = 1$, and $c = -2$.
   Then find the discriminant.
   $$b^2 - 4ac = 1^2 - 4(3)(-2)$$
   $$= 1 + 24$$
   $$= 25, \text{ or } 5^2$$
   Since a, b, and c are integers and the

Now Try:

4. Find the discriminant. Use it to predict the number and type of solutions for each equation. Then tell whether the equation can be solved by factoring or whether the quadratic formula should be used.

   a. $10x^2 + 21x + 9 = 0$

   _____

Name: Date:
Instructor: Section:

discriminant 25 is a perfect square, there will be two rational solutions. The equation can be solved by factoring.

b. $16x^2 + 25 = 40x$

Write in standard form: $16x^2 - 40x + 25 = 0$.
$a = 16$, $b = -40$, and $c = 25$
$$b^2 - 4ac = (-40)^2 - 4(16)(25)$$
$$= 1600 - 1600$$
$$= 0$$

Because the discriminant is 0, this quadratic equation will have one distinct rational solution. The equation can be solved by factoring.

c. $5y^2 - 5y + 2 = 0$

$a = 5$, $b = -5$, and $c = 2$
$$b^2 - 4ac = (-5)^2 - 4(5)(2)$$
$$= 25 - 40$$
$$= -15$$

Because the discriminant is negative and a, b, and c are integers, this quadratic equation will have two nonreal complex solutions. The quadratic equation should be used to solve it.

b. $25x^2 + 9 = 30x$

_____

c. $2y^2 + 4y + 8 = 0$

_____

Objective 3 Practice Exercises

For extra help, see Example 4 on pages 742–743 of your text.

Use the discriminant to determine whether the solutions for each equation are
  A. two rational numbers   B. one rational number,
  C. two irrational numbers   D. two imaginary numbers.

Do not actually solve.

4. $m^2 - 4m + 4 = 0$ 4. _____

5. $z^2 + 6z + 3 = 0$ 5. _____

6. $16x^2 - 12x + 9 = 0$ 6. _____

Name: Date:
Instructor: Section:

Chapter 11 QUADRATIC EQUATIONS, INEQUALITIES, AND FUNCTIONS

11.4 Solving Equations Quadratic in Form

Learning Objectives
1. Solve an equation with fractions by writing it in quadratic form.
2. Use quadratic equations to solve applied problems.
3. Solve an equation with radicals by writing it in quadratic form.
4. Solve an equation that is quadratic in form by substitution.

Key Terms

Use the vocabulary terms listed below to complete each statement in exercises 1–2.

quadratic in form   standard form

1. A quadratic equation written in the form $ax^2 + bx + c = 0$, $a \neq 0$ is written in _____.

2. A nonquadratic equation that can be written as a quadratic equation is called _____.

Objective 1 Solve an equation with fractions by writing it in quadratic form.

Video Examples

Review this example for Objective 1:

1. Solve $5 + \dfrac{6}{m+1} = \dfrac{14}{m}$.

   Multiply each side by the least common denominator, $m(m+1)$. The domain must be restricted to $m \neq 0$, $m \neq -1$.

   $$5 + \dfrac{6}{m+1} = \dfrac{14}{m}$$
   $$m(m+1)\left(5 + \dfrac{6}{m+1}\right) = m(m+1)\dfrac{14}{m}$$
   $$m(m+1)(5) + m(m+1)\dfrac{6}{m+1} = m(m+1)\dfrac{14}{m}$$
   $$5m^2 + 5m + 6m = 14m + 14$$
   $$5m^2 + 11m = 14m + 14$$
   $$5m^2 - 3m - 14 = 0$$
   $$(5m+7)(m-2) = 0$$
   $$5m + 7 = 0 \quad \text{or} \quad m - 2 = 0$$
   $$m = -\dfrac{7}{5} \quad \text{or} \quad m = 2$$

   The solution set is $\left\{-\dfrac{7}{5},\, 2\right\}$.

Now Try:

1. Solve $4 - \dfrac{8}{x-1} = -\dfrac{35}{x}$.

   _____

Name: Date:
Instructor: Section:

Objective 1 Practice Exercises

For extra help, see Example 1 on page 745 of your text.

Solve each equation. Check your solutions.

1. $\dfrac{5}{x} + \dfrac{1}{2x+7} = -\dfrac{2}{3}$

   1. _____

2. $\dfrac{2m}{m-5} + \dfrac{7}{m+1} = 0$

   2. _____

3. $1 + \dfrac{49}{2x} = \dfrac{15}{x+1}$

   3. _____

Objective 2 Use quadratic equations to solve applied problems.

Video Examples

Review this example for Objective 2:

2. Amy rows her boat 6 miles upstream and then returns in $2\dfrac{6}{7}$ hours. The speed of the current is 2 miles per hour. How fast can she row?

   Step 1 Read the problem carefully.

   Step 2 Assign a variable. Let x = the rate that Amy rows in still water. The current slows Amy when she is going upstream, so Amy's rate going upstream is her rate in still water less the rate of

Now Try:

2. Mike can row 3 miles per hour in still water. It takes him 3 hours and 36 minutes to row 3 miles upstream and return. Find the speed of the current.

   _____

the current, or $x - 2$. Similarly, the current makes Amy row faster when she is going downstream, so her downstream rate is $x + 2$.

Complete a table. Recall that $d = rt$.

	d	r	t
Upstream	6	$x - 2$	$\dfrac{6}{x-2}$
Downstream	6	$x + 2$	$\dfrac{6}{x+2}$

Step 3 Write an equation. The time upstream plus the time downstream equals the total time, $2\dfrac{6}{7}$ hours, or $\dfrac{20}{7}$ hours.

$$\frac{6}{x-2} + \frac{6}{x+2} = \frac{20}{7}$$

Step 4 Solve the equation. The LCD is $7(x-2)(x+2)$.

$$7(x-2)(x+2)\left(\frac{6}{x-2} + \frac{6}{x+2}\right)$$
$$= 7(x-2)(x+2)\left(\frac{20}{7}\right)$$
$$7(x+2)6 + 7(x-2)6 = (x-2)(x+2)20$$
$$42(x+2) + 42(x-2) = 20(x^2 - 4)$$
$$42x + 84 + 42x - 84 = 20x^2 - 80$$
$$20x^2 - 84x - 80 = 0$$
$$4(5x^2 - 21x - 20) = 0$$
$$4(5x+4)(x-5) = 0$$
$$5x + 4 = 0 \quad \text{or} \quad x - 5 = 0$$
$$x = -\frac{4}{5} \quad \text{or} \quad x = 5$$

Step 5 State the answer. The rate cannot be $-\dfrac{4}{5}$ mph, so the answer is Amy rows at 5 mph.

Step 6 Check that this value satisfies the original equation.

Name: Date:
Instructor: Section:

Objective 2 Practice Exercises

For extra help, see Examples 2–3 on pages 745–748 of your text.

Solve each problem. Round answers to the nearest tenth, if necessary.

4. Two pipes together can fill a large tank in 10 hours. One of the pipes, used alone, takes 15 hours longer than the other to fill the tank. How long would each pipe used alone take to fill the tank?

4. pipe 1 _____

   pipe 2 _____

5. A jet plane traveling at a constant speed goes 1200 miles with the wind, then turns around and travels for 1000 miles against the wind. If the speed of the wind is 50 miles per hour and the total flight takes 4 hours, find the speed of the plane.

5. _____

6. A man rode a bicycle for 12 miles and then hiked an additional 8 miles. The total time for the trip was 5 hours. If his rate when he was riding the bicycle was 10 miles per hour faster than his rate walking, what was each rate?

6. bike _____

   hike _____

Name: Date:
Instructor: Section:

Objective 3 Solve an equation with radicals by writing it in quadratic form.

Video Examples

Review these examples for Objective 3:

4. Solve each equation.

   a. $y = \sqrt{y+42}$

   Start by squaring each side.
   $$y = \sqrt{y+42}$$
   $$y^2 = \left(\sqrt{y+42}\right)^2$$
   $$y^2 = y + 42$$
   $$y^2 - y - 42 = 0$$
   $$(y+6)(y-7) = 0$$
   $$y+6 = 0 \quad \text{or} \quad y-7 = 0$$
   $$y = -6 \quad \text{or} \quad y = 7$$

   We must check all proposed solutions in the original equation because squaring each side of an equation can introduce extraneous solutions.

   Check
   $$y = \sqrt{y+42} \qquad y = \sqrt{y+42}$$
   $$-6 \stackrel{?}{=} \sqrt{-6+42} \qquad 7 \stackrel{?}{=} \sqrt{7+42}$$
   $$-6 \stackrel{?}{=} \sqrt{36} \qquad 7 \stackrel{?}{=} \sqrt{49}$$
   $$-6 = 6 \quad \text{False} \qquad 7 = 7 \quad \text{True}$$

   The solution set is $\{7\}$.

   b. $\sqrt{x} + 2 = x$
   $$\sqrt{x} = x - 2$$
   $$\left(\sqrt{x}\right)^2 = (x-2)^2$$
   $$x = x^2 - 4x + 4$$
   $$0 = x^2 - 5x + 4$$
   $$0 = (x-1)(x-4)$$
   $$x-1 = 0 \quad \text{or} \quad x-4 = 0$$
   $$x = 1 \quad \text{or} \quad x = 4$$

   Check each proposed solution in the original equation.

Now Try:

4. Solve each equation.

   a. $x = \sqrt{x+2}$

   _____

   b. $\sqrt{2x} + 4 = x$

   _____

Name: Date:
Instructor: Section:

$$\sqrt{x}+2=x \qquad \sqrt{x}+2=x$$
$$\sqrt{1}+2\overset{?}{=}1 \qquad \sqrt{4}+2\overset{?}{=}4$$
$$1+2\overset{?}{=}1 \qquad 2+2\overset{?}{=}4$$
$$3=1 \quad \text{False} \qquad 4=4 \quad \text{True}$$

The solution set is {4}.

Objective 3 Practice Exercises

For extra help, see Example 4 on page 748 of your text.

Solve each equation. Check your solutions.

7. $\sqrt{7y-10}=y$ 7. _____

8. $x=\sqrt{\dfrac{x+3}{2}}$ 8. _____

9. $\sqrt{4x}+3=x$ 9. _____

Name:
Instructor:
Date:
Section:

Objective 4 Solve an equation that is quadratic in form by substitution.

Video Examples

Review these examples for Objective 4:

6. Solve the equation.

$$c^4 - 20c^2 + 64 = 0$$

Write this equation in quadratic form by substituting u for c^2.

$$u^2 - 20u + 64 = 0$$
$$(u-4)(u-16) = 0$$

$u - 4 = 0$ or $u - 16 = 0$
$u = 4$ or $u = 16$
$c^2 = 4$ or $c^2 = 16$
$c = \pm 2$ or $c = \pm 4$

The solution set is $\{-4, -2, 2, 4\}$.

7. Solve each equation.

a. $(m+5)^2 + 6(m+5) + 8 = 0$

Step 1 Substitute u for $m+5$.

$$(m+5)^2 + 6(m+5) + 8 = 0$$
$$u^2 + 6u + 8 = 0$$

Step 2 $(u+4)(u+2) = 0$

$u + 4 = 0$ or $u + 2 = 0$
$u = -4$ or $u = -2$

Step 3 $m + 5 = -4$ or $m + 5 = -2$

Step 4 $m = -9$ or $m = -7$

Step 5 Check that the solution set of the original equation is $\{-9, -7\}$.

b. $x^{4/3} - 20x^{2/3} + 36 = 0$

Step 1 Substitute u for $x^{2/3}$.

$$x^{4/3} - 20x^{2/3} + 36 = 0$$
$$u^2 - 20u + 36 = 0$$

Step 2 $(u-2)(u-18) = 0$

$u - 2 = 0$ or $u - 18 = 0$
$u = 2$ or $u = 18$

Now Try:

6. Solve the equation.

$$x^4 - 5x^2 + 4 = 0$$

_____

7. Solve each equation.

a. $(x-5)^2 + 2(x-5) - 35 = 0$

_____

b. $x^{2/3} - 2x^{1/3} = 3$

_____

Name: Date:
Instructor: Section:

Step 3 $\quad x^{2/3} = 2 \quad$ or $\quad x^{2/3} = 18$

Step 4 $\quad (x^{2/3})^{3/2} = (2)^{3/2} \quad$ or $\quad (x^{2/3})^{3/2} = (18)^{3/2}$

$\quad\quad\quad\quad x = 2\sqrt{2} \quad$ or $\quad x = 54\sqrt{2}$

Step 5 Check that the solution set of the original equation is $\{2\sqrt{2},\ 54\sqrt{2}\}$.

Objective 4 Practice Exercises

For extra help, see Examples 5–7 on pages 749–752 of your text.

Solve each equation. Check your solutions.

10. $\quad 4t^4 = 21t^2 - 5$

10. _____

11. $\quad p^{4/3} - 12p^{2/3} + 27 = 0$

11. _____

12. $\quad (t^2 - 3t)^2 = 14(t^2 - 3t) - 40$

12. _____

Name: Date:
Instructor: Section:

Chapter 11 QUADRATIC EQUATIONS, INEQUALITIES, AND FUNCTIONS

11.5 Formulas and Further Applications

Learning Objectives
1. Solve formulas for variables involving squares and square roots.
2. Solve applied problems using the Pythagorean theorem.
3. Solve applied problems using area formulas.
4. Solve applied problems using quadratic functions as models.

Key Terms

Use the vocabulary terms listed below to complete each statement in exercises 1–2.

quadratic function **Pythagorean theorem**

1. A function defined by $f(x) = ax^2 + bx + c$, for real numbers a, b, and c, with $a \neq 0$, is a _____.

2. The _____ states that the sum of the squares of the lengths of the legs of a right triangle equals the square of the length of the hypotenuse.

Objective 1 Solve formulas for variables involving squares and square roots.

Video Examples

Review these examples for Objective 1:	Now Try:
1. Solve the formula for the given variable. $y = \frac{1}{2}gt^2$ for t The goal is to isolate t on one side. $y = \frac{1}{2}gt^2$ $2y = gt^2$ $\frac{2y}{g} = t^2$ $t = \pm\sqrt{\frac{2y}{g}}$ $t = \pm\frac{\sqrt{2y}}{\sqrt{g}} \cdot \frac{\sqrt{g}}{\sqrt{g}}$ $t = \pm\frac{\sqrt{2yg}}{g}$	1. Solve the formula for the given variable. $F = \frac{mx}{t^2}$ for t _____

426   Copyright © 2016 Pearson Education, Inc.

Name: Date:
Instructor: Section:

2. Solve $rk^2 - 3k = -s$ for k.

Write the equation in standard form and then use the quadratic formula to solve for k.

$$rk^2 - 3k = -s$$
$$rk^2 - 3k + s = 0$$

Let $a = r$, $b = -3$, and $c = s$.

$$k = \frac{-(-3) \pm \sqrt{(-3)^2 - 4(r)(s)}}{2r}$$

$$k = \frac{3 \pm \sqrt{9 - 4rs}}{2r}$$

The solutions are $\frac{3 + \sqrt{9 - 4rs}}{2r}$ and $\frac{3 - \sqrt{9 - 4rs}}{2r}$.

2. Solve $p^2q^2 + pkq = k^2$ for q.

2. _____

Objective 1 Practice Exercises

For extra help, see Examples 1–2 on pages 757–758 of your text.

Solve each equation for the indicated variable. (Leave ± in your answers.)

1. $F = \frac{kl}{\sqrt{d}}$ for d

1. _____

2. $p = \sqrt{\frac{kl}{g}}$ for k

2. _____

3. $b^2a^2 + 2bca = c^2$ for a

3. _____

Name: Date:
Instructor: Section:

Objective 2 Solve applied problems using the Pythagorean theorem.

Video Examples

Review this example for Objective 2:

3. A 13-foot ladder is leaning against a building. The distance from the bottom of the ladder to the building is 2 feet more than twice the distance from the top of the ladder to the ground. How far is the bottom of the ladder from the building?

    Step 1 Read the problem carefully.

    Step 2 Assign a variable. Let x = the distance from the top of the ladder to the ground. Then $2x + 2$ = the distance from the bottom of the ladder to the building. Draw a picture to represent the problem.

    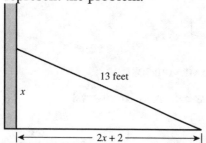

    Step 3 Write an equation. Use the Pythagorean theorem.
    $$a^2 + b^2 = c^2$$
    $$x^2 + (2x+2)^2 = 13^2$$

    Step 4 Solve.
    $$x^2 + 4x^2 + 8x + 4 = 169$$
    $$5x^2 + 8x - 165 = 0$$
    $$(5x + 33)(x - 5) = 0$$
    $5x + 33 = 0$ or $x - 5 = 0$
    $x = -\dfrac{33}{5}$ or $x = 5$

    Step 5 State the answer. Length cannot be negative, so discard the negative solution. The distance from the top of the ladder to the ground is 5 feet. However, we are asked to find the distance from the bottom of the ladder to the building. This distance is $2(5) + 2 = 12$ feet.

    Step 6 Check. Since $5^2 + 12^2 = 13^2$, the answer is correct.

Now Try:

3. Two cars left an intersection at the same time, one heading south, the other heading east. Sometime later, the car traveling south had gone 18 miles farther than the car headed east. At that time they were 90 miles apart. How far had each car traveled?

    south _____

    east _____

Name: Date:
Instructor: Section:

Objective 2 Practice Exercises

For extra help, see Example 3 on page 758 of your text.

Solve each problem.

4. A child flying a kite has let out 45 feet of string to the kite. The distance from the kite to the ground is 9 feet more than the distance from the child to a point directly below the kite. How high up is the kite?

   4. _____

5. A ladder is leaning against a building so that the top is 8 feet above the ground. The length of the ladder is 2 feet less than twice the distance of the bottom of the ladder from the building. Find the length of the ladder.

   5. _____

6. Two cars left an intersection at the same time, one heading north, the other heading west. Later they were exactly 95 miles apart. The car headed west had gone 38 miles less than twice as far as the car headed north. How far had each car traveled?

   6. north _____

      west _____

Copyright © 2016 Pearson Education, Inc.

Name: Date:
Instructor: Section:

Objective 3 Solve applied problems using area formulas.

Video Examples

Review this example for Objective 3:

4. A fish pond is 3 feet by 4 feet. How wide a strip of concrete can be poured around the pond if there is enough concrete for 44 square feet?

Step 1 Read the problem carefully.

Step 2 Assign a variable. Let x = the width of the strip of concrete. Then $2x + 3$ = the width of the fish pond with the two strips of concrete and $2x + 4$ = the length of the fish pond with the two strips of concrete.

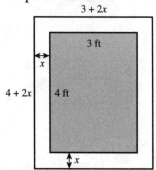

Step 3 Write an equation. The area of the strip is 44 sq ft and the area of the fish pond is $3(4) = 12$ sq ft, so the total area of the outer rectangle is $44 + 12 = 56$ sq ft.
$$(3+2x)(4+2x) = 56$$

Step 4 Solve.
$$(3+2x)(4+2x) = 56$$
$$12 + 14x + 4x^2 = 56$$
$$4x^2 + 14x - 44 = 0$$
$$2(2x^2 + 7x - 22) = 0$$
$$2(2x+11)(x-2) = 0$$
$$2x + 11 = 0 \quad \text{or} \quad x - 2 = 0$$
$$x = -\frac{11}{2} \quad \text{or} \quad x = 2$$

Step 5 State the answer. Since length cannot be negative, we disregard the negative solution. The concrete strip should be 2 ft wide.

Step 6 Check. If $x = 2$, then the area of the large rectangle is $(3 + 2 \cdot 2)(4 + 2 \cdot 2) = 7(8) = 56$ sq ft. The area of the fish pond is $3(4) = 12$ sq ft, so the area of the concrete strip is $56 - 12 = 44$ sq ft.

Now Try:

4. A picture 9 inches by 12 inches is to be mounted on a piece of mat board so that there is an even width of mat all around the picture. How wide will the matted border be if the area of the mounted picture is 238 square inches?

_____

Name: Date:
Instructor: Section:

Objective 3 Practice Exercises

For extra help, see Example 4 on page 759 of your text.

Solve each problem.

7. A rug is to fit in a room so that a border of even width is left on all four sides. If the room is 16 feet by 20 feet and the area of the rug is 165 square feet, how wide to the nearest tenth of a foot will the border be?

7. _____

8. A rectangular garden has an area of 12 feet by 5 feet. A gravel path of equal width is to be built around the garden. How wide can the path be if there is enough gravel for 138 square feet?

8. _____

9. A doghouse 2 feet by 4 feet is to be built with a cement path around it of equal width on all sides. The area available for the doghouse and path is 120 square feet. How wide will the path be?

9. _____

Name: Date:
Instructor: Section:

Objective 4 Solve applied problems using quadratic functions as models.

Video Examples

Review this example for Objective 4:

5. A certain projectile is located at a distance of $d(t) = 3t^2 - 6t + 1$ feet from its starting point after t seconds. How many seconds will it take the projectile to travel 10 feet?

   Let $d = 10$ in the formula and solve for t.
   $$d = 3t^2 - 6t + 1$$
   $$10 = 3t^2 - 6t + 1$$
   $$9 = 3t^2 - 6t$$
   $$3 = t^2 - 2t$$
   $$3 + 1 = t^2 - 2t + 1$$
   $$4 = (t-1)^2$$
   $$\sqrt{4} = t - 1 \quad \text{or} \quad -\sqrt{4} = t - 1$$
   $$2 = t - 1 \quad \text{or} \quad -2 = t - 1$$
   $$3 = t \quad \text{or} \quad -1 = t$$

   Since t represents time, we reject the negative solution. It will take 3 seconds for the projectile to travel 10 feet.

Now Try:

5. A baseball is thrown upward from a building 20 m high with a velocity of 15 m/sec. Its distance from the ground after t seconds is modeled by the function
$f(t) = -4.9t^2 + 15t + 20$.
When will the ball hit the ground? Round your answer to the nearest tenth.

_____

For extra help, see Examples 5–6 on pages 760–761 of your text.

Solve each problem. Round answers to the nearest tenth.

10. A population of microorganisms grows according to the function $p(x) = 100 + 0.2x + 0.5x^2$, where x is given in hours. How many hours does it take to reach a population of 250 microorganisms?

10. _____

Name: Date:
Instructor: Section:

11. An object is thrown downward from a tower 280 feet high. The distance the object has fallen at time t in seconds is given by $s(t) = 16t^2 + 68t$. How long will it take the object to fall 100 feet?

11. _____

12. A widget manufacturer estimates that her monthly revenue can be modeled by the function $R(x) = -0.006x^2 + 32x - 10,000$. What is the minimum number of items that must be sold for the revenue to equal $30,000?

12. _____

Name: Date:
Instructor: Section:

Chapter 11 QUADRATIC EQUATIONS, INEQUALITIES, AND FUNCTIONS

11.6 Graphs of Quadratic Functions

Learning Objectives
1. Graph a quadratic function.
2. Graph parabolas with horizontal and vertical shifts.
3. Use the coefficient of x^2 to predict the shape and direction in which a parabola opens.
4. Find a quadratic function to model data.

Key Terms

Use the vocabulary terms listed below to complete each statement in exercises 1–4.

parabola   vertex   axis   quadratic function

1. The vertical (or horizontal) line through the vertex of a vertical (or horizontal) parabola is its _____.

2. The point on a parabola that has the least y-value (if the parabola opens up) or the greatest y-value (if the parabola opens down) is called the _____ of the parabola.

3. A function defined by $f(x) = ax^2 + bx + c$, for real numbers a, b, and c, with $a \neq 0$, is a _____.

4. The graph of a quadratic function is a _____.

Objective 1   Graph a quadratic function.

For extra help, see page 766 of your text.

Objective 2   Graph parabolas with horizontal and vertical shifts.

Video Examples

Review these examples for Objective 2:	Now Try:
1. Graph $g(x) = x^2 + 2$. Give the vertex, axis, domain, and range. The graph of $g(x)$ has the same shape as that of $f(x) = x^2$ but shifted 2 units up with vertex (0, 2). Every function value is 2 more than the corresponding function value of $f(x) = x^2$.	1. Graph $f(x) = x^2 - 1$. Give the vertex, axis, domain, and range. Vertex _____ Axis _____ Domain _____ Range _____

Name: _____ Date: _____
Instructor: _____ Section: _____

x	$f(x)=x^2$	$g(x)=x^2+2$
-2	4	6
-1	1	3
0	0	2
1	1	3
2	4	6

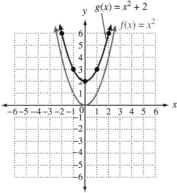

The vertex is (0, 2). The axis is $x = 0$. The domain is $(-\infty, \infty)$. The range is $[2, \infty)$.

2. Graph $g(x)=(x-1)^2$. Give the vertex, axis, domain, and range.

The graph of $g(x)$ has the same shape as that of $f(x)=x^2$ but shifted 1 unit right with vertex (1, 0).

x	$f(x)=x^2$	$g(x)=(x-1)^2$
-2	4	9
-1	1	4
0	0	1
1	1	0
2	4	1
3	9	4

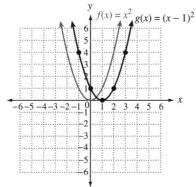

The vertex is (1, 0). The axis is $x = 1$. The domain is $(-\infty, \infty)$. The range is $[0, \infty)$.

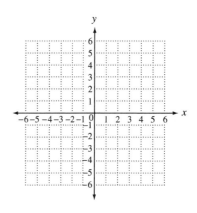

2. Graph $f(x)=(x-3)^2$. Give the vertex, axis, domain, and range.

Vertex _____

Axis _____

Domain _____

Range _____

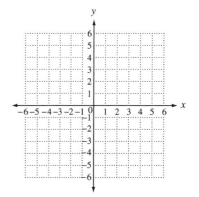

Copyright © 2016 Pearson Education, Inc.

3. Graph $g(x)=(x-1)^2-2$. Give the vertex, axis, domain, and range.

The graph of $g(x)$ has the same shape as that of $f(x)=x^2$ but shifted 1 unit right (since $x-1=0$ if $x=1$) and 2 units down (because of the -2).

x	$g(x)=(x-1)^2-2$
-2	7
-1	2
0	-1
1	-2
2	-1
3	2
4	7

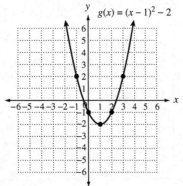

The vertex is $(1, -2)$. The axis is $x=1$. The domain is $(-\infty, \infty)$. The range is $[-2, \infty)$.

3. Graph $f(x)=(x+2)^2-1$. Give the vertex, axis, domain, and range.

Vertex _____

Axis _____

Domain _____

Range _____

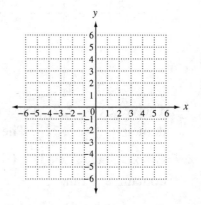

Name: Date:
Instructor: Section:

Objective 2 Practice Exercises

For extra help, see Examples 1–3 on pages 767–768 of your text.

Sketch the graph of each parabola. Give the vertex, axis, domain, and range.

1. $f(x) = x^2 - 4$

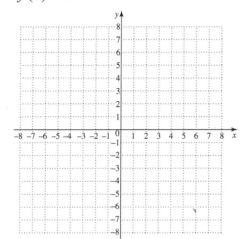

1. vertex_____

   axis_____

   domain_____

   range _____

2. $f(x) = (x-3)^2$

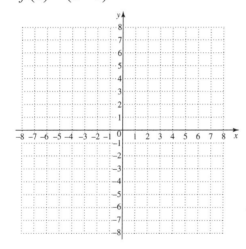

2. vertex_____

   axis_____

   domain_____

   range _____

Copyright © 2016 Pearson Education, Inc.

Name: Date:
Instructor: Section:

3. $f(x) = (x-3)^2 - 1$

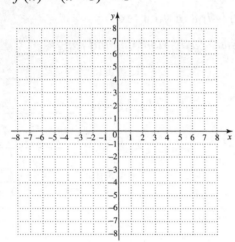

3. vertex _____

axis _____

domain _____

range _____

Objective 3 Use the coefficient of x^2 to predict the shape and direction in which a parabola opens.

Video Examples

Review these examples for Objective 3:

4. Graph $g(x) = -2x^2$. Give the vertex, axis, domain, and range.

The graph of $g(x)$ has the same shape as that of $f(x) = x^2$ but is narrower and opens downward.

x	$g(x) = -2x^2$
-2	-8
-1	-2
0	0
1	-2
2	-8

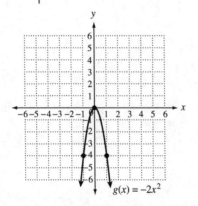

The vertex is $(0, 0)$. The axis is $x = 0$. The domain is $(-\infty, \infty)$. The range is $(-\infty, 0]$.

Now Try:

4. Graph $g(x) = -\frac{1}{4}x^2$. Give the vertex, axis, domain, and range.

Vertex _____

Axis _____

Domain _____

Range _____

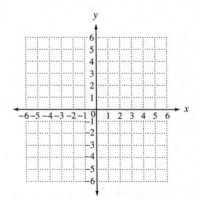

Name: Date:
Instructor: Section:

5. Graph $g(x) = -\frac{1}{2}(x+1)^2 - 2$. Give the vertex, axis, domain, and range.

The parabola opens down because $a < 0$ and is wider than the graph of $f(x) = x^2$. The parabola has vertex $(-1, -2)$.

x	$g(x) = -\frac{1}{2}(x+1)^2 - 2$
-3	-4
-2	-2.5
-1	-2
0	-2.5
1	-4

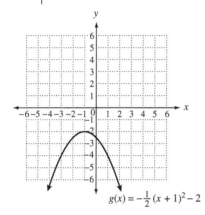

$g(x) = -\frac{1}{2}(x+1)^2 - 2$

The vertex is $(-1, -2)$. The axis is $x = -1$. The domain is $(-\infty, \infty)$. The range is $(-\infty, -2]$.

5. Graph $f(x) = 3(x-1)^2 + 1$. Give the vertex, axis, domain, and range.

Vertex   _____

Axis   _____

Domain   _____

Range   _____

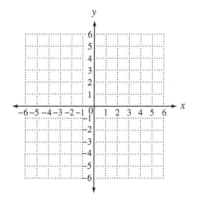

Objective 3 Practice Exercises

For extra help, see Examples 4–5 on page 769 of your text.

For each quadratic function, tell whether the graph opens up or down and whether the graph is wider, narrower, or the same shape as the graph of $f(x) = x^2$. Then give the vertex, domain, and range.

4. $f(x) = -\frac{4}{3}x^2 - 1$

4. _____

vertex _____

domain _____

range _____

Name: Date:
Instructor: Section:

5. $f(x) = -2(x+1)^2$

5. _____

vertex _____

domain _____

range _____

6. $f(x) = \frac{5}{4}(x-1)^2 + 7$

6. _____

vertex _____

domain _____

range _____

Objective 4 Find a quadratic function to model data.

Video Examples

Review this example for Objective 4:

6. The number of ice cream cones sold by an ice cream parlor from 2003–2009 is shown in the following table.

Year	Years since 2003, x	Number of cones sold
2003	0	1775
2004	1	4194
2005	2	5063
2006	3	5161
2007	4	4663
2008	5	4639
2009	6	3710

Use the ordered pairs (x, number of cones sold) to make a scatter diagram of the data. Determine a quadratic function that models these data by using a system of equations. Use the ordered pairs (0, 1775), (3, 5161), and (6, 3710). Round the values of a, b, and c in your model to the nearest tenth, as necessary.

Now Try:

6. The table lists the average price of a Major League Baseball ticket.

Year	Years since 1990, x	Price
1991	1	$9.14
1994	4	$10.60
1997	7	$12.49
2000	10	$16.81
2004	14	$19.82
2010	20	$26.74

Use the ordered pairs (x, price) to make a scatter diagram of the data. Determine a quadratic function that models these data by using a system of equations. Use the ordered pairs (1, 9.14), (10, 16.81), and (20, 26.74). Round the values of a, b, and c in your model to the nearest hundredth, as necessary.

_____

Name: Date:
Instructor: Section:

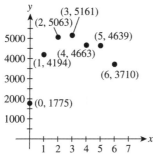

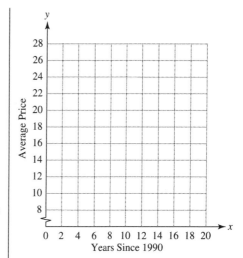

It appears that the parabola opens down, so the coefficient a is negative.

Using the chosen ordered pairs, we substitute the x- and y-values into the quadratic form $y = ax^2 + bx + c$ to obtain the three equations.

$$a(0)^2 + b(0) + c = 1775 \quad (1)$$

$$a(3)^2 + b(3) + c = 5161 \quad (2)$$

$$a(6)^2 + b(6) + c = 3710 \quad (3)$$

Equation (1) simplifies to $c = 1775$, so substitute 1775 for c in equation (2) and (3).

$$9a + 3b + 1775 = 5161 \quad (2)$$

$$36a + 6b + 1775 = 3710 \quad (3)$$

Subtract 1775 from each side of both equations.

$$9a + 3b = 3386 \quad (2)$$

$$36a + 6b = 1935 \quad (3)$$

Solve by elimination. Multiply equation (2) by -2 and add to equation (3).

$$-18a - 6b = -6772 \quad \text{Multiply (2) by } -2.$$

$$\underline{36a + 6b = 1935 \quad (3)}$$

$$18a = -4837$$

$$a \approx -268.7 \quad \text{Round to the tenth.}$$

Substitute this value for a into equation (2) and solve for b.

$$9a + 3b = 3386 \quad (2)$$

$$9(-268.7) + 3b = 3386$$

$$-2418.3 + 3b = 3386$$

$$3b = 5804.3$$

$$b \approx 1934.8$$

Therefore, the model is $y = -268.7x^2 + 1934.8x + 1775$.

Name: Date:
Instructor: Section:

Objective 4 Practice Exercises

For extra help, see Example 6 on pages 770–771 of your text.

Tell whether a linear or quadratic function would be a more appropriate model for each set of graphed data. If linear, tell whether the slope should be positive or negative. If quadratic, tell whether the coefficient a of x^2 should be positive or negative.

7.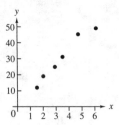

7. _____

8.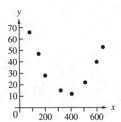

8. _____

Solve the problem.

9. The number of publicly traded companies filing for bankruptcy for selected years between 1990 and 2000 are shown in the table, with 0 representing 1990, 2 representing 1992, etc.

Year	Number of Bankruptcies
0	115
2	91
4	70
6	84
8	120
10	176

Use the ordered pairs to make a scatter diagram of the data.

Use the ordered pairs (0, 115), (4, 70), and (8, 120) to find a function that models the data. Round the values of *a*, *b*, and *c* to three decimal places, if necessary.

Source: Lial, Margaret L., John Hornsby, Terry McGinnis, *Intermediate Algebra* Eighth Edition. Boston: Pearson Education, 2006.

9.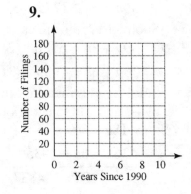

_____

Name: Date:
Instructor: Section:

Chapter 11 QUADRATIC EQUATIONS, INEQUALITIES, AND FUNCTIONS

11.7 More about Parabolas and Their Applications

Learning Objectives
1. Find the vertex of a vertical parabola.
2. Graph a quadratic function.
3. Use the discriminant to find the number of x-intercepts of a parabola with a vertical axis.
4. Use quadratic functions to solve problems involving maximum or minimum value.
5. Graph parabolas with horizontal axes.

Key Terms

Use the vocabulary terms listed below to complete each statement in exercises 1–2.

   discriminant   **vertex**

1. The _____ of a quadratic function is found by using the formula $b^2 - 4ac$.

2. The maximum or minimum value of a quadratic function occurs at the _____ of its graph.

Objective 1 Find the vertex of a vertical parabola.

Video Examples

Review these examples for Objective 1:
1. Find the vertex of the graph of $f(x) = x^2 + 6x + 10$.

   We can express $x^2 + 6x + 10$ in the form $(x-h)^2 + k$ by completing the square on $x^2 + 6x$. Because we want to keep $f(x)$ alone on one side of the equation, we add and subtract the appropriate number on just one side.

   $f(x) = x^2 + 6x + 10 \qquad \left[\frac{1}{2}(6)\right]^2 = 9$

   $f(x) = (x^2 + 6x + 9 - 9) + 10$

   $f(x) = (x^2 + 6x + 9) + 10 - 9$

   $f(x) = (x+3)^2 + 1$

   The vertex of the parabola is (–3, 1).

Now Try:
1. Find the vertex of the graph of $f(x) = x^2 - 6x + 4$.

   _____

Name: Date:
Instructor: Section:

2. Find the vertex of the graph of $f(x) = 3x^2 + 6x + 10$.

    Because the x^2-term has a coefficient other than 1, we factor that coefficient out of the first two terms before completing the square.

    $f(x) = 3x^2 + 6x + 10$

    $f(x) = 3(x^2 + 2x) + 10 \qquad \left[\frac{1}{2}(2)\right]^2 = 1$

    $f(x) = 3(x^2 + 2x + 1 - 1) + 10$

    $f(x) = 3(x^2 + 2x + 1) + 3(-1) + 10$

    $f(x) = 3(x + 1)^2 + 7$

    The vertex is (–1, 7).

2. Find the vertex of the graph of $f(x) = -2x^2 + 4x - 1$.

    _____

3. Use the vertex formula to find the vertex of the graph of $f(x) = -4x^2 + 5x + 3$.

    The x-coordinate of the vertex of the parabola is given by $\frac{-b}{2a}$.

    $\frac{-b}{2a} = \frac{-5}{2(-4)} = \frac{5}{8}$

    The y-coordinate is $f\left(\frac{-b}{2a}\right) = f\left(\frac{5}{8}\right)$.

    $f\left(\frac{5}{8}\right) = -4\left(\frac{5}{8}\right)^2 + 5\left(\frac{5}{8}\right) + 3 = \frac{73}{16}$

    The vertex is $\left(\frac{5}{8}, \frac{73}{16}\right)$.

3. Use the vertex formula to find the vertex of the graph of $f(x) = 2x^2 - 6x + 5$.

    _____

Objective 1 Practice Exercises

For extra help, see Examples 1–3 on pages 775–777 of your text.

Find the vertex of each parabola.

1. $f(x) = x^2 - 2x + 4$

    1. _____

2. $f(x) = 5x^2 - 4x + 1$

    2. _____

3. $f(x) = -\frac{1}{4}x^2 - 3x - 9$

    3. _____

Name: Date:
Instructor: Section:

Objective 2   Graph a quadratic function.

Video Examples

Review this example for Objective 2:

4. Graph the quadratic function defined by $f(x) = x^2 - 3x + 2$. Give the vertex, axis, domain, and range.

Step 1 From the equation, $a = 1$, so the graph opens up.

Step 2 The x-coordinate of the vertex is $\frac{3}{2}$. The y-coordinate of the vertex is

$f\left(\frac{3}{2}\right) = \left(\frac{3}{2}\right)^2 - 3\left(\frac{3}{2}\right) + 2 = -\frac{1}{4}$. The vertex is $\left(\frac{3}{2}, -\frac{1}{4}\right)$.

Step 3 Find any intercepts. Since the vertex is in quadrant IV and the graph opens up, there will be two x-intercepts. Let $f(x) = 0$ and solve.

$x^2 - 3x + 2 = 0$
$(x-1)(x-2) = 0$
$x - 1 = 0 \quad \text{or} \quad x - 2 = 0$
$x = 1 \quad \text{or} \quad x = 2$

The x-intercepts are (1, 0) and (2, 0). Find the y-intercept by evaluating $f(0)$.

$f(0) = 0^2 - 3(0) + 2$

The y-intercept is (0, 2).

Step 4 Plot the points found so far and additional points as needed using symmetry about the axis, $x = \frac{3}{2}$.

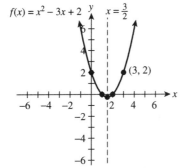

The domain is $(-\infty, \infty)$. The range is $\left[-\frac{1}{4}, \infty\right)$.

Now Try:

4. Graph the quadratic function defined by $f(x) = x^2 + 4x + 5$. Give the vertex, axis, domain, and range.

Vertex   _____

Axis   _____

Domain   _____

Range   _____

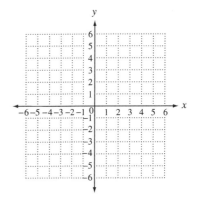

Name: Date:
Instructor: Section:

Objective 2 Practice Exercises

For extra help, see Example 4 on pages 777–778 of your text.

Sketch the graph of each parabola. Give the vertex, axis, domain, and range.

4. $f(x) = -x^2 + 8x - 10$

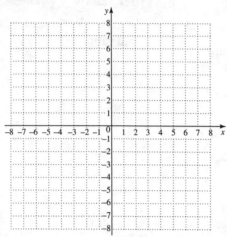

4. vertex _____

   axis _____

   domain _____

   range _____

5. $f(x) = 3x^2 + 6x + 2$

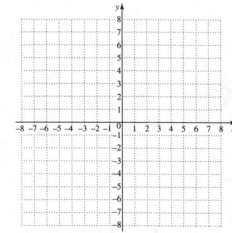

5. vertex _____

   axis _____

   domain _____

   range _____

6. $f(x) = -2x^2 + 4x + 1$

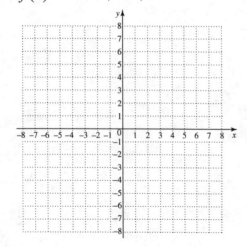

6. vertex _____

   axis _____

   domain _____

   range _____

Name: Date:
Instructor: Section:

Objective 3 Use the discriminant to find the number of x-intercepts of a parabola with a vertical axis.

Video Examples

Review this example for Objective 3:

5. Use the discriminant to determine the number of x-intercepts of the graph of the quadratic function.

$f(x) = 4x^2 + 12x + 9$

$b^2 - 4ac = 12^2 - 4(4)(9) = 0$

Since the discriminant is zero, the graph has only one x-intercept, its vertex.

Now Try:

5. Use the discriminant to determine the number of x-intercepts of the graph of the quadratic function.

$f(x) = 9x^2 - 24x + 16$

_____

Objective 3 Practice Exercises

For extra help, see Example 5 on pages 778–779 of your text.

Use the discriminant to determine the number of x-intercepts of the graph of each function.

7. $f(x) = 2x^2 - 3x + 2$

7. _____

8. $f(x) = -3x^2 - x + 5$

8. _____

9. $f(x) = 3x^2 - 6x + 3$

9. _____

Objective 4 Use quadratic functions to solve problems involving maximum or minimum value.

Video Examples

Review this example for Objective 4:

6. A farmer has 1000 yards of fencing to enclose a rectangular field. What is the largest area that the farmer can enclose? What are the dimensions of the field when the area is maximized?

If the length of the field is represented by *l* and the width of the field is represented by *w*, the perimeter is given by $2l + 2w = 1000$ or $l + w = 500$ or $w = 500 - l$.

Now Try:

6. A farmer has 1000 yards of fencing to enclose a rectangular field next to a building. What is the largest area that the farmer can enclose? What are the dimensions of the field when the area is maximized?

_____

Copyright © 2016 Pearson Education, Inc.

Name: Date:
Instructor: Section:

The area is given by $A = lw$ or
$A(l) = l(500 - l) = 500l - l^2 = -l^2 + 500l$.
This is a quadratic equation, so its maximum occurs at the vertex of its graph. The
x-coordinate is given by $\frac{-b}{2a} = \frac{-500}{2(-1)} = 250$.

The y-coordinate is
$A(250) = -250^2 + 500(250) = 62{,}500$.
If $l = 250$, then $w = 500 - 250 = 250$.
Therefore, the maximum area is 62,500 sq yd when the length of the field is 250 yd and the width is 250 yd.

Objective 4 Practice Exercises

For extra help, see Examples 6–7 on pages 779–780 of your text.

Solve each problem.

10. Jean sells ceramic pots. She has weekly costs of $C(x) = x^2 - 100x + 2700$, where x is the number of pots she sells each week. How many pots should she sell to minimize her costs? What is the minimum cost?

10. units _____

 cost _____

11. The length and width of a rectangle have a sum of 48. What width will produce the maximum area?

11. _____

Name: Date:
Instructor: Section:

12. A projectile is fired upward so that its distance (in feet) above the ground t seconds after firing is given by $s(t) = -16t^2 + 80t + 156$. Find the maximum height it reaches and the number of seconds it takes to reach that height.

12. height _____

time _____

Objective 5 Graph parabolas with horizontal axes.

Video Examples

Review this example for Objective 5:

9. Graph $x = -y^2 + 6y - 9$. Give the vertex, axis, domain, and range.

We must complete the square in order to write the equation in $x = (y-k)^2 + h$ form.

$x = -(y^2 - 6y) - 9$

$x = -(y^2 - 6y + 9 - 9) - 9$

$x = -(y^2 - 6y + 9) - 1(-9) - 9$

$x = -(y-3)^2$

The vertex is (0, 3). The axis is $y = 3$.

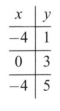

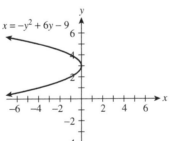

domain: $(-\infty, 0]$
range: $(-\infty, \infty)$

Now Try:

9. Graph $x = -y^2 + 4y - 4$. Give the vertex, axis, domain, and range.

Vertex _____

Axis _____

Domain _____

Range _____

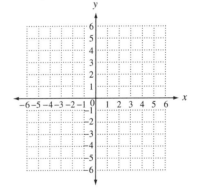

Copyright © 2016 Pearson Education, Inc.

Name: Date:
Instructor: Section:

Objective 5 Practice Exercises

For extra help, see Examples 8–9 on page 781 of your text.

Sketch the graph of each parabola. Give the vertex, axis, domain, and range.

13. $x = -y^2 + 2$

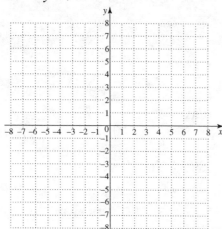

13. vertex _____

   axis _____

   domain _____

   range _____

14. $x = y^2 - 3$

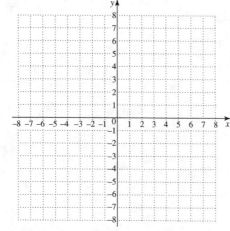

14. vertex _____

   axis _____

   domain _____

   range _____

15. $x = -y^2 - 6y - 10$

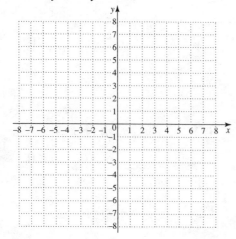

15. vertex _____

   axis _____

   domain _____

   range _____

Name: Date:
Instructor: Section:

Chapter 11 QUADRATIC EQUATIONS, INEQUALITIES, AND FUNCTIONS

11.8   Polynomial and Rational Inequalities

Learning Objectives
1. Solve quadratic inequalities.
2. Solve polynomial inequalities of degree 3 or greater.
3. Solve rational inequalities.

Key Terms

Use the vocabulary terms listed below to complete each statement in exercises 1–2.

quadratic inequality   **rational inequality**

1. An inequality that involves a rational expression is a _____.

2. An inequality that can be written in the form $ax^2 + bx + c < 0$ or $ax^2 + bx + c > 0$, where *a*, *b*, and *c* are real numbers with $a \neq 0$ is called a _____.

Objective 1   Solve quadratic inequalities.

Video Examples

Review these examples for Objective 1:	Now Try:
1. Solve each inequality.	1. Solve each inequality.

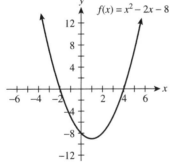
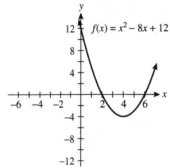

a. $x^2 - 2x - 8 > 0$ **a.** $x^2 - 8x + 12 > 0$

From the graph, we see that the *y*-values are greater than 0 when the *x*-values are less than –2 or greater than 4. Therefore, the solution set of $x^2 - 2x - 8 > 0$ is $(-\infty, -2) \cup (4, \infty)$.

b. $x^2 - 2x - 8 < 0$ **b.** $x^2 - 8x + 12 < 0$

From the graph, we see that the *y*-values are less than 0 when the *x*-values are greater than –2 and less than 4. Therefore, the solution set of $x^2 - 2x - 8 < 0$ is $(-2, 4)$.

Copyright © 2016 Pearson Education, Inc.

Name: Date:
Instructor: Section:

2. Solve and graph the solution set of $x^2 + 5x + 4 \geq 0$.

Solve the quadratic equation by factoring.
$$(x+1)(x+4) = 0$$
$$x+1 = 0 \quad \text{or} \quad x+4 = 0$$
$$x = -1 \quad \text{or} \quad x = -4$$

The numbers −4 and −1 divide a number line into intervals A, B, and C, as shown below.

Since the numbers −4 and −1 are the only numbers that make the quadratic expression $x^2 + 5x + 4$ equal to 0, all other numbers make the expression either positive or negative. If one number in an interval satisfies the inequality, then all the numbers in that interval will satisfy the inequality.

Choose any number in interval A as a test number; we will choose −5.
$$x^2 + 5x + 4 \geq 0$$
$$(-5)^2 + 5(-5) + 4 \overset{?}{\geq} 0$$
$$4 \geq 0 \quad \text{True}$$

Because −5 satisfies the inequality, all numbers from interval A are solutions.

Now try −2 from interval B.
$$x^2 + 5x + 4 \geq 0$$
$$(-2)^2 + 5(-2) + 4 \overset{?}{\geq} 0$$
$$-2 \geq 0 \quad \text{False}$$

The numbers in interval B are not solutions.

Finally, try 0 from interval C.
$$x^2 + 5x + 4 \geq 0$$
$$0^2 + 5(0) + 4 \overset{?}{\geq} 0$$
$$4 \geq 0 \quad \text{True}$$

Because 0 satisfies the inequality, all numbers from interval C are solutions.

Because the inequality is greater than or equal to zero, we include the endpoints of the intervals in the solution set. Thus, the solution set is $(-\infty, -4] \cup [-1, \infty)$.

2. Solve and graph the solution set of $x^2 - x - 2 < 0$.

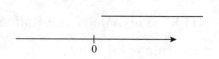

452 Copyright © 2016 Pearson Education, Inc.

Name: Date:
Instructor: Section:

4. Solve each inequality.

   a. $(2k+5)^2 \geq -1$

   Because $(2k+5)^2$ is never negative, it is always greater than –1. The solution set is $(-\infty, \infty)$.

   b. $(2k+5)^2 \leq -1$

   Because $(2k+5)^2$ is never negative, there is no solution. The solution set is $\emptyset$.

4. Solve each inequality.

   a. $(4m+1)^2 \geq -3$

   _____

   b. $(4m+1)^2 \leq -3$

   _____

Objective 1 Practice Exercises

For extra help, see Examples 1–4 on pages 787–790 of your text.

Solve each inequality, and graph the solution set.

1. $a^2 - a - 2 \leq 0$

   1. _____

2. $8k^2 + 10k > 3$

   2. _____

3. $(3x-2)^2 < -1$

   3. _____

Name: Date:
Instructor: Section:

Objective 2 Solve polynomial inequalities of degree 3 or greater.

Video Examples

Review this example for Objective 2:

5. Solve and graph the solution set of $(x+1)(x-2)(x+4) \leq 0$.

   Set the factored polynomial equal to 0, then use the zero-factor property.

   $x+1=0$ or $x-2=0$ or $x+4=0$
   $x=-1$ or $x=2$ or $x=-4$

   Locate –4, –1, and 2 on a number line to determine the intervals A, B, C, and D.

   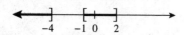

   Substitute a test number from each interval in the original inequality to determine which intervals satisfy the inequality.

Interval	Test Number	Test of inequality	True or False?
A	-5	$-28 \leq 0$	T
B	-2	$8 \leq 0$	F
C	0	$-8 \leq 0$	T
D	5	$162 \leq 0$	F

   The numbers in intervals A and C are in the solution set. The three endpoints are included in the solution set since the inequality symbol, $\leq$, includes equality. Thus, the solution set is $(-\infty, -4] \cup [-1, 2]$.

Now Try:

5. Solve and graph the solution set of $(2x-1)(2x+3)(3x+1) \leq 0$.

   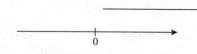

Objective 2 Practice Exercises

For extra help, see Example 5 on pages 790–791 of your text.

Solve each inequality, and graph the solution set.

4. $(y+2)(y-1)(y-2) < 0$

4. _____

Name: Date:
Instructor: Section:

5. $(k+5)(k-1)(k+3) \leq 0$

5. _____

6. $(x-1)(x-3)(x+2) \geq 0$

6. _____

Objective 3 Solve rational inequalities.

Video Examples

Review these examples for Objective 3:

6. Solve and graph the solution set of $\dfrac{7}{x-1} < 1$.

Write the inequality so that 0 is on one side.

$$\dfrac{7}{x-1} - 1 < 0$$

$$\dfrac{7}{x-1} - \dfrac{x-1}{x-1} < 0 \quad \text{The LCD is } x-1.$$

$$\dfrac{7-x+1}{x-1} < 0$$

$$\dfrac{8-x}{x-1} < 0$$

The sign of $\dfrac{8-x}{x-1}$ will change from positive to negative or negative to positive only at those numbers that make the numerator or denominator 0. These two numbers, 1 and 8, divide a number line into three intervals.

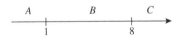

Test a number in each interval using the original inequality.

Interval	Test Number	Test of inequality	True or False?
A	0	$-7 < 1$	T
B	2	$7 < 1$	F
C	10	$\dfrac{7}{9} < 1$	T

The solution set is $(-\infty, 1) \cup (8, \infty)$. This interval does not include 1 because it would

Now Try:

6. Solve and graph the solution set of $\dfrac{y}{y+1} > 3$.

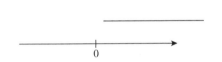

make the denominator of the original inequality 0. The number 8 is not included because the inequality symbol, <, does not include equality.

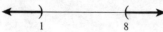

7. Solve and graph the solution set of $\frac{x+1}{x-5} \geq 3$.

    Write the inequality so that 0 is on one side.
    $$\frac{x+1}{x-5} - 3 \geq 0$$
    $$\frac{x+1}{x-5} - \frac{3(x-5)}{x-5} \geq 0$$
    $$\frac{x+1}{x-5} - \frac{3x-15}{x-5} \geq 0$$
    $$\frac{x+1-3x+15}{x-5} \geq 0$$
    $$\frac{-2x+16}{x-5} \geq 0$$

    The sign of $\frac{-2x+16}{x-5}$ will change from positive to negative or negative to positive only at those numbers that make the numerator or denominator 0. These two numbers, 5 and 8, divide a number line into three intervals.

    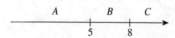

    Test a number in each interval using original inequality.

Interval	Test Number	Test of inequality	True or False?
A	0	$-\frac{1}{5} \geq 3$	F
B	6	$7 \geq 3$	T
C	10	$\frac{11}{5} \geq 3$	F

    The solution set is (5, 8]. This interval does not include 5 because it would make the denominator of the original inequality 0. The number 8 is included because the inequality symbol, $\geq$, does includes equality.

    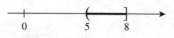

7. Solve and graph the solution set of $\frac{z+2}{z-3} \leq 2$.

    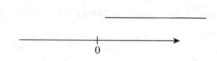

Name: Date:
Instructor: Section:

For extra help, see Examples 6–7 on pages 791–793 of your text.

Solve each inequality, and graph the solution set.

7. $\dfrac{7}{x-1} \le 1$

7. _____

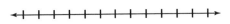

8. $\dfrac{2p-1}{3p+1} \le 1$

8. _____

9. $\dfrac{5}{x-3} \le -1$

9. _____

Name: Date:
Instructor: Section:

Chapter 12 INVERSE, EXPONENTIAL, AND LOGARITHMIC FUNCTIONS

12.1 Inverse Functions

Learning Objectives	
1	Decide whether a function is one-to-one and, if it is, find its inverse.
2	Use the horizontal line test to determine whether a function is one-to-one.
3	Find the equation of the inverse of a function.
4	Graph f^{-1} from the graph of f.

Key Terms

Use the vocabulary terms listed below to complete each statement in exercises 1–2.

one-to-one function **inverse of a function** f

1. A function in which each x-value corresponds to just one y-value and each y-value corresponds to just one x-value is a(n) _____.

2. If f is a one-to-one function, the _____ is the set of all ordered pairs of the form (y, x) where (x, y) belongs to f.

Objective 1 Decide whether a function is one-to-one and, if it is, find its inverse.

Video Examples

Review these examples for Objective 1:

1. Find the inverse of each function that is one-to-one.

   a. $F = \{(2, 1), (-1, 1), (0, 0), (1, 1)\}$

   Every x-value in F corresponds to only one y-value. However, the y-value 1 corresponds to two x-values, so F is not a one-to-one function.

   b. $G = \{(-3,-1), (-2, 0), (-1, 1), (0, 2)\}$

   Every x-value in G corresponds to only one y-value, and every y-value corresponds to only one x-value, so G is a one-to-one function.
   The inverse function is found by interchanging the x- and y-values in each ordered pair.
   $G^{-1} = \{(-1,-3), (0,-2), (1,-1), (2, 0)\}$

Now Try:

1. Find the inverse of each function that is one-to-one.

   a. $F = \{(2, 4), (-1, 1), (0, 0), (1, 1), (2, 6)\}$

   _____

   b. $G = \{(3, 2), (-3,-2), (2, 3), (-2,-3)\}$

   _____

Name: Date:
Instructor: Section:

c.

State	Number of National Parks
AK	8
AZ	3
CA	8
CO	4
FL	3
HI	2
UT	5

c.

State	Number of representatives
AK	1
AZ	8
CA	53
FL	25
NY	29
DE	1

Let N be the function defined in the table, with the states forming the domain and the number of national parks forming the range. Then, N is not one-to-one, because two different states have the same number of national parks.

Objective 1 Practice Exercises

For extra help, see Example 1 on page 807 of your text.

If the function is one-to-one, find its inverse.

1. $\{(-3,-1), (-2, 2), (-1, 3), (0, 4)\}$

   1. _____

2. $\{(1, 0), (2, 0), (3, 5), (4, 1)\}$

   2. _____

3. $\{(0, 0), (1, 1), (-1, -1), (2, 2), (-2, -2)\}$

   3. _____

Name: Date:
Instructor: Section:

Objective 2 Use the horizontal line test to determine whether a function is one-to-one.

Video Examples

Review these examples for Objective 2:

2. Use the horizontal line test to determine whether each graph is the graph of a one-to-one function.

   a.

   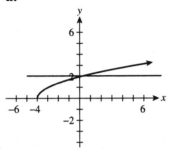

   Every horizontal line will intersect the graph in exactly one point. The function is one-to-one.

   b.

   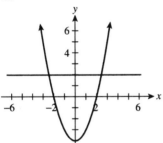

   Because a horizontal line intersects the graph in more than one point, the function is not one-to-one.

Now Try:

2. Use the horizontal line test to determine whether each graph is the graph of a one-to-one function.

   a.

   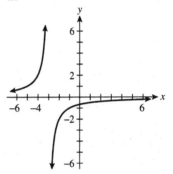

   _____

   b.

   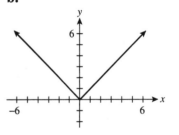

   _____

Objective 2 Practice Exercises

For extra help, see Example 2 on page 808 of your text.

Use the horizontal line test to determine whether each function is one-to-one.

4.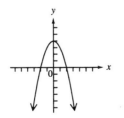

4. _____

Name: Date:
Instructor: Section:

5.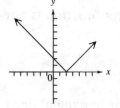

5. _____

6.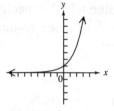

6. _____

Objective 3 Find the equation of the inverse of a function.

Video Examples

Review these examples for Objective 3:

3. Decide whether each equation represents a one-to-one function. If so, find the equation for the inverse.

   a. $f(x) = 3x - 5$

   The graph of $y = 3x - 5$ is a nonvertical line, so by the horizontal line test, f is a one-to-one function. To find the inverse, let $y = f(x)$, interchange x and y, then solve for y.
   $$y = 3x - 5$$
   $$x = 3y - 5 \quad \text{Interchange } x \text{ and } y.$$
   $$x + 5 = 3y$$
   $$\frac{x+5}{3} = y$$
   $$f^{-1}(x) = \frac{x+5}{3} = \frac{x}{3} + \frac{5}{3}$$
   $$f^{-1}(x) = \frac{1}{3}x + \frac{5}{3}$$

   b. $f(x) = 2x^2 + 3$

   The graph of $y = 2x^2 + 3$ is a vertical parabola, so by the horizontal line test, f is not a one-to-one function and does not have an inverse.

Now Try:

3. Decide whether each equation represents a one-to-one function. If so, find the equation for the inverse.

   a. $f(x) = 4x - 1$

   _____

   b. $f(x) = -\frac{3}{2}x^2$

   _____

c. $f(x) = x^3 + 1$

The graph of $y = x^3 + 1$ is a cubing function. The function is one-to-one and has an inverse.

$y = x^3 + 1$

$x = y^3 + 1$ Interchange x and y.

$x - 1 = y^3$

$\sqrt[3]{x-1} = y$

$f^{-1}(x) = \sqrt[3]{x-1}$

c. $f(x) = 2x^3 - 3$

Objective 3 Practice Exercises

For extra help, see Examples 3–4 on pages 808–810 of your text.

If the function is one-to-one, find its inverse.

7. $f(x) = 2x - 5$

7. _____

8. $f(x) = x^3 - 1$

8. _____

9. $f(x) = x^2 - 1$

9. _____

Objective 4   Graph f^{-1} from the graph of f.

Video Examples

Review this example for Objective 4:
5. Use the given graph to graph the inverse of f.

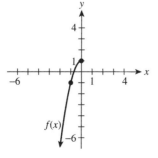

We can find the graph of f^{-1} from the graph of f by locating the mirror image of each point in f

Now Try:
5. Use the given graph to graph the inverse of f.

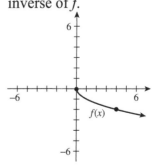

with respect to the line $y = x$.

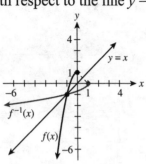

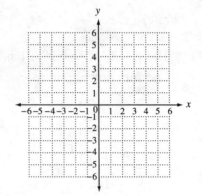

Objective 4 Practice Exercises

For extra help, see Example 5 on pages 810–811 of your text.

If the function is one-to-one, graph the function f and its inverse f^{-1} on the same set of axes.

10.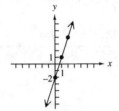

10. _____

11.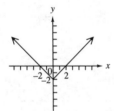

11. _____

12.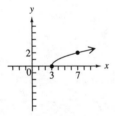

12. _____

Name: Date:
Instructor: Section:

Chapter 12 INVERSE, EXPONENTIAL, AND LOGARITHMIC FUNCTIONS

12.2 Exponential Functions

Learning Objectives
1   Use a calculator to find approximations of exponentials.
2   Define and graph exponential functions.
3   Solve exponential equations of the form $a^x = a^k$ for x.
4   Use exponential functions in applications involving growth or decay.

Key Terms

Use the vocabulary terms listed below to complete each statement in exercises 1–2.

exponential equation   **inverse**

1. If f is a one-to-one function, then the _____ of f is the set of all ordered pairs formed by interchanging the coordinates of the ordered pairs of f.

2. An equation that has a variable as an exponent, is an _____.

Objective 1   Use a calculator to find approximations of exponentials.

Video Examples

Review these examples for Objective 1:	Now Try:

1. Use a calculator to find an approximation to three decimal places for each exponential expression.
 	
    a. $3^{1.8}$
 	
    $3^{1.8} \approx 7.225$
 	
    b. $3^{-1.4}$
 	
    $3^{-1.4} \approx 0.215$
 	
    c. $3^{1/4}$
 	
    $3^{1/4} \approx 1.316$
	1. Use a calculator to find an approximation to three decimal places for each exponential expression.
	
	    a. $3^{1.9}$
	
	    _____
	
	    b. $3^{-1.6}$
	
	    _____
	
	    c. $3^{1/5}$
	
	    _____

Name: Date:
Instructor: Section:

Objective 1 Practice Exercises

For extra help, see Example 1 on page 815 of your text.

Use a calculator to find an approximation to three decimal places for each exponential expression.

1. $3^{1.2}$ 1. _____

2. $3^{-1.2}$ 2. _____

3. $3^{1/3}$ 3. _____

Objective 2 Define and graph exponential functions.

Video Examples

Review these examples for Objective 2:   **Now Try:**

2. Graph $f(x) = 6^x$.   2. Graph $f(x) = 3^x$.

Create a table of values, then plot the points and draw a smooth curve through them.

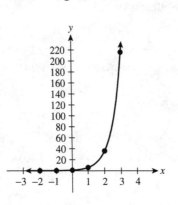

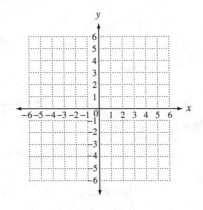

3. Graph $f(x) = \left(\dfrac{1}{6}\right)^x$.   3. Graph $f(x) = \left(\dfrac{1}{3}\right)^x$.

Create a table of values, then plot the points and draw a smooth curve through them.

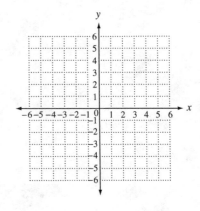

Name: Date:
Instructor: Section:

4. Graph $f(x) = 3^{2x-1}$.

Create a table of values, then plot the points and draw a smooth curve through them.

x	$2x-1$	$f(x) = 3^{2x-1}$
-1	-3	$\frac{1}{27}$
0	-1	$\frac{1}{3}$
1	1	3
2	3	27

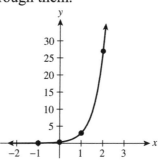

4. Graph $f(x) = 2^{1-x}$.

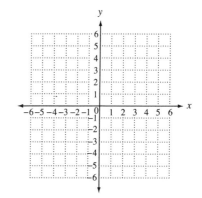

Objective 2 Practice Exercises

For extra help, see Examples 2–4 on pages 816–817 of your text.

Graph each exponential function.

4. $f(x) = 2^{-x}$

4.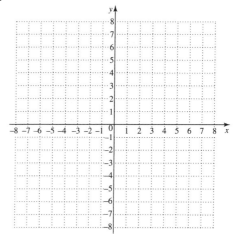

5. $f(x) = \left(\frac{1}{8}\right)^x$

5.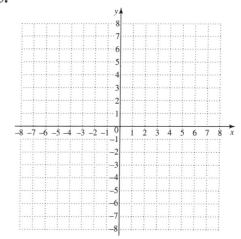

Name: Date:
Instructor: Section:

6. $f(x) = 4^{2x-3}$

6.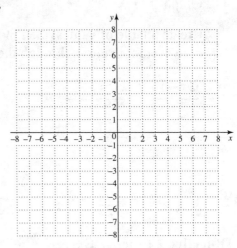

Objective 3 Solve exponential equations of the form $a^x = a^k$ for x.

Video Examples

Review these examples for Objective 3:

5. Solve the equation $16^x = 64$.

$$16^x = 64$$
$$(2^4)^x = 2^6 \quad \text{Write with the same base.}$$
$$2^{4x} = 2^6 \quad \text{Power rule for exponents}$$
$$4x = 6 \quad \text{If } a^x = a^y, \text{ then } x = y.$$
$$x = \frac{6}{4} = \frac{3}{2} \quad \text{Solve for } x; \text{ simplify.}$$

Check: Substitute 3/2 for x.
$$16^{3/2} = (16^{1/2})^3 = 4^3 = 64$$

The solution set is $\left\{\frac{3}{2}\right\}$.

6. Solve each equation.

   a. $16^{x-2} = 64^x$

   $$16^{x-2} = 64^x$$
   $$(2^4)^{x-2} = (2^6)^x \quad \text{Write with the same base.}$$
   $$2^{4x-8} = 2^{6x} \quad \text{Power rule for exponents}$$
   $$4x - 8 = 6x \quad \text{If } a^x = a^y, \text{ then } x = y.$$
   $$-8 = 2x \quad \text{Solve for } x.$$
   $$-4 = x$$

   The solution set is $\{-4\}$.

Now Try:

5. Solve the equation $25^x = 125$.

   _____

6. Solve each equation.

   a. $4^{x-1} = 8^x$

   _____

468 Copyright © 2016 Pearson Education, Inc.

Name: Date:
Instructor: Section:

b. $4^x = \dfrac{1}{64}$

$4^x = \dfrac{1}{64}$

$4^x = \dfrac{1}{4^3}$ $64 = 4^3$

$4^x = 4^{-3}$ Write with the same base.

$x = -3$ Set exponents equal.

The solution set is $\{-3\}$.

c. $\left(\dfrac{2}{5}\right)^x = \dfrac{125}{8}$

$\left(\dfrac{2}{5}\right)^x = \dfrac{125}{8}$

$\left(\dfrac{2}{5}\right)^x = \left(\dfrac{8}{125}\right)^{-1}$

$\left(\dfrac{2}{5}\right)^x = \left[\left(\dfrac{2}{5}\right)^3\right]^{-1}$ Write with the same base.

$\left(\dfrac{2}{5}\right)^x = \left(\dfrac{2}{5}\right)^{-3}$ Power rule for exponents

$x = -3$ Set exponents equal.

The solution set is $\{-3\}$.

b. $3^x = \dfrac{1}{243}$

_____

c. $\left(\dfrac{3}{2}\right)^x = \dfrac{16}{81}$

_____

Objective 3 Practice Exercises

For extra help, see Examples 5–6 on pages 818–819 of your text.

Solve each equation.

7. $25^{1-t} = 5$ 7. _____

8. $8^{2x+1} = 4^{4x}$ 8. _____

9. $\left(\dfrac{3}{4}\right)^x = \dfrac{16}{9}$ 9. _____

Name:　　　　　　　　　　　　　Date:
Instructor:　　　　　　　　　　　Section:

Objective 4 Practice Exercises

Video Examples

Review these examples for Objective 4:

7. Suppose the number of bacteria present in a certain culture after t minutes is given by the equation $Q(t) = 2500(2^{0.05t})$,

   How many bacteria were present after 20 minutes?

   Start with the given function. Replace t with 20.
   $$Q(t) = 2500(2^{0.05t})$$
   $$Q(20) = 2500(2^{0.05 \times 20})$$
   $$Q(20) = 2500(2^1)$$
   $$Q(20) = 5000$$
   There were 5000 bacteria present after 20 minutes.

8. The amount of radioactive material in a sample is given by the function $A(t) = 90\left(\dfrac{1}{2}\right)^{t/18}$, where $A(t)$ is the amount present, in grams, t days after the initial measurement.

   How many grams will be present after 3 days? Round to the nearest hundredth.

   Start with the given function. Replace t with 3.
   $$A(t) = 90\left(\dfrac{1}{2}\right)^{t/18}$$
   $$A(3) = 90\left(\dfrac{1}{2}\right)^{3/18}$$
   $$A(3) \approx 80.18$$
   After 3 days, there were about 80.18 grams in the sample.

Now Try:

7. The population of Evergreen Park is now 16,000. The population t years from now is given by the formula
$$P = 16,000(2^{t/10}).$$
Using the model, what will be the population 40 years from now?

_____

8. An industrial city in Ohio has found that its population is declining according to the equation $y = 70,000(2)^{-0.01x}$, where x is the time in years from 1910.
According to the model, what will the city's population be in the year 2020?

_____

Name: Date:
Instructor: Section:

Objective 4   Use exponential functions in applications involving growth or decay.

For extra help, see Examples 7–8 on pages 819–820 of your text.

Solve each problem.

10. The population of Canadian geese that spend the summer at Gemini Lake each year has been growing according to the function $f(x) = 56(2)^{0.2x}$, where x is the time in years from 1990. Find the number of geese in 2010.

10. _____

11. A sample of a radioactive substance with mass in grams decays according to the function $f(x) = 100(10)^{-0.2x}$, where x is the time in hours after the original measurement. Find the mass of the substance after 10 hours.

11. _____

12. A culture of a certain kind of bacteria grows according to $f(x) = 7750(x)^{0.75x}$, where x is the number of hours after 12 noon. Find the number of bacteria in the culture at 12 noon.

12. _____

Chapter 12 INVERSE, EXPONENTIAL, AND LOGARITHMIC FUNCTIONS

12.3 Logarithmic Functions

Learning Objectives
1. Define a logarithm.
2. Convert between exponential and logarithmic forms, and evaluate logarithms.
3. Solve logarithmic equations of the form $\log_a b = k$ for a, b, or k.
4. Use the definition of logarithm to simplify logarithmic expressions.
5. Define and graph logarithmic functions.
6. Use logarithmic functions in applications involving growth or decay.

Key Terms

Use the vocabulary terms listed below to complete each statement in exercises 1−2.

    logarithm   **logarithmic equation**

1. The _____ of a positive number is the exponent indicating the power to which it is necessary to raise a given number (the base) to give the original number.

2. An equation with a logarithm in at least one term is a _____.

Objective 1   Define a logarithm.

For extra help, see pages 823–824 of your text.

Objective 2   Convert between exponential and logarithmic forms, and evaluate logarithms.

Video Examples

Review these examples for Objective 2:
1.
    a. Write $5^3 = 125$ in logarithmic form.

    $\log_5 125 = 3$

    b. Write $\log_{16} 4 = \frac{1}{2}$ in exponential form.

    $16^{1/2} = 4$

Now Try:
1.
    a. Write $8^2 = 64$ in logarithmic form.

    _____

    b. Write $\log_{16} \frac{1}{4} = -\frac{1}{2}$ in exponential form.

    _____

Name: Date:
Instructor: Section:

2. Use a calculator to approximate each logarithm to four decimal places.

   a. $\log_2 9$

   $\log_2 9 \approx 3.1699$

   b. $\log_7 15$

   $\log_7 15 \approx 1.3917$

   c. $\log_{1/3} 20$

   $\log_{1/3} 20 \approx -2.7268$

   d. $\log_{10} 17$

   $\log_{10} 17 \approx 1.2304$

2. Use a calculator to approximate each logarithm to four decimal places.

   a. $\log_2 6$

   _____

   b. $\log_4 21$

   _____

   c. $\log_{1/4} 25$

   _____

   d. $\log_{10} 12$

   _____

Objective 2 Practice Exercises

For extra help, see Examples 1–2 on pages 824–825 of your text.

Write in exponential form.

1. $\log_{10} 0.001 = -3$

1. _____

Write in logarithmic form.

2. $2^{-7} = \dfrac{1}{128}$

2. _____

Use a calculator to approximate the logarithm to four decimal places.

3. $\log_3 25$

3. _____

Objective 3 Solve logarithmic equations of the form $\log_a b = k$ for a, b, or k.

Video Examples

Review these examples for Objective 3:
3. Solve each equation.

   a. $\log_{3/2} x = -2$

   By definition, $\log_{3/2} x = -2$ is equivalent to $x = \left(\dfrac{3}{2}\right)^{-2}$, and $\left(\dfrac{3}{2}\right)^{-2} = \left(\dfrac{2}{3}\right)^2 = \dfrac{4}{9}$. The solution set is $\left\{\dfrac{4}{9}\right\}$.

Now Try:
2. Solve each equation.

   a. $\log_4 x = -3$

   _____

Name: Date:
Instructor: Section:

b. $\log_5(3x+1) = 2$

$\log_5(3x+1) = 2$

$3x+1 = 5^2$ Write in exponential form.

$3x = 24$ Apply the exponent; subtract 1.

$x = 8$ Divide by 3.

The solution set is $\{8\}$.

c. $\log_x 6 = 2$

$\log_x 6 = 2$

$x^2 = 6$ Write in exponential form.

$x = \pm\sqrt{6}$ Take square root.

Only the principal square root satisfies the equation since the base must be a positive number. The solution set is $\{\sqrt{6}\}$.

d. $\log_{64} \sqrt[4]{8} = x$

$\log_{64} \sqrt[4]{8} = x$

$64^x = \sqrt[4]{8}$ Write in exponential form.

$(8^2)^x = 8^{1/4}$ Write with the same base.

$8^{2x} = 8^{1/4}$ Power rule for exponents

$2x = \dfrac{1}{4}$

$x = \dfrac{1}{8}$

The solution set is $\left\{\dfrac{1}{8}\right\}$.

b. $\log_9(2x+1) = 2$

_____

c. $\log_x 12 = 2$

_____

d. $\log_{81} \sqrt[3]{9} = x$

_____

Objective 3 Practice Exercises

For extra help, see Example 3 on pages 825–826 of your text.

Solve each equation.

4. $x = \log_{32} 8$ 4. _____

5. $\log_{1/3} r = -4$ 5. _____

Name: Date:
Instructor: Section:

6. $\log_a 4 = \frac{1}{2}$

6. _____

Objective 4 Use the definition of logarithm to simplify logarithmic expressions.

Video Examples

Review these examples for Objective 4:

4. Use special properties to evaluate each expression.

  a. $\log_8 8$

  $\log_8 8 = 1$

  b. $\log_{64} 1$

  $\log_{64} 1 = 0$

  c. $\log_{0.2} 1$

  $\log_{0.2} 1 = 0$

  d. $\log_4 4^{11}$

  $\log_4 4^{11} = 11$

  e. $8^{\log_8 5}$

  $8^{\log_8 5} = 5$

  f. $\log_4 64$

  $\log_4 64 = \log_4 4^3 = 3$

Now Try:

4. Use special properties to evaluate each expression.

  a. $\log_4 4$

  _____

  b. $\log_{100} 1$

  _____

  c. $\log_{1/3} 1$

  _____

  d. $\log_6 6^3$

  _____

  e. $6^{\log_6 9}$

  _____

  f. $\log_5 125$

  _____

Objective 4 Practice Exercises

For extra help, see Example 4 on page 826 of your text.

Use the special properties to evaluate each expression.

7. $\log_{3.4} 1$

7. _____

8. $\log_8 8^3$

8. _____

Name:
Instructor:
Date:
Section:

9. $2^{\log_2 5}$

9. _____

Objective 5 Define and graph logarithmic functions.

Video Examples

Review these examples for Objective 5:

5. Graph $f(x) = \log_5 x$.

   Begin by writing $y = \log_5 x$ in exponential form as $x = 5^y$. Then, create a table of values, plot the points and draw a smooth curve through them.

$x = 5^y$	y
$\frac{1}{5}$	-1
1	0
5	1
25	2

Now Try:

5. Graph $f(x) = \log_3 x$.

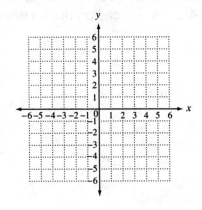

6. Graph $f(x) = \log_{1/3} x$.

   Begin by writing $y = \log_{1/3} x$. in exponential form. Then, create a table of values, plot the points and draw a smooth curve through them.

$x = \left(\frac{1}{3}\right)^y$	y
$\frac{1}{3}$	1
1	0
3	-1
9	-2

6. Graph $f(x) = \log_{1/4} x$.

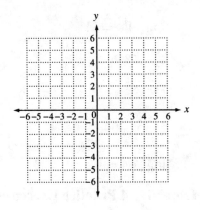

476 Copyright © 2016 Pearson Education, Inc.

Name: Date:
Instructor: Section:

Objective 5 Practice Exercises

For extra help, see Examples 5–6 on page 827 of your text.

Graph each logarithmic function.

10. $y = \log_9 x$

10.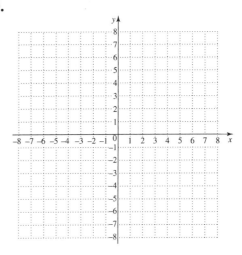

11. $y = \log_{1/4} x$

11.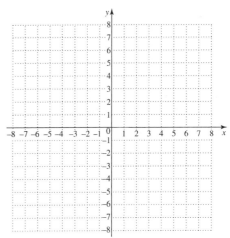

Name: Date:
Instructor: Section:

Objective 6   Use logarithmic functions in applications involving growth or decay.

Video Examples

Review this example for Objective 6:

6. A company analyst has found that total sales in thousands of dollars after a major advertising campaign are given by $S(x) = 100\log_2(x+2)$, where x is time in weeks after the campaign was introduced. Find the amount of sales two weeks after the campaign was introduced.

   Two weeks after the campaign, $x = 2$, so we have
   $S(2) = 100\log_2(2+2)$
   $S(2) = 100\log_2(2)^2$
   $S(2) = 100 \cdot 2$
   $S(2) = 200$

   Two weeks after the campaign, sales were $200,000.

Now Try:

6. The number of fish in an aquarium is given by the function $f(t) = 8\log_5(2t+5)$, where t is time in months. Find the number of fish present after 10 months.

   _____

Objective 6 Practice Exercises

For extra help, see Example 7 on page 828 of your text.

Solve each problem.

12. The population of foxes in an area t months after the foxes were introduced there is approximated by the function $F(t) = 500\log_{10}(2t+10)$. Find the number of foxes in the area when the foxes were first introduced into the area.

    12. _____

13. A population of mites in a laboratory is growing according to the function
    $p = 50\log_3(20t+7) - 25\log_9(80t+1)$, where t is the number of days after a study is begun. Find the number of mites present 1 day after the beginning of the study.

    13. _____

Name: Date:
Instructor: Section:

14. Sales (in thousands) of a new product are approximated by
$S = 125 + 20\log_2(30t + 4) + 30\log_4(35t - 6),$ where t is the number of years after the product is introduced. Find the total sales 2 years after the product is introduced.

14. _____

Name: Date:
Instructor: Section:

Chapter 12 INVERSE, EXPONENTIAL, AND LOGARITHMIC FUNCTIONS

12.4 Properties of Logarithms

Learning Objectives
1. Use the product rule for logarithms.
2. Use the quotient rule for logarithms.
3. Use the power rule for logarithms.
4. Use properties to write alternative forms of logarithmic expressions.

Key Terms

Use the vocabulary terms listed below to complete each statement in exercises 1–4.

    product rule for logarithms   **quotient rule for logarithms**

    power rule for logarithms   **special properties**

1. The equations $b^{\log_b x} = x$, $x > 0$ and $\log_b b^x = x$ are referred to as _____ of logarithms.

2. The equation $\log_b \frac{x}{y} = \log_b x - \log_b y$ is referred to as the _____.

3. The equation $\log_b xy = \log_b x + \log_b y$ is referred to as the _____.

4. The equation $\log_b x^r = r \log_b x$ is referred to as the _____.

Objective 1   Use the product rule for logarithms.

Video Examples

Review these examples for Objective 1:
1. Use the product rule to rewrite each logarithm. Assume $x > 0$.

    a. $\log_4 (6 \cdot 11)$

Use the product rule.
$\log_4 (6 \cdot 11) = \log_4 6 + \log_4 11$

    b. $\log_3 8 + \log_3 2$

Use the product rule.
$\log_3 8 + \log_3 2 = \log_3 (8 \cdot 2) = \log_3 16$

    c. $\log_2 (2x)$

$\log_2 (2x) = \log_2 2 + \log_2 x$ Product rule
$ = 1 + \log_2 x$ $\log_2 2 = 1$

Now Try:
1. Use the product rule to rewrite each logarithm. Assume $x > 0$.

    a. $\log_6 (5 \cdot 3)$

_____

    b. $\log_5 7 + \log_5 3$

_____

    c. $\log_4 (4x)$

_____

Name: Date:
Instructor: Section:

d. $\log_3 x^4$

$\log_3 x^4 = \log_3(x \cdot x \cdot x \cdot x)$
$= \log_3 x + \log_3 x + \log_3 x + \log_3 x$
$= 4\log_3 x$

d. $\log_5 x^4$

_____

Objective 1 Practice Exercises

For extra help, see Example 1 on page 832 of your text.

Use the product rule to express each logarithm as a sum of logarithms.

1. $\log_7 5m$

1. _____

2. $\log_2 6xy$

2. _____

Use the product rule to express the sum as a single logarithm.

3. $\log_4 7 + \log_4 3$

3. _____

Objective 2 Use the quotient rule for logarithms.

Video Examples

Review these examples for Objective 2:
2. Use the quotient rule to rewrite each logarithm. Assume $x > 0$.

a. $\log_4 \frac{8}{7}$

$\log_4 \frac{8}{7} = \log_4 8 - \log_4 7$

b. $\log_3 8 - \log_3 x$

$\log_3 8 - \log_3 x = \log_3 \frac{8}{x}$

c. $\log_2 \frac{16}{11}$

$\log_2 \frac{16}{11} = \log_2 16 - \log_2 11 = 4 - \log_2 11$

Now Try:
2. Use the quotient rule to rewrite each logarithm. Assume $x > 0$.

a. $\log_5 \frac{4}{9}$

_____

b. $\log_6 x - \log_6 3$

_____

c. $\log_4 \frac{16}{11}$

_____

Name: Date:
Instructor: Section:

Objective 2 Practice Exercises

For extra help, see Example 2 on page 833 of your text.

Use the quotient rule for logarithms to express each logarithm as a difference of logarithms, or as a single number if possible.

4. $\log_2 \dfrac{5}{m}$ 4. _____

5. $\log_6 \dfrac{k}{3}$ 5. _____

Use the quotient rule for logarithms to express the difference as a single logarithm.

6. $\log_2 7q^4 - \log_2 5q^2$ 6. _____

Objective 3   Use the power rule for logarithms.

Video Examples

Review these examples for Objective 3:
3. Use the power rule to rewrite each logarithm. Assume that $b > 0$, $x > 0$, and $b \neq 1$.

  a. $\log_4 3^5$

  $\log_4 3^5 = 5 \log_4 3$

  b. $\log_b \sqrt{11}$

  $\log_b \sqrt{11} = \log_b 11^{1/2} = \dfrac{1}{2} \log_b 11$

  c. $\log_2 \sqrt[5]{x^4}$

  $\log_2 \sqrt[5]{x^4} = \log_2 x^{4/5} = \dfrac{4}{5} \log_2 x$

  d. $\log_3 \dfrac{1}{x^5}$

  $\log_3 \dfrac{1}{x^5} = \log_3 x^{-5} = -5 \log_3 x$

Now Try:
3. Use the power rule to rewrite each logarithm. Assume that $b > 0$, $x > 0$, and $b \neq 1$.

  a. $\log_6 4^3$

  _____

  b. $\log_b \sqrt{13}$

  _____

  c. $\log_3 \sqrt[4]{x^3}$

  _____

  d. $\log_3 \dfrac{1}{x^7}$

  _____

Name: Date:
Instructor: Section:

Objective 3 Practice Exercises

For extra help, see Example 3 on page 834 of your text.

Use the power rule for logarithms to rewrite each logarithm or as a single number if possible.

7. $\log_m 2^7$

7. _____

8. $\log_3 \sqrt[3]{5}$

8. _____

9. $3^{\log_3 \sqrt[3]{7}}$

9. _____

Objective 4 Use properties to write alternative forms of logarithmic expressions.

Video Examples

Review these examples for Objective 4:

4. Use the properties of logarithms to rewrite each expression if possible. Assume that all variables represent positive real numbers.

   a. $\log_5 6x^3$

   $\log_5 6x^3 = \log_5 6 + \log_5 x^3$
   $= \log_5 6 + 3\log_5 x$

   b. $\log_b \sqrt{\dfrac{5}{x}}$

   $\log_b \sqrt{\dfrac{5}{x}} = \log_b \left(\dfrac{5}{x}\right)^{1/2}$
   $= \dfrac{1}{2}\log_b \dfrac{5}{x}$
   $= \dfrac{1}{2}(\log_b 5 - \log_b x)$

Now Try:

4. Use the properties of logarithms to rewrite each expression if possible. Assume that all variables represent positive real numbers.

   a. $\log_6 36x^3$

   _____

   b. $\log_b \sqrt{\dfrac{x}{3}}$

   _____

c. $3\log_b x - \left(2\log_b y + \frac{1}{2}\log_b z\right)$

$3\log_b x - \left(2\log_b y + \frac{1}{2}\log_b z\right)$

$= \log_b x^3 - (\log_b y^2 + \log_b z^{1/2})$

$= \log_b x^3 - \log_b y^2\sqrt{z}$

$= \log_b \dfrac{x^3}{y^2\sqrt{z}}$

d. $2\log_3 x + \log_3(x-1) - \dfrac{1}{2}\log_3(x+1)$

$2\log_3 x + \log_3(x-1) - \dfrac{1}{2}\log_3(x+1)$

$= \log_3 x^2 + \log_3(x-1) - \log_3(x+1)^{1/2}$

$= \log_3\left(x^2(x-1)\right) - \log_3\sqrt{x+1}$

$= \log_3 \dfrac{x^3 - x^2}{\sqrt{x+1}}$

e. $\log_2(2x+3y)$

$\log_2(2x+3y)$ cannot be rewritten using the properties of logarithms.

5. Given that $\log_2 9 \approx 3.1699$ and $\log_2 11 \approx 3.4594$, use properties of logarithms to evaluate each expression.

a. $\log_2 99$

$\log_2 99 = \log_2(9 \cdot 11)$

$\quad\quad\quad = \log_2 9 + \log_2 11$

$\quad\quad\quad = 3.1699 + 3.4594$

$\quad\quad\quad = 6.6293$

b. $\log_2 \dfrac{1}{9}$

$\log_2 \dfrac{1}{9} = \log_2 1 - \log_2 9$

$\quad\quad\quad = 0 - 3.1699$

$\quad\quad\quad = -3.1699$

c. $\log_b x + 4\log_b y - \log_b z$

_____

d.

$2\log_2 x + \log_2(x-1) - \dfrac{1}{3}\log_2(x^2+1)$

_____

e. $\log_3(3x-y)$

_____

5. Given that $\log_2 9 \approx 3.1699$ and $\log_2 11 \approx 3.4594$, use properties of logarithms to evaluate each expression.

a. $\log_2 198$

_____

b. $\log_2 \dfrac{1}{11}$

_____

Name: Date:
Instructor: Section:

c. $\log_2 121$

$$\begin{aligned}\log_2 121 &= \log_2 11^2 \\ &= 2\log_2 11 \\ &= 2(3.4594) \\ &= 6.9188\end{aligned}$$

c. $\log_2 729$

_____

6. Decide whether each statement is *true* or *false*.

    a. $\log_2 32 - \log_2 16 = \log_2 16$

Evaluate each side.
Left side:
$$\begin{aligned}\log_2 32 - \log_2 16 &= \log_2 2^5 - \log_2 2^4 \\ &= 5 - 4\log_2 2 \\ &= 1\end{aligned}$$

Right side:
$$\log_2 16 = \log_2 2^4 = 4$$
The statement is false because $1 \neq 4$.

    b. $\log_3(\log_4 64) = \dfrac{\log_{12} 144}{\log_6 36}$

Evaluate each side.
Left side:
$$\begin{aligned}\log_3(\log_4 64) &= \log_3(\log_4 4^3) \\ &= \log_3 3 \\ &= 1\end{aligned}$$

Right side:
$$\frac{\log_{12} 144}{\log_6 36} = \frac{\log_{12} 12^2}{\log_6 6^2} = \frac{2}{2} = 1$$
The statement is true because $1 = 1$.

6. Decide whether each statement is *true* or *false*.

    a. $\log_3 27 + \log_3 9 = \log_5 5$

_____

    b. $(\log_2 8)(\log_2 4) = \log_2 32$

_____

For extra help, see Examples 4–6 on pages 835–836 of your text.

Use the properties of logarithms to express the sum or difference of logarithms as a single logarithm, or as a single number if possible.

10. $\log_4 10y + \log_4 3y - \log_4 6y^3$ **10.** _____

Name: Date:
Instructor: Section:

Given that $\log_2 6 \approx 2.5850$ *and* $\log_2 12 \approx 3.5850,$ *use properties of logarithms to evaluate the expression.*

11. $\log_2 72$

11. _____

Decide whether the statement is true *or* false.

12. $\log_2 4p^3 = 6 + 3\log_2 p$

12. _____

Name: Date:
Instructor: Section:

Chapter 12 INVERSE, EXPONENTIAL, AND LOGARITHMIC FUNCTIONS

12.5 Common and Natural Logarithms

Learning Objectives
1. Evaluate common logarithms using a calculator.
2. Use common logarithms in applications.
3. Evaluate natural logarithms using a calculator.
4. Use natural logarithms in applications.
5. Use the change-of-base rule.

Key Terms

Use the vocabulary terms listed below to complete each statement in exercises 1–2.

common logarithm   **natural logarithm**

1. A logarithm to the base e is a _____.

2. A logarithm to the base 10 is a _____.

Objective 1 Evaluate common logarithms using a calculator.

Video Examples

Review this example for Objective 1:
1. Evaluate the logarithm to four decimal places using a calculator.

   $\log 436.2$

   $\log 436.2 \approx 2.6397$

Now Try:
1. Evaluate the logarithm to four decimal places using a calculator.
   $\log 983.5$

   _____

Objective 1 Practice Exercises

For extra help, see Example 1 on pages 838–839 of your text.

Use a calculator to find each logarithm. Give an approximation to four decimal places.

1. $\log 57.23$ 1. _____

2. $\log 0.0914$ 2. _____

3. $\log 87{,}123$ 3. _____

Name: Date:
Instructor: Section:

Objective 2 Use common logarithms in applications.

Video Examples

Review these examples for Objective 2:

2. Wetlands are classified as bogs, fens, marshes, and swamps, on the basis of pH values. A pH value between 6.0 and 7.5 indicates that the wetland is a "rich fen." When the pH is between 3.0 and 6.0, the wetland is a "poor fen," and if the pH falls to 3.0 or less, it is a "bog." Suppose that the hydronium ion concentration of a sample of water from a wetland is 5.4×10^{-4}. Find the pH value for the water and determine how the wetland should be classified.

$$pH = -\log(5.4 \times 10^{-4}) \quad \text{Definition of pH}$$
$$= -(\log 5.4 + \log 10^{-4}) \quad \text{Product rule}$$
$$= -(0.7324 - 4) \quad \text{Use a calculator to find log 5.4.}$$
$$= 3.2676$$

Since the pH is between 3.0 and 6.0, the wetland is a poor fen.

3. Find the hydronium ion concentration of a solution with pH 5.4.

$$pH = -\log[H_3O^+] \quad \text{Definition of pH}$$
$$5.4 = -\log[H_3O^+]$$
$$\log[H_3O^+] = -5.4 \quad \text{Multiply by } -1.$$
$$H_3O^+ = 10^{-5.4} \quad \text{Write in exponential form.}$$
$$\approx 4.0 \times 10^{-6} \quad \text{Use a calculator.}$$

4. Find the decibel level to the nearest whole number of the sound with intensity I of $5.012 \times 10^{10} I_0$.

$$D = 10\log\left(\frac{I}{I_0}\right) = 10\log\left(\frac{5.012 \times 10^{10} I_0}{I_0}\right)$$
$$= 10\log(5.012 \times 10^{10})$$
$$\approx 107 \text{ db}$$

Now Try:

2. Suppose that the hydronium ion concentration of a sample of water from a wetland is 6.2×10^{-8}. Find the pH value for the water and determine how the wetland should be classified.

_____

3. Find the hydronium ion concentration of a solution with pH 3.6.

_____

4. Find the decibel level to the nearest whole number of the sound with intensity I of $3.16 \times 10^8 I_0$.

_____

Name: Date:
Instructor: Section:

Objective 2 Practice Exercises

For extra help, see Examples 2–4 on pages 839–840 of your text.

Solve each problem.

4. Find the pH of a solution with the given hydronium ion concentration. Round the answer to the nearest tenth.

    a. 4.3×10^{-9} b. 2.8×10^{-6}

4. a._____

   b._____

5. Find the decibel level to the nearest whole number of the sound with intensity I of $2.5 \times 10^{13} I_0$.

5. _____

6. Find the hydronium ion concentration of a solution with the given pH value.

    a. 5.2   b. 1.3

6. a._____

   b._____

Objective 3 Evaluate natural logarithms using a calculator.

Video Examples

Review this example for Objective 3:

5. Using a calculator, evaluate the logarithm to four decimal places.

    $\ln 436.2$

    $\ln 436.2 \approx 6.0781$

Now Try:

5. Using a calculator, evaluate the logarithm to four decimal places.

    $\ln 98$

    _____

Objective 3 Practice Exercises

For extra help, see Example 5 on page 842 of your text.

Find each natural logarithm. Give an approximation to four decimal places.

7. $\ln 76.3$

7. _____

8. $\ln 0.102$

8. _____

9. $\ln 50$

9. _____

Name:
Instructor:
Date:
Section:

Objective 4 Use natural logarithms in applications.

Video Examples

Review this example for Objective 4:
6. The time t in years for an investment increasing at a rate of r percent (in decimal form) to double is given by
$$t = \frac{\ln 2}{\ln(1+r)}.$$
This is called the doubling time. Find the doubling time to the nearest tenth for an investment at 4%.

$4\% = 0.04$, so $t = \frac{\ln 2}{\ln(1+0.04)} = \frac{\ln 2}{\ln 1.04} \approx 17.7$

The doubling time for the investment is about 17.7 years.

Now Try:
6. Use the formula at the left to find the doubling time to the nearest tenth for an investment at 6%.

_____

Objective 4 Practice Exercises

For extra help, see Example 6 on page 842 of your text.

The time t in years for an amount increasing at a rate of r (in decimal form) to double (the doubling time) is given by $t = \frac{\ln 2}{\ln(1+r)}$. *Find the doubling time for an investment at the interest rate. Round to the nearest whole number.*

10. 3%   10. _____

The half-life of a radioactive substance is the time it takes for half of the material to decay. The amount A in pounds of substance remaining after t years is given by $\ln \frac{A}{C} = -\frac{t}{h} \ln 2$, *where C is the initial amount in pounds, and h is its half-life in years. Use the formula to solve the following problems. Round to the nearest whole number.*

11. The half-life of radium-226 is 1620 years. How long, to the nearest year, will it take for 100 pounds to decay to 25 pounds?   11. _____

490

Name: Date:
Instructor: Section:

Newton's Law of Cooling describes the cooling of a warmer object to the cooler temperature of the surrounding environment. The formula can be given as
$t = \frac{1}{k} \ln \frac{T_s - T_1}{T_s - T_2}$, *where t is the elapsed time, T_1 is the initial temperature measurement of the object, T_2 is the second temperature measurement of the object, and T_s is the temperature of the surrounding environment. Use this formula to solve the problem. Round to the nearest tenth.*

12.   A corpse was discovered in a motel room at midnight and its temperature was 80°F. The temperature in the room was 60°F. Assuming that the person's temperature at the time of death was 98.6° F and using $k = 0.1438$, determine t and the time of death.

12. t _____

      time _____

Objective 5   Use the change-of-base rule.

Video Examples

Review this example for Objective 5:

7.   Evaluate $\log_7 28$ to four decimal places.

$$\log_7 28 = \frac{\log 28}{\log 7} = 1.7124$$

Now Try:

7.   Evaluate $\log_5 180$ to four decimal places.

_____

Objective 5 Practice Exercises

For extra help, see Example 7 on page 843 of your text.

Use the change-of-base rule to find each logarithm. Give approximations to four decimal places.

13. $\log_{16} 27$ 13. _____

14. $\log_6 0.25$ 14. _____

15. $\log_{1/2} 5$ 15. _____

Name: Date:
Instructor: Section:

Chapter 12 INVERSE, EXPONENTIAL, AND LOGARITHMIC FUNCTIONS

12.6 Exponential and Logarithmic Equations; Further Applications

Learning Objectives
1. Solve equations involving variables in the exponents.
2. Solve equations involving logarithms.
3. Solve applications involving compound interest.
4. Solve applications involving base *e* exponential growth and decay.

Key Terms

Use the vocabulary terms listed below to complete each statement in exercises 1–2.

 compound interest   continuous compounding

1. The formula for _____ is $A = Pe^{rt}$.

2. The formula for _____ is $A = P\left(1+\frac{r}{n}\right)^{nt}$.

Objective 1 Solve equations involving variables in the exponents.

Video Examples

Review these examples for Objective 1:

1. Solve $4^x = 30$. Approximate the solution to three decimal places.

   $4^x = 30$

   $\log 4^x = \log 30$ If $x = y$, and $x > 0$, $y > 0$, then $\log_b x = \log_b y$.

   $x \log 4 = \log 30$ Power rule

   $x = \dfrac{\log 30}{\log 4}$ Divide by log 4.

   $x \approx 2.453$ Use a calculator.

   Check

   $4^x = 4^{2.453} \approx 30$

   The solution set is $\{2.453\}$.

Now Try:

1. Solve $3^x = 15$. Approximate the solution to three decimal places.

   _____

Name: Date:
Instructor: Section:

2. Solve $e^{0.005x} = 9$. Approximate the solution to three decimal places.

$$e^{0.005x} = 9$$
$$\ln e^{0.005x} = \ln 9 \quad \text{If } x = y, \text{ and } x > 0, y > 0, \text{ then } \ln x = \ln y.$$
$$0.005x \ln e = \ln 9 \quad \text{Power rule}$$
$$0.005x = \ln 9 \quad \ln e = 1$$
$$x = \frac{\ln 9}{0.005} \quad \text{Divide by 0.005.}$$
$$x \approx 439.445 \text{ Use a calculator.}$$

The solution set is {439.445}.

2. Solve $e^{0.4x} = 15$. Approximate the solution to three decimal places.

_____

Objective 1 Practice Exercises

For extra help, see Examples 1–3 on pages 847–848 of your text.

Solve each equation. Give solutions to three decimal places.

1. $25^{x+2} = 125^{3-x}$

1. _____

2. $4^{x-1} = 3^{2x}$

2. _____

3. $e^{-0.45x} = 7$

3. _____

Name: Date:
Instructor: Section:

Objective 2   Solve equations involving logarithms.

Video Examples

Review these examples for Objective 2:

5. Solve $\log_3 (x-1)^3 = 5$. Give the exact solution.

$$\log_3 (x-1)^3 = 5$$
$$(x-1)^3 = 3^5 \quad \text{Write in exponential form.}$$
$$(x-1)^3 = 243$$
$$x - 1 = \sqrt[3]{243} \quad \text{Take the cube root of each side.}$$
$$x - 1 = 3\sqrt[3]{9} \quad \text{Simplify the cube root.}$$
$$x = 1 + 3\sqrt[3]{9} \quad \text{Add 1.}$$

Check:
$$\log_3 (x-1)^3 = 5$$
$$\log_3 \left(1 + 3\sqrt[3]{9} - 1\right)^3 \stackrel{?}{=} 5$$
$$\log_3 \left(\sqrt[3]{243}\right)^3 \stackrel{?}{=} 5$$
$$\log_3 (243) \stackrel{?}{=} 5$$
$$\log_3 3^5 \stackrel{?}{=} 5$$
$$5 = 5$$

The solution set is $\{1 + 3\sqrt[3]{9}\}$.

6. Solve $\log_3 (5x+42) - \log_3 x = \log_3 26$.

$$\log_3 (5x+42) - \log_3 x = \log_3 26$$
$$\log_3 \frac{5x+42}{x} = \log_3 26$$
$$\frac{5x+42}{x} = 26 \quad \text{If } \log_b x = \log_b y \text{ then } x = y.$$
$$5x + 42 = 26x \quad \text{Multiply by } x.$$
$$42 = 21x \quad \text{Subtract } 5x.$$
$$2 = x \quad \text{Divide by 21.}$$

Now Try:

5. Solve $\log_6 (x+1)^3 = 2$. Give the exact solution.

_____

6. Solve
$\log_6 (2x+7) - \log_6 x = \log_6 16$.

_____

Name: Date:
Instructor: Section:

Check:
$$\log_3(5x+42) - \log_3 x = \log_3 26$$
$$\log_3(5\cdot 2+42) - \log_3 2 \overset{?}{=} \log_3 26$$
$$\log_3 52 - \log_3 2 \overset{?}{=} \log_3 26$$
$$\log_3 \tfrac{52}{2} \overset{?}{=} \log_3 26$$
$$\log_3 26 = \log_3 26$$

The solution set is $\{2\}$.

7. Solve $\log_2(x+7) + \log_2(x+3) = \log_2 77$.

$$\log_2(x+7) + \log_2(x+3) = \log_2 77$$
$$\log_2[(x+7)(x+3)] = \log_2 77$$
$$\quad\text{Product rule}$$
$$(x+7)(x+3) = 77$$
$$\quad\text{If } \log_b x = \log_b y \text{ then } x = y.$$
$$x^2 + 10x + 21 = 77 \quad \text{Multiply.}$$
$$x^2 + 10x - 56 = 0 \quad \text{Subtract 77.}$$
$$(x-4)(x+14) = 0 \quad \text{Factor.}$$
$$x-4 = 0 \quad \text{or} \quad x+14 = 0$$
$$x = 4 \qquad\qquad x = -14$$

The value -14 must be rejected since it leads to the logarithm of a negative number in the original equation.
A check shows that the only solution is 4.
The solution set is $\{4\}$.

7. Solve.
$$\log_4(4x-3) + \log_4 x = \log_4(2x-1)$$

_____

Objective 2 Practice Exercises

For extra help, see Examples 4–7 on pages 848–850 of your text.

Solve each equation. Give exact solution.

4. $\log(-a) + \log 4 = \log(2a+5)$

4. _____

Name: Date:
Instructor: Section:

5. $\log_3(x^2 - 10) - \log_3 x = 1$

5. _____

6. $\ln(x+4) + \ln(x-2) = \ln 7$

6. _____

Objective 3   Solve applications involving compound interest.

Video Examples

Review these examples for Objective 3:

8. How much money will be in an account at the end of 5 years if $5000 is deposited at 4% compounded monthly?

   Because interest is compounded monthly, $n = 12$. The other given values are $P = 5000$, $r = 0.04$, and $t = 5$.

   $$A = P\left(1 + \frac{r}{n}\right)^{nt}$$

   $$A = 5000\left(1 + \frac{0.04}{12}\right)^{12 \cdot 5}$$

   $$A = 5000(1.0033)^{60}$$

   $$A = 6104.98$$

   There will be $6104.98 in the account at the end of 5 years.

9. Approximate the time it would take for money deposited in an account paying 5% interest compounded quarterly to double. Round to the nearest hundredth.

   We want the number of years t for P dollars to grow to $2P$ dollars at a rate of 5% per year. In the compound interest formula, we substitute $2P$ for A, and let $r = 0.05$ and $n = 4$.

Now Try:

8. How much money will be in an account at the end of 5 years if $10,000 is deposited at 4% compounded quarterly?

   _____

9. Approximate the time it would take for money deposited in an account paying 5% interest compounded monthly to double. Round to the nearest hundredth.

   _____

Name: Date:
Instructor: Section:

$$2P = P\left(1 + \frac{0.05}{4}\right)^{4t}$$

$$2 = 1.0125^{4t}$$

$$\log 2 = \log 1.025^{4t}$$

$$\log 2 = 4t \log 1.0125$$

$$t = \frac{\log 2}{4 \log 1.0125}$$

$$t \approx 13.95$$

It will take about 13.95 years for the investment to double.

10. Suppose that $5000 is invested at 4% interest for 3 years.

   a. How much will the investment be worth if it is compounded continuously?

   $A = Pe^{rt}$

   $A = 5000e^{0.04 \cdot 3}$

   $A = 5000e^{0.12}$

   $A = 5637.48$

   The investment will be worth $5637.48.

   b. Approximate the amount of time it would take for the investment to double. Round to the nearest tenth.

   Find the value of t that will cause A to be 2($5000) = $10,000.

   $A = Pe^{rt}$

   $10,000 = 5000e^{0.04t}$

   $2 = e^{0.04t}$ Divide by 5000.

   $\ln 2 = \ln e^{0.04t}$ If $x = y$, then $\ln x = \ln y$.

   $\ln 2 = 0.04t$ $\ln e^k = k$

   $\frac{\ln 2}{0.04} = t$ Divide by 0.04.

   $t \approx 17.3$

   It will take about 17.3 years for the amount to double.

10. Suppose that $5000 is invested at 2% interest for 3 years.

   a. How much will the investment be worth if it is compounded continuously?

   _____

   b. Approximate the amount of time it would take for the investment to double. Round to the nearest tenth.

   _____

Name: Date:
Instructor: Section:

Objective 3 Practice Exercises

For extra help, see Examples 8–10 on pages 851–852 of your text.

Solve each problem.

7. How much will be in an account after 10 years if $25,000 is invested at 8% compounded quarterly? Round to the nearest cent.

7. _____

8. How much will be in an account after 5 years if $10,000 is invested at 4.5% compounded continuously? Round to the nearest cent.

8. _____

9. How long will it take an investment to double if it is placed in an account paying 9% interest compounded continuously? Round to the nearest tenth.

9. _____

Objective 4 Solve applications involving base *e* exponential growth and decay.

Video Examples

Review these examples for Objective 4:

11. A sample of 500 g of lead-210 decays according to the function $y = y_0 e^{-0.032t}$, where *t* is the time in years, *y* is the amount of the sample at time *t*, and y_0 is the initial amount present at $t = 0$.

  a. How much lead will be left in the sample after 20 years? Round to the nearest tenth of a gram.

  Let $t = 20$ and $y_0 = 500$.

  $y = 500 e^{-0.032 \cdot 20} \approx 263.6$

  There will be about 263.6 grams after 20 years.

Now Try:

11. Cesium-137, a radioactive isotope used in radiation therapy, decays according to the function $y = y_0 e^{-0.0231t}$, where *t* is the time in years and y_0 is the initial amount present at $t = 0$.

  a. If an initial sample contains 36 mg of cesium-137, how much cesium-137 will be left in the sample after 50 years? Round to the nearest tenth.

  _____

Name: Date:
Instructor: Section:

b. Approximate the half-life of lead-210 to the nearest tenth.

Let $y = \frac{1}{2}(500) = 250$.

$$250 = 500e^{-0.032t}$$
$$0.5 = e^{-0.032t}$$
$$\ln 0.5 = \ln e^{-0.032t}$$
$$\ln 0.5 = -0.032t$$
$$t = \frac{\ln 0.5}{-0.032} \approx 21.7$$

The half-life of lead-210 is about 21.7 years.

b. Approximate the half-life of cesium-137 to the nearest tenth.

_____

Objective 4 Practice Exercises

For extra help, see Example 11 on page 853 of your text.

Solve each problem.

10. Radioactive strontium decays according to the function $y = y_0 e^{-0.0239t}$, where t is the time in years. If an initial sample contains $y_0 = 15$ g of radioactive strontium, how many grams will be present after 25 years? Round to the nearest hundredth of a gram.

10. _____

11. How long will it take the initial sample of strontium in exercise 19 to decay to half of its original amount?

11. _____

12. The concentration of a drug in a person's system decreases according to the function $C(t) = 2e^{-0.2t}$, where $C(t)$ is given in mg and t is in hours. How much of the drug will be in the person's system after one hour? Approximate answer to the nearest hundredth.

12. _____

Answers

Chapter R PREALGEBRA REVIEW

R.1   Fractions

Key Terms
1. equivalent fractions
2. improper fraction
3. numerator
4. proper fraction
5. denominator
6. composite number
7. prime factorization
8. prime number
9. lowest terms

Objective 1
Now Try
1. $2 \cdot 3 \cdot 5 \cdot 7$

Practice Exercises
1. $2 \cdot 7 \cdot 7$
3. $2 \cdot 3 \cdot 7 \cdot 13$

Objective 2
Now Try
2. $\dfrac{3}{5}$

Practice Exercises
5. $\dfrac{5}{6}$

Objective 3
Now Try
3. $14\dfrac{4}{5}$
4. $\dfrac{86}{7}$

Practice Exercises
7. $21\dfrac{2}{5}$
9. $\dfrac{244}{11}$

Objective 4
Now Try
5. $\dfrac{1}{10}$
6. $\dfrac{16}{21}$

Practice Exercises
11. $\dfrac{7}{5}$ or $1\dfrac{2}{5}$

Objective 5
Now Try
7. $\dfrac{3}{4}$
8. $\dfrac{23}{24}$
9. $\dfrac{2}{9}$

Copyright © 2016 Pearson Education, Inc.

Answers

Practice Exercises

13. $\frac{256}{225}$ or $1\frac{31}{225}$ 15. $\frac{119}{24}$ or $4\frac{23}{24}$

Objective 6
Now Try

10. $8\frac{3}{4}$ cups

Practice Exercises

17. $11\frac{3}{4}$ yards

Objective 7
Now Try

11a. Home; $\frac{1}{50}$ 11b. 104 workers

Practice Exercises

19. Lunch Room   21. 754 workers

R.2   Decimals and Percents

Key Terms

1. decimals   2. place value   3. percent

Objective 1
Now Try

1a. $\frac{72}{100}$ 1b. $\frac{53}{1000}$ 1c. $\frac{37,058}{10,000}$

Practice Exercises

1. $\frac{7}{1000}$ 3. $\frac{300,005}{10,000}$

Objective 2
Now Try

2a. 37.871   2b. 27.282

Practice Exercises

5. 755.098

Objective 3
Now Try

3a. 251.116   3b. 0.028   4a. 32.4

4b. 1.43

Answers

Practice Exercises
7. 2.3424   9. 0.4292

Objective 4
  Now Try
6a. 0.35   6b. $3.\overline{5}$ or 3.555...

  Practice Exercises
11. $0.\overline{4}$ or 0.444...

Objective 5
  Now Try
9a. 0.91   9b. 0.06   9c. 43%
9d. 520%

  Practice Exercises
13. 3.62   15. 8.4%

Objective 6
  Now Try
10a. $\frac{3}{10}$ 10b. $1\frac{1}{4}$ 11a. 15%
11b. $5\frac{5}{9}\%$ exact, or 5.6% rounded

  Practice Exercises
17. $\frac{139}{250}$

Objective 7
  Now Try
12. discount: $10.64, sale price: $8.36

  Practice Exercises
19. $14; sale price: $56   21. $42; sale price: $14

Answers

Chapter 1 THE REAL NUMBER SYSTEM

1.1 Exponents, Order of Operations, and Inequality

Key Terms
1. exponential expression
2. base
3. exponent

Objective 1
 Now Try
 1. 49

 Practice Exercises
 1. 27
 3. 0.16

Objective 2
 Now Try
 2. 23

 Practice Exercises
 5. 45

Objective 3
 Now Try
 3. 215

 Practice Exercises
 7. −8
 9. 48

Objective 4
 Now Try
 4. true

 Practice Exercises
 11. false

Objective 5
 Now Try
 5. $19 \leq 11 + 8$

 Practice Exercises
 13. $7 = 13 - 6$
 15. $20 \geq 2 \cdot 7$

Objective 6
 Now Try
 6. $11 < 15$

 Practice Exercises
 17. $8 \leq 12$

Answers

1.2 Variables, Expressions, and Equations

Key Terms

1. equation
2. elements
3. variable
4. algebraic expression
5. set
6. Solution
7. constant

Objective 1
Now Try
1. 252, 567
2. 65

Practice Exercises
1. 8
3. $\dfrac{28}{13}$

Objective 2
Now Try
3. $20x$

Practice Exercises
5. $8x - 11$

Objective 3
Now Try
4. no

Practice Exercises
7. no
9. yes

Objective 4
Now Try
5. $x + 7 = 11$; 4

Practice Exercises
11. $5 + x = 14$; 9

Objective 5
Now Try
6. expression

Practice Exercises
13. expression
15. equation

Answers

1.3 Real Numbers and the Number Line

Key Terms

1. whole numbers
2. additive inverse
3. integers
4. natural numbers
5. absolute value
6. number line
7. irrational number
8. coordinate
9. negative number
10. positive number
11. real numbers
12. set-builder notation
13. rational number
14. signed numbers

Objective 1
  Now Try
  1. −282
  2. (number line with points at −3, −2, 1, 2)
  3a. 0, 7
  3b. −10, 0, 7
  3c. $-10, -\frac{5}{8}, 0, 0.\overline{4}, 5\frac{1}{2}, 7, 9.9$
  3d. $\sqrt{5}$

  Practice Exercises
  1. −75 pounds
  3. (number line with points at −4, −2, 1, 3)

Objective 2
  Now Try
  4. true

  Practice Exercises
  5. false

Objective 3
  Practice Exercises
  7. 25
  9. −4.5

Objective 4
  Now Try
  5a. 10
  5b. −10
  5c. 3

  Practice Exercises
  11. 1.22

Objective 5
  Now Try
  6. Milk and Electricity

  Practice Exercises
  13. Gasoline
  15. Eggs in 2012 to 2013

Answers

1.4 Adding and Subtracting Real Numbers

Key Terms
1. minuend
2. subtrahend
3. difference
4. sum
5. addends

Objective 1
Now Try
1. −5
2. −12

Practice Exercises
1. −18
3. $-5\frac{5}{8}$

Objective 2
Now Try
3. 3
4. $-\frac{7}{10}$

Practice Exercises
5. $-\frac{1}{6}$

Objective 3
Now Try
6a. −3
6b. 2

Practice Exercises
7. −25
9. 0

Objective 4
Now Try
7. 1

Practice Exercises
11. −6

Objective 5
Now Try
8. −10 + 11 + 2; 3
9. −17 − 9; −26
10. 5464 ft

Practice Exercises
13. −4 − 4; −8
15. −51.2°C

Objective 6
Now Try
11. −$0.006

Practice Exercises
17. −$0.061

Answers

1.5 Multiplying and Dividing Real Numbers

Key Terms
1. quotient
2. reciprocals
3. product

Objective 1
Now Try
1. −56

Practice Exercises
1. −28
3. −13.12

Objective 2
Now Try
2a. 30

Practice Exercises
5. $\frac{4}{5}$

Objective 3
Now Try
2b. −12, −6, −4, −3, −2, −1, 1, 2, 3, 4, 6, and 12

Practice Exercises
7. −8, −4, −2, −1, 1, 2, 4, 8
9. −42, −21, −14, −7, −6, −3, −2, −1, 1, 2, 3, 6, 7, 14, 21, 42

Objective 4
Now Try
3a. 2
3b. 3

Practice Exercises
11. 0

Objective 5
Now Try
4a. 28
4b. $-\frac{5}{4}$

Practice Exercises
13. 64
15. $\frac{16}{21}$

Objective 6
Now Try
5. −432

Practice Exercises
17. 25

Answers

Objective 7
Now Try
6. $\frac{5}{6}[-8+(-4)]$; -10

Practice Exercises
19. $(-7)(3)+(-7)$; -28
21. $-12+\frac{49}{-7}$; -19

Objective 8
Now Try
8. $\frac{36}{x}=-4$

Practice Exercises
23. $-8x=72$

1.6 Properties of Real Numbers

Key Terms
1. identity element for addition
2. identity element for multiplication

Objective 1
Now Try
1a. -12
1b. 2

Practice Exercises
1. 4
3. $(4+z)$

Objective 2
Now Try
2a. 4
2b. $[(-3)\cdot 4]$
3a. associative
3b. commutative
3c. both
4. $39x+20$

Practice Exercises
5. $[(-4+3y)]$

Objective 3
Now Try
5a. 0
5b. 1
6a. $\frac{7}{9}$
6b. $\frac{2}{3}$

Practice Exercises
7. 4
9. $\frac{6}{7}$

Answers

Objective 4
  Now Try
  7a. $\dfrac{5}{8}$ 7b. -8 7c. -10
  7d. 11 8. 6

  Practice Exercises
  11. 0; identity

Objective 5
  Now Try
  9a. $17x - 102$ 9b. $-8x + 20$ 9c. $3(11+7)$
  9d. $12(y+6+x)$ 10a. $-3x-4$ 10b. $4x+5y-z$
  10c. $3(x+y+1)$

  Practice Exercises
  13. $2an - 4bn + 6cn$ 15. $2k - 7$

1.7 Simplifying Expressions

Key Terms
  1. numerical coefficient   2. term   3. like terms

Objective 1
  Now Try
  1a. $35x - 21y$ 1b. $11 - 7x$

  Practice Exercises
  1. $8x + 27$ 3. $10x + 3$

Objective 2
  Practice Exercises
  5. $\dfrac{7}{9}$

Objective 3
  Practice Exercises
  7. like   9. unlike

Objective 4
  Now Try
  2a. $23r$ 2b. $19x$ 2c. $8x^2$
  3a. $37k - 24$ 3b. $-\dfrac{5}{4}x + 6$

Answers

Practice Exercises
11. $2x-14$

Objective 5
Now Try
4. $11+10x+8x+4x$; $11+22x$

Practice Exercises
13. $6x+12+4x=10x+12$
15. $4(2x-6x)+6(x+9)=-10x+54$

Chapter 2 LINEAR EQUATIONS AND INEQUALITIES IN ONE VARIABLE

2.1 The Addition Property of Equality

Key Terms
1. equivalent equations
2. linear equation
3. solution set

Objective 1
Practice Exercises
1. no
3. yes

Objective 2
Now Try
1. $\{21\}$
3. $\{-19\}$
5. $\{13\}$
6. $\{-3\}$

Practice Exercises
5. $\dfrac{1}{2}$

Objective 3
Now Try
7. $\{29\}$
8. $\{8\}$

Practice Exercises
7. 7
9. 7.2

Answers

2.2 The Multiplication Property of Equality

Key Terms
1. multiplication property of equality
2. addition property of equality

Objective 1
Now Try
1. {14}
3. {5.7}
4. {24}
5. {36}
6. {−3}

Practice Exercises
1. {−17}
3. {6.4}

Objective 2
Now Try
7. {4}

Practice Exercises
5. {−5}

2.3 More on Solving Linear Equations

Key Terms
1. contradiction
2. conditional equation
3. identity

Objective 1
Now Try
1. {−6}
2. {8}
4. $\left\{\dfrac{15}{2}\right\}$

Practice Exercises
1. $\left\{\dfrac{5}{2}\right\}$
3. $\left\{-\dfrac{1}{5}\right\}$

Objective 2
Now Try
6. {all real numbers}
7. ∅

Practice Exercises
5. infinitely many

Objective 3
Now Try
9. {2}
10. {2}

Practice Exercises
7. {2}
9. {10}

Answers

Objective 4
Now Try
11. $67 - t$

Practice Exercises
11. $\dfrac{17}{p}$

2.4 Applications of Linear Equations

Key Terms
1. supplementary angles
2. complementary angles
3. right angle
4. straight angle
5. consecutive integers

Objective 1
Practice Exercises
1. Read the problem; assign a variable to represent the unknown; write an equation; solve the equation; state the answer; check the answer.

Objective 2
Now Try
2. 16

Practice Exercises
3. $-2(4 - x) = 24$; 16

Objective 3
Now Try
3. 52
6. 8 feet

Practice Exercises
5. $x + (x + 5910) = 34,730$; Mt. Rainier: 14,410 ft; Mt. McKinley: 20,320 feet
7. $x + (5 + 3x) + 4x = 29$; Mark: 3 laps; Pablo: 14 laps; Faustino: 12 laps

Objective 4
Now Try
8. −2, 0, 2, 4

Practice Exercises
9. 27, 28

Objective 5
Now Try
10. 66°

Practice Exercises
11. 133°
13. 27°

Answers

2.5 Formulas and Additional Applications from Geometry

Key Terms
1. vertical angles   2. formula   3. perimeter
4. area

Objective 1
  Now Try
  1. $W = 5.5$

  Practice Exercises
  1. $a = 36$ 3. $h = 12$

Objective 2
  Now Try
  2. 12 ft   3. 15 ft, 20 ft, 30 ft

  Practice Exercises
  5. 1.5 years

Objective 3
  Now Try
  5. 54°, 126°

  Practice Exercises
  7. 35°, 35°   9. 129°, 51°

Objective 4
  Now Try
  6. $r = \dfrac{d}{t}$ 7. $a = P - b - c$ 8. $h = \dfrac{2A}{b+B}$

  Practice Exercises
  11. $n = \dfrac{S}{180} + 2$ or $n = \dfrac{S+360}{180}$

2.6 Ratio, Proportion, and Percent

Key Terms
1. proportion   2. ratio   3. terms
4. cross products

Objective 1
  Now Try
  1a. $\dfrac{11}{17}$ 1b. $\dfrac{4}{15}$ 2. 24-ounce jar, $0.054 per oz

  Practice Exercises
  1. $\dfrac{8}{3}$ 3. 45-count box

Answers

Objective 2
Now Try
3. True 5. $\left\{-\dfrac{3}{4}\right\}$

Practice Exercises
5. $\left\{\dfrac{10}{3}\right\}$

Objective 3
Now Try
6. $259.20

Practice Exercises
7. 15 inches 9. $135

Objective 4
Now Try
7a. 140 7b. 950 7c. 85%

8. 87 male students

Practice Exercises
11. 2%

2.7 Further Applications of Linear Equations

Objective 1
Now Try
1. $117

Practice Exercises
1. 1100 students 3. 15,000 students

Objective 2
Now Try
2. 8 oz

Practice Exercises
5. 60 lb

Objective 3
Now Try
4. 5%: $1600; 7%: $3500

Practice Exercises
7. 7%: $800; 9%: $300

Answers

Objective 4
  Now Try
  5. 32 quarters; 18 nickels

  Practice Exercises
  9. 25 large jars; 55 small jars

  11. 1500 general admission; 750 reserved

Objective 5
  Now Try
  6. 270 miles   7. 5 hours

  Practice Exercises
  13. 450 miles

2.8 Solving Linear Inequalities

Key Terms
  1. three-part inequality   2. interval
  3. linear inequality   4. inequalities   5. interval notation

Objective 1
  Now Try

  1.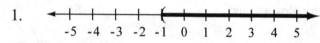

  Practice Exercises

  1. $(3, \infty)$;

  3. $(-\infty, -4)$;

Objective 2
  Now Try
  2. $[-3, \infty)$

  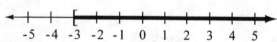

  Practice Exercises

  5. $(2, \infty)$;

516   Copyright © 2016 Pearson Education, Inc.

Answers

Objective 3
 Now Try
 3a. $(-\infty, -5]$
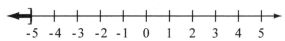

 3b. $(-\infty, -4)$
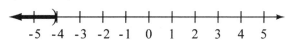

 Practice Exercises

 7. $(-2, \infty)$;

 9. $(-\infty, 4]$
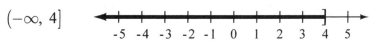

Objective 4
 Now Try
 4. $[2, \infty)$
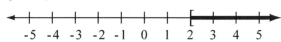

 Practice Exercises
 11. $(-\infty, 5]$;

Objective 5
 Now Try
 7. 10 feet

 Practice Exercises
 13. 89   15. all numbers greater than 5

Objective 6
 Now Try
 9. $[2, 4)$
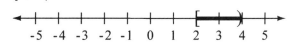

 Practice Exercises
 17. $[-5, -3)$;

Copyright © 2016 Pearson Education, Inc.

Answers

Chapter 3 LINEAR EQUATIONS AND INEQUALITIES IN TWO VARIABLES

3.1 Linear Equations and Rectangular Coordinates

Key Terms
1. line graph
2. linear equation in two variables
3. coordinates
4. x-axis
5. y-axis
6. ordered pair
7. rectangular (Cartesian) coordinate system
8. quadrants
9. origin
10. Plot
11. scatter diagram
12. table of values
13. plane

Objective 1
  Now Try
  1a. 2000-2001, 2002-2003, 2004-2005   1b. 300 degrees

  Practice Exercises
  1. 2001-2002, 2003-2004

Objective 2
  Practice Exercises
  3. (4, 7)   5. (0.2, 0.3)

Objective 3
  Now Try
  2a. yes   2b. no

  Practice Exercises
  7. not a solution

Objective 4
  Now Try
  3. (5, 13)

  Practice Exercises
  9. (a) (2, −1); (b) (0, −5); (c) (4, 3); (d) (−1, −7); (e) (7, 9)

Objective 5
  Now Try
  4. (1, −4), (5, 12), (2, 0), (3, 4)

x	y
1	−4
5	12
2	0
3	4

Answers

Practice Exercises

11. $(1, -3), (1, 0), (1, 5)$

13. $(0, 4), (3, 0), \left(\dfrac{15}{4}, -1\right)$

Objective 6
Now Try
5.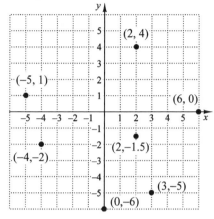

Practice Exercises

15.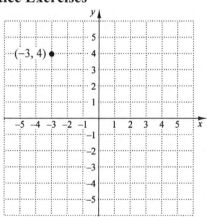

Answers

3.2 Graphing Linear Equations in Two Variables

Key Terms
1. *y*-intercept
2. *x*-intercept
3. graphing
4. graph

Objective 1
Now Try
2.

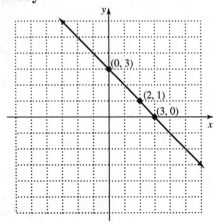

Practice Exercises

1. $(0,-2), \left(\dfrac{2}{3}, 0\right), (2, 4)$

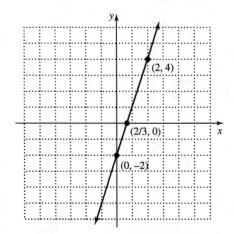

3. $\left(0, -\dfrac{1}{2}\right), (1, 0), (-3, -2)$

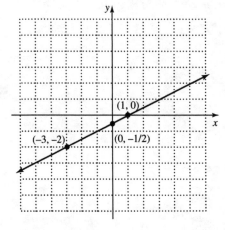

Answers

Objective 2
Now Try
3.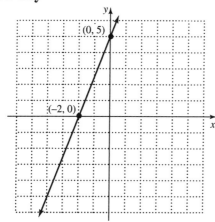
4.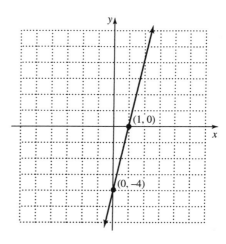

Practice Exercises
5.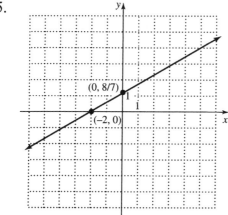

Objective 3
Now Try
5.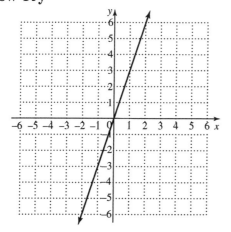

Copyright © 2016 Pearson Education, Inc. 521

Answers

Practice Exercises
7. $x + y = 0$

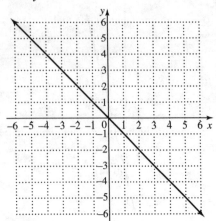

Objective 4
Now Try
6.

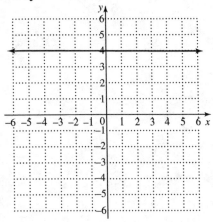

7.

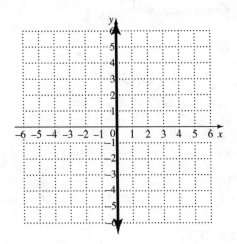

Practice Exercises
9. $x - 1 = 0$

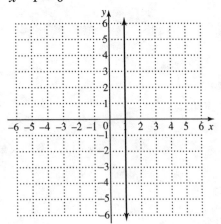

Answers

Objective 5
Now Try
8a. 0 calculators, $45
5000 calculators, $42
20,000 calculators, $33
45,000 calculators, $18

8b. (0, $45), (5, $42), (20, $33), (45, $18)

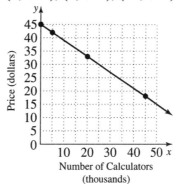

8c. 30,000 calculators, $27

Practice Exercises
11. 2004, 4.9 million;
2005, 5.53 million;
2006, 6.16 million;
2007, 6.79 million

13. 48, 20°C;
54, 22°C;
60, 24°C;
66, 26°C

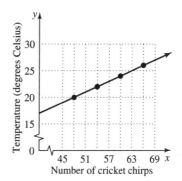

3.3 The Slope of a Line

Key Terms
1. perpendicular lines
2. slope
3. rise
4. parallel lines
5. run

Objective 1
Now Try
1. $\frac{4}{1}$, or 4
2. $-\frac{16}{9}$
3. 0
4. undefined slope

Answers

Practice Exercises
1. −2
3. 0

Objective 2
  Now Try
5. $\dfrac{7}{4}$

  Practice Exercises
5. $\dfrac{2}{3}$

Objective 3
  Now Try
6a. parallel   6b. perpendicular

  Practice Exercises
7. 1; 1; parallel   9. −3; $\dfrac{1}{3}$; perpendicular

3.4 Slope-Intercept Form of a Linear Equation

Key Terms
1. point-slope form   2. standard form   3. slope-intercept form

Objective 1
  Now Try
1a. slope: −12; y-intercept: (0, 6)   1b. slope: $-\dfrac{1}{7}$; y-intercept: $\left(0, -\dfrac{7}{5}\right)$

  Practice Exercises
1. slope: $\dfrac{3}{2}$; y-intercept: $\left(0, -\dfrac{2}{3}\right)$

Objective 2
  Now Try
2.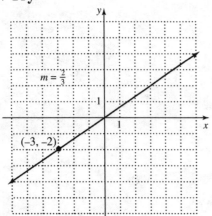

524

Answers

Practice Exercises

3.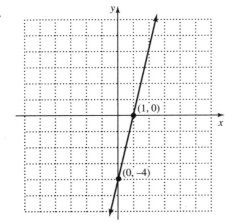

5.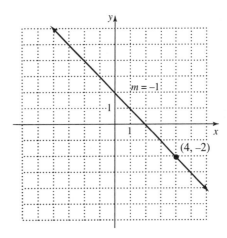

Objective 3
Now Try

4a. $y = \dfrac{3}{4}x - 5$

4b. $y = 6x + 10$

Practice Exercises

7. $y = -2x + 12$

Objective 4
Now Try

5a.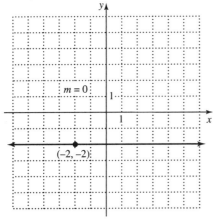

5b.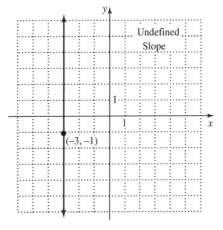

6a. $y = 5$ 6b. $x = -5$

Answers

Practice Exercises

9.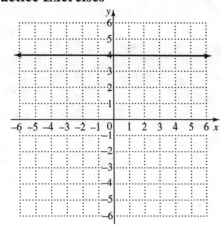

11. $x = -3$

3.5 Point-Slope Form of a Linear Equation and Modeling

Key Terms

1. standard form
2. slope-intercept form
3. point-slope form

Objective 1

  Now Try

  1. $y = -\frac{2}{3}x + 15$

  Practice Exercises

  1. $y = -\frac{3}{5}x + \frac{11}{5}$
  3. $y = -\frac{3}{2}x + 5$

Objective 2

  Now Try

  2. $y = -\frac{3}{4}x + \frac{81}{4}$; $3x + 4y = 81$

  Practice Exercises

  5. $3x - 2y = 0$

Objective 3

  Now Try

  3.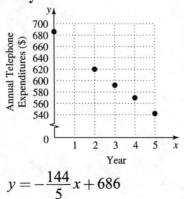

  $y = -\frac{144}{5}x + 686$

Practice Exercises

7.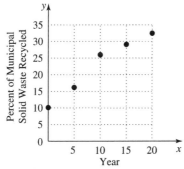

$y = \dfrac{28}{25}x + 10.1$

Chapter 4 EXPONENTS AND POLYNOMIALS

4.1 The Product Rule and Power Rules for Exponents

Key Terms
1. power
2. exponential expression
3. base

Objective 1
  Now Try
  1. 4^5
  2a. base: 2; exponent: 6; value: 64

2b. base: –2; exponent: 6; value 64

  Practice Exercises
  1. $\dfrac{1}{243}$
  3. -6561; base: 3; exponent: 8

Objective 2
  Now Try
  3a. 9^{13}
  3b. m^{27}
  3c. $18x^{11}$

  3d. 108

  Practice Exercises
  5. $8c^{15}$

Objective 3
  Now Try
  4a. 7^8
  4b. x^{30}

  Practice Exercises
  7. 7^{12}
  9. $(-3)^{21}$

Answers

Objective 4
　Now Try
　5. $64a^3b^3$

　Practice Exercises
　11. $-0.008a^{12}b^3$

Objective 5
　Now Try
　6. $\dfrac{1}{1024}$

　Practice Exercises
　13. $-\dfrac{8x^3}{125}$　　　15. $-\dfrac{128a^7}{b^{14}}$

Objective 6
　Now Try
　7a. $\dfrac{5^5}{2^3}$, or $\dfrac{3125}{8}$　　　7b. $-x^{27}y^{13}$

　Practice Exercises
　17. $32a^9b^{14}c^5$

Objective 7
　Now Try
　8. $28x^5$

　Practice Exercises
　19. $36x^5$　　　21. $28q^{11}$

4.2　Integer Exponents and the Quotient Rule

Key Terms
1. power rule for exponents
2. base; exponent
3. product rule for exponents

Objective 1
　Now Try
　1a. 1　　　1b. −1　　　1c. 1　　　1d. 0

　Practice Exercises
　1. −1　　　3. 0

Answers

Objective 2
Now Try
2a. $\dfrac{1}{27}$ 2b. 25 2c. $\dfrac{8}{27}$ 2d. $\dfrac{1}{8}$ 2e. $\dfrac{1}{p^5}$

3a. $\dfrac{125}{36}$ 3b. $\dfrac{y^2}{x^7}$ 3c. $\dfrac{qr^5}{4p^3}$

Practice Exercises
5. $\dfrac{1}{m^{18}n^9}$

Objective 3
Now Try
4a. 9 4b. z^{10} 4c. $(a-b)^2$ 4d. $\dfrac{36b^7}{a^8}$

Practice Exercises
7. $\dfrac{k^4 m^5}{2}$ 9. $\dfrac{p^8}{3^5 m^3}$ or $\dfrac{p^8}{243 m^3}$

Objective 4
Now Try
5a. 6 5b. $3125b^5$ 5c. $\dfrac{243}{32p^{20}}$ 5d. $\dfrac{x^{13}y^2}{343z}$

Practice Exercises
11. $a^{16}b^{22}$

4.3   Scientific Notation

Key Terms
1. scientific notation   2. power rule   3. quotient rule

Objective 1
Now Try
1b. 4.771×10^{10} 1c. 4.63×10^{-2}

Practice Exercises
1. 2.3651×10^4 3. -2.208×10^{-4}

Objective 2
Now Try
2a. $27{,}960{,}000$ 2b. 0.000164

Practice Exercises
5. 0.0064

Answers

Objective 3
  Now Try
  3a. 2.7×10^8, or 270,000,000   3b. 3×10^{-8}, or 0.00000003
  4. 9×10^{23} grains of sand

Practice Exercises
  7. 2.53×10^2 9. 4.86×10^{19} atoms

4.4   Adding, Subtracting, and Graphing Polynomials

Key Terms
1. degree of a term   2. descending powers   3. term
4. trinomial   5. polynomial   6. monomial
7. degree of a polynomial   8. binomial   9. like terms
10. line of symmetry   11. vertex   12. axis
13. parabola

Objective 1
  Now Try
  1.   1, 7, –2; three terms

Practice Exercises
  1.   three terms; 3, –2, 1   3.   three terms; 8, –1, –1

Objective 2
  Now Try
  2. $15m^3 + 29m^2$

Practice Exercises
  5. $2.6z^8 - 0.9z^7$

Objective 3
  Now Try
  3a. $3x^3 - 7x^2 + 4x$; degree 3; trinomial
  3b. $4w^5 - 3w^2$; degree 5; binomial

Practice Exercises
  7. $n^8 - n^2$; degree 8; binomial
  9. $5c^5 + 3c^4 - 10c^2$; degree 5; trinomial

Objective 4
  Now Try
  4.   1285

Practice Exercises

11. a. 71; b. −19

Objective 5
Now Try

6a. $4x^3 + x + 12$ 6b. $7x^3 + 8x^2 - 14x - 1$ 7. $-11x^3 - 7x + 1$

9. $7 + 8x - 4x^2$ 10. $x^2y + 4xy$

Practice Exercises

13. $3r^3 + 7r^2 - 5r - 2$ 15. $-7x^2y + 3xy + 5xy^2$

Objective 6
Now Try

11.
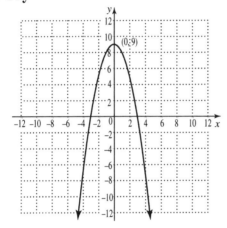

Practice Exercises

17.
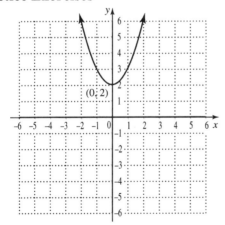

Vertex: (0, 2)

Answers

4.5 Multiplying Polynomials

Key Terms
1. inner product   2. FOIL   3. outer product

Objective 1
  Now Try
  1. $32x^4 + 64x^3$

  Practice Exercises
  1. $35z^4 + 14z$ 3. $-6y^5 - 9y^4 + 12y^3 - 33y^2$

Objective 2
  Now Try
  2. $4x^7 - 2x^5 + 37x^4 - 18x^2 + 9x$
  3. $28x^4 - 33x^3 + 51x^2 + 17x - 15$
  4. $2x^5 - 6x^4 + 10x^3 - 29x^2 + 5$

  Practice Exercises
  5. $6m^5 + 4m^4 - 5m^3 + 2m^2 - 4m$

Objective 3
  Now Try
  5. $x^2 + 3x - 54$ 6. $16xy + 72y - 14x - 63$
  7. $15k^2 + 44kn + 32n^2$

  Practice Exercises
  7. $20a^2 + 11ab - 3b^2$ 9. $-6m^2 - mn + 12n^2$

4.6 Special Products

Key Terms
  1. binomial   2. conjugate

Objective 1
  Now Try
  2a. $4a^2 + 36ak + 81k^2$ 2b. $9p^2 + p + \frac{1}{36}$

  Practice Exercises
  1. $49 + 14x + x^2$ 3. $16y^2 - 5.6y + 0.49$

Answers

Objective 2
Now Try

3a. $x^2 - 81$ 3b. $\dfrac{25}{36} - a^2$ 4a. $121x^2 - y^2$

4b. $4p^5 - 144p$

Practice Exercises

5. $64k^2 - 25p^2$

Objective 3
Now Try

5a. $x^3 + 18x^2 + 108x + 216$

5b. $81x^4 - 540x^3 + 1350x^2 - 1500x + 625$

Practice Exercises

7. $a^3 - 9a^2 + 27a - 27$ 9. $256s^4 + 768s^3t + 864s^2t^2 + 432st^3 + 81t^4$

4.7   Dividing Polynomials

Key Terms

1. dividend   2. quotient   3. divisor

Objective 1
Now Try

1. $10x^3 - 5x$ 2. $3n^2 - 4n - \dfrac{2}{n}$ 3. $\dfrac{7z^4}{2} - 4z^3 - \dfrac{5}{z} - \dfrac{3}{z^2}$

4. $-2a^3b^2 - 4a^2b + 3$

Practice Exercises

1. $2a^3 - 3a$ 3. $-13m^2 + 4m - \dfrac{5}{m^2}$

Objective 2
Now Try

5. $4x + 3$ 6. $2x^2 - 2x - 2 + \dfrac{-5}{5x-1}$ 7. $x^2 + 10x + 100$

8. $3x^2 + 5x + 5 + \dfrac{8x+29}{x^2-4}$

Practice Exercises

5. $3x^2 - 6x + 2 + \dfrac{13x-7}{2x^2+3}$

Answers

Objective 3
Now Try
10. $L = 2r^2 - r + 5$ units

Practice Exercises
7. $4y^2 + 24y + 100$ units

Chapter 5 FACTORING AND APPLICATIONS

5.1 The Greatest Common Factor; Factoring by Graphing

Key Terms
1. factoring
2. factored form
3. greatest common factor
4. factor

Objective 1
Now Try
1a. 6
1b. 8
1c. 1
2. $6x^4$

Practice Exercises
1. 28
3. $9xy^2$

Objective 2
Now Try
3. $4y^2(5y^2 - 3y + 1)$
5a. $(y+8)(y+4)$
5b. $(z+5)(z^2-11)$

Practice Exercises
5. $(x-2y)(2a+9b)$

Objective 3
Now Try
6a. $(9+t)(4x+1)$
6b. $(x-7)(4x+5y)$
6c. $(x+7)(x^2-2)$
7. $(8x-3y)(7x+4)$

Practice Exercises
7. $(5-y)(3-x)$
9. $(r^2+s^2)(3r-2s)$

Answers

5.2 Factoring Trinomials

Key Terms
1. factoring
2. greatest common factor
3. prime polynomial

Objective 1
Now Try
1. $(x+3)(x+8)$
2. $(y-7)(y-5)$
3. $(p+9)(p-3)$
5. prime
6. $(p-7q)(p+2q)$

Practice Exercises
1. prime
3. $(x-11)(x+3)$

Objective 2
Now Try
7. $7x^4(x-5)(x-2)$

Practice Exercises
5. $2ab(a-3b)(a-2b)$

5.3 More on Factoring Trinomials

Key Terms
1. coefficient
2. trinomial
3. inner product
4. FOIL
5. outer product

Objective 1
Now Try
1. $(5x+2)(x+3)$
2a. $(7x-5)(2x+1)$
2b. $(3m-7)(m+2)$
2c. $(5x+3y)(2x-y)$
3. $3x^3(5x-3)(2x+7)$

Practice Exercises
1. $(4b+3)(2b+3)$
3. $(5c-7t)(2c-3t)$

Objective 2
Now Try
5. $(3x+1)(5x+7)$
6. $(4x-1)(5x-2)$
7. $(4x+7)(2x-3)$
8. $(6x-5y)(4x+3y)$
9. $-6a(3a-5)(a-2)$

Practice Exercises
5. $(a+2b)(3a+2b)$

Answers

5.4 Special Factoring Techniques

Key Terms
1. difference
2. perfect square trinomial

Objective 1
Now Try
1. $(z+6)(z-6)$
2a. $(2x+9)(2x-9)$
2b. $(5t+7)(5t-7)$
3a. $10(3x+7)(3x-7)$
3b. $(p^2+16)(p+4)(p-4)$

Practice Exercises
1. $(x-7)(x+7)$
3. prime

Objective 2
Now Try
4. $(p+8)^2$
5a. prime
5b. $5x(2x+5)^2$
5c. $(8m+3)^2$

Practice Exercises
5. $(3j+2)^2$

Objective 3
Now Try
6a. $(t-6)(t^2+6t+36)$
6b. $(3k-y)(9k^2+3ky+y^2)$
6c. $(3x+7y^2)(9x^2-21xy^2+49y^4)$

Practice Exercises
7. $(2a-5b)(4a^2+10ab+25b^2)$
9. $2n(3m^2+n^2)$

Objective 4
Now Try
7a. $(6x+1)(36x^2-6x+1)$
7b. $6(x+2y)(x^2-2xy+4y^2)$

Practice Exercises
11. $8(a+2b)(a^2-2ab+4b^2)$

Answers

5.5 Solving Quadratic Equations Using the Zero-Factor Property

Key Terms
1. standard form
2. double solution
3. quadratic equation

Objective 1

Now Try

1a. $\left\{-12, \frac{7}{4}\right\}$
1b. $\left\{0, \frac{11}{6}\right\}$
3. $\left\{\frac{4}{5}, 3\right\}$
4. $\left\{-\frac{3}{10}, \frac{3}{10}\right\}$

Practice Exercises

1. $\left\{-\frac{5}{2}, 4\right\}$
3. $\left\{-4, \frac{3}{5}\right\}$

Objective 2

Now Try

6a. $\{-5, 0, 5\}$
6b. $\left\{\frac{2}{5}, 2, 9\right\}$
7. $\{-4, 3\}$

Practice Exercises

5. $\{-9, 0, 1\}$

5.6 Applications of Quadratic Equations

Key Terms
1. legs
2. hypotenuse

Objective 1

Now Try
1. width: 3 m, length: 5 m

Practice Exercises
1. width: 8 in., length: 24 in.
3. height: 4 ft, width: 6 ft

Objective 2

Now Try
2. 0, 1, 2, or 5, 6, 7

Practice Exercises
5. 6, 8

Objective 3

Now Try
3. 16 ft

Practice Exercises
7. 45 m, 60 m, 75 m
9. 20 mi

Answers

Objective 4
  Now Try
  4. 1 sec

  Practice Exercises
  11. 40 items or 110 items

Chapter 6 RATIONAL EXPRESSIONS AND APPLICATIONS

6.1   The Fundamental Property of Rational Expressions

Key Terms
  1. rational expression   2. lowest terms

Objective 1
  Now Try
  1. 14

  Practice Exercises
  1. a. $-\frac{11}{9}$; b. -4 3. a. $-\frac{1}{6}$; b. $-\frac{7}{2}$

Objective 2
  Now Try
  2a. $y \neq \frac{1}{7}$ 2b. $m \neq 5, m \neq -4$ 2c. never undefined

  Practice Exercises
  5. none

Objective 3
  Now Try
  3. $\frac{3}{k^3}$ 4a. $\frac{7}{9}$ 4b. $\frac{m+6}{2m+3}$
  5. -1

  Practice Exercises
  7. $\frac{-5b}{8c}$ 9. $\frac{9(x+3)}{2}$

Objective 4
  Now Try
  7. $\frac{-(10x-7)}{4x-3}$, $\frac{-10x+7}{4x-3}$, $\frac{10x-7}{-(4x-3)}$, $\frac{10x-7}{-4x+3}$

  Practice Exercises
  11. $\frac{-(2p-1)}{-(1-4p)}$; $\frac{1-2p}{4p-1}$; $\frac{-(2p-1)}{4p-1}$; $\frac{1-2p}{-(1-4p)}$

Answers

6.2 Multiplying and Dividing Rational Expressions

Key Terms
1. reciprocal 2. rational expression 3. lowest terms

Objective 1
Now Try

1a. $\dfrac{4}{15}$ 1b. $\dfrac{4}{3x}$ 2. $\dfrac{s^2}{6(r-s)}$

3. $\dfrac{35}{x}$

Practice Exercises

1. $\dfrac{10m^3n}{3}$ 3. $\dfrac{x+4}{2x-8}$

Objective 2
Now Try

4a. $\dfrac{15}{2}$ 4b. $\dfrac{y-2}{6(y+2)}$ 6. $\dfrac{8x}{(x-3)^2}$

7. $\dfrac{-(m-8)}{5m(m+9)}$

Practice Exercises

5. $\dfrac{(m-1)(m+n)}{m(m-n)}$

6.3 Least Common Denominators

Key Terms
1. equivalent expressions   2. least common denominator

Objective 1
Now Try

1. 72   2. $120a^4$ 3a. $9w(w-2)$

3b. $(b+4)(b+1)(b-4)^2$ 3c. $p-14$ or $14-p$

Practice Exercises

1. $108b^4$ 3. $w(w+3)(w-3)(w-2)$

Objective 2
Now Try

4. $\dfrac{65}{30}$ 5a. $\dfrac{76}{24c-20}$

Copyright © 2016 Pearson Education, Inc.

Answers

5b. $\dfrac{3(z+2)}{z(z-7)(z+2)}$ or $\dfrac{3z+6}{z^3-5z^2-14z}$

Practice Exercises
5. $30r$

6.4 Adding and Subtracting Rational Expressions

Key Terms
1. greatest common factor
2. least common multiple

Objective 1
 Now Try
1a. $\dfrac{4}{5}$
1b. $2x$

 Practice Exercises
1. $\dfrac{4}{w^2}$
3. $\dfrac{1}{x-2}$

Objective 2
 Now Try
2a. $\dfrac{137}{315}$
2b. $\dfrac{46}{63y}$
4. $\dfrac{7x^2+11x+8}{(x+2)(x+1)(x-4)}$

 Practice Exercises
5. $\dfrac{7z^2-z-6}{(z+2)(z-2)^2}$

Objective 3
 Now Try
9. $\dfrac{8x^2+37x+15}{(x+5)(x-5)^2}$
8. 7

 Practice Exercises
7. $\dfrac{8z}{(z-2)(z+2)}$ or $\dfrac{8z}{z^2-4}$
9. $\dfrac{2m^2-m+2}{(m-2)(m+2)^2}$

6.5 Complex Fractions

Key Terms
1. complex fraction
2. LCD

Answers

Objective 1
Objective 2
Now Try

1. $\dfrac{90}{x}$
2. $\dfrac{bc^2}{a^2}$
3. $\dfrac{-9x+65}{2x-5}$

Practice Exercises

1. $\dfrac{7m^2}{2n^3}$
3. $\dfrac{9s+12}{6s^2+2s}$ or $\dfrac{3(3s+4)}{2s(3s+1)}$

Objective 3
Now Try

4. $\dfrac{12}{x}$
5. $\dfrac{5n-9}{3(6n+5)}$

Practice Exercises

5. $\dfrac{(x-2)^2}{x(x+2)}$

Objective 4
Now Try

7. $\dfrac{(x+y)(x+1)}{x}$

Practice Exercises

7. $\dfrac{2z^3 + xy^2z^3}{x}$
9. $\dfrac{2y^3}{x^2y^3 + 3x^2}$

6.6 Solving Equations with Rational Expressions

Key Terms
1. extraneous solution
2. proposed solution

Objective 1
Now Try

1a. equation; {14}
1b. expression; $\dfrac{1}{2}x$

Practice Exercises

1. equation; {−2}
3. expression; $\dfrac{41x}{15}$

Objective 2
Now Try
4. ∅
7. {−4}

Answers

Practice Exercises
5. $\{-4, 16\}$

Objective 3
Now Try

9a. $b = aq + c$ 9b. $y = \dfrac{x - wz}{w}$, or $\dfrac{x}{w} - z$ 10. $x = \dfrac{yz}{y+z}$

Practice Exercises
7. $f = \dfrac{d_0 d_1}{d_0 + d_1}$ 9. $q = \dfrac{2pf - Ab}{Ab}$ or $\dfrac{2pf}{Ab} - 1$

6.7   Applications of Rational Expressions
Key Terms
1. numerator   2. denominator   3. reciprocal

Objective 1
Now Try
1. 3

Practice Exercises
1. $-\dfrac{2}{3}$ or 1   3. $\dfrac{3}{5}$

Objective 2
Now Try
2. 3 miles per hour

Practice Exercises
5. 24 miles per hour

Objective 3
Now Try
3. $\dfrac{2}{5}$ hour

Practice Exercises
7. $2\dfrac{2}{5}$ hr

Answers

Chapter 7 GRAPHS, LINEAR EQUATIONS, AND SYSTEMS

7.1 Linear Equations in Two Variables

Key Terms
1. y-intercept
2. linear equation in two variables
3. coordinate
4. origin
5. x-intercept
6. x-axis
7. quadrant
8. ordered pair
9. y-axis
10. components
11. plot
12. rectangular (Cartesian) coordinate system
13. slope
14. graph of an equation
15. linear equation in two variables
16. rise
17. run

Objective 2
Now Try

1.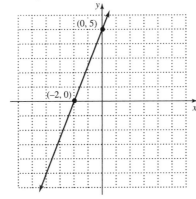

2.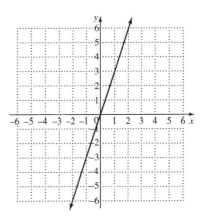

Practice Exercises

1.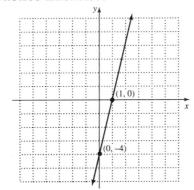

Answers

Objective 3
 Now Try
3a. 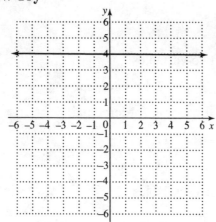 3b.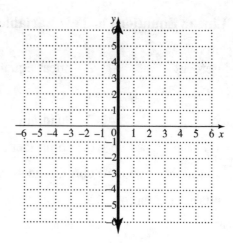

Practice Exercises

3. $x - 1 = 0$

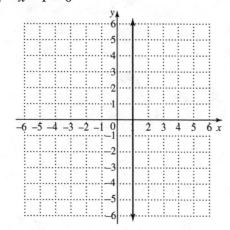

Objective 4
 Now Try
 4. $(5, -2)$

 Practice Exercises
 5. $(2, 2)$ 7. $(-5.3, -1.8)$

Objective 5
 Now Try
 5. $-\dfrac{16}{9}$

 Practice Exercises
 9. -3

Answers

Objective 6
Now Try
8.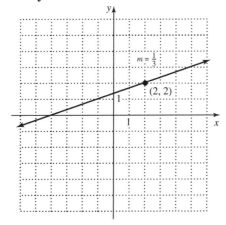

Practice Exercises
11.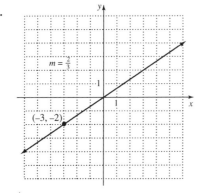

Objective 7
Now Try
9a. parallel   9b. perpendicular

Practice Exercises
13. neither   15. perpendicular

Objective 8
Now Try
10. 113.5 ft/min   11. an increase of 5 employees/yr

Practice Exercises
17. 4500 people/yr

Answers

7.2 Review of Equations of Lines; Linear Models

Key Terms
1. point-slope form
2. standard form
3. slope-intercept form

Objective 1
  Now Try
  1. $y = \dfrac{7}{9}x + 8$

  Practice Exercises
  1. $y = \dfrac{3}{2}x - \dfrac{2}{3}$
  3. $y = -\dfrac{6}{5}x + \dfrac{2}{5}$

Objective 2
  Now Try
  2.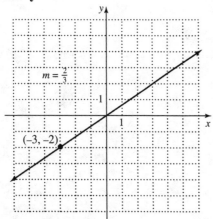

  Practice Exercises
  5.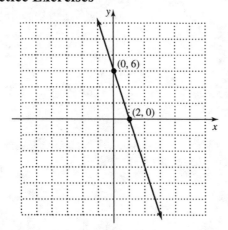

Objective 3
  Now Try
  3. $y = \dfrac{4}{5}x - \dfrac{44}{5}$

  Practice Exercises
  7. $3x + 5y = 11$
  9. $2x - 3y = -8$

Answers

Objective 4
Now Try
4. $y = -\dfrac{3}{4}x + \dfrac{81}{4}$; $3x + 4y = 81$

Practice Exercises
11. $x + 3y = -1$

Objective 5
Now Try
5a. $y = 5$ 5b. $x = -5$

Practice Exercises
13. $x = -1$ 15. $y = -5$

Objective 6
Now Try
6a. $y = -\dfrac{3}{4}x + \dfrac{7}{2}$ 6b. $y = \dfrac{4}{3}x + 16$

Practice Exercises
17. $4x - 3y = -17$

Objective 7
Now Try
7a. $y = -28.8x + 686$ 7b. $(1, 657.2)$

Practice Exercises
19. (a) $y = 1.25x + 25$; (b) $(0, 25)$, $(5, 31.25)$, $(10, 37.50)$

7.3 Solving Systems of Linear Equations by Graphing

Key Terms
1. independent equations
2. consistent system
3. solution set of the system
4. solution of the system
5. dependent equations
6. inconsistent system
7. system of linear equations

Objective 1
Now Try
1. no

Practice Exercises
1. no   3. no

Answers

Objective 2
  Now Try
2.
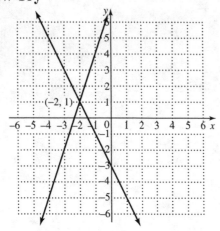

  Practice Exercises
5.
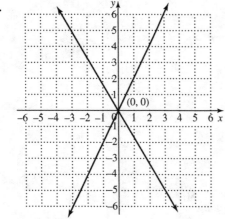

Objective 3
  Now Try
  3a. ∅   3b. $\{(x, y)\mid 4x - 2y = 8\}$

  Practice Exercises
  7. no solution

Objective 4
  Now Try
  4a. neither   4b. intersecting lines   4c. exactly one solution

  Practice Exercises
  9. (a) neither   (b) intersecting lines   (c) one solution

  11. (a) dependent   (b) one line   (c) infinitely many solutions

Answers

7.4 Solving Systems of Linear Equations by Substitution

Key Terms
1. ordered pair
2. substitution
3. dependent system
4. inconsistent system

Objective 1
  Now Try
  1. $\{(1, 6)\}$
  2. $\{(9, -4)\}$
  3. $\{(8, -2)\}$

  Practice Exercises
  1. $(2, 4)$
  3. $(4, -9)$

Objective 2
  Now Try
  4. $\varnothing$
  5. $\{(x, y)\mid 5x + 4y = 20\}$

  Practice Exercises
  5. $\{(x, y) \mid x - 2y = -6\}$

Objective 3
  Now Try
  6. $\{(-2, 5)\}$
  7. $\{(3, -5)\}$

  Practice Exercises
  7. $(-9, -11)$
  9. $\{(x, y) \mid 0.3x + 0.4y = 0.5\}$

7.5 Solving Systems of Linear Equations by Elimination

Key Terms
1. elimination method
2. addition property of equality
3. substitution

Objective 1
  Now Try
  1. $\{(8, 3)\}$

  Practice Exercises
  1. $(8, 3)$
  3. $(5, 0)$

Objective 2
  Now Try
  3. $\{(-4, 9)\}$

  Practice Exercises
  5. $\left(\dfrac{1}{2}, 1\right)$

Answers

Objective 3
  Now Try
  4. $\left\{\left(\frac{9}{17}, \frac{11}{17}\right)\right\}$

  Practice Exercises
  7. $(2, -4)$ 9. $(3, -2)$

Objective 4
  Now Try
  5a. $\{(x, y) \mid 9x - 7y = 5\}$ 5b. $\varnothing$

  Practice Exercises
  11. $\{(x, y) \mid -x - 2y = 3\}$

7.6 Systems of Linear Equations in Three Variables

Key Terms
  1. ordered triple   2. dependent system   3. inconsistent system

Objective 1
  Practice Exercises
  1. The planes intersect in one point.

Objective 2
  Now Try
  1. $\{(-3, 1, 2)\}$

  Practice Exercises
  3. $\{(-1, 2, 1)\}$ 5. $\{(4, -4, 1)\}$

Objective 3
  Now Try
  2. $\{(2, -5, 3)\}$

  Practice Exercises
  7. $\{(4, 2, -1)\}$

Objective 4
  Now Try
  4. $\{(x, y, z) \mid x - 5y + 2z = 0\}$ 3. $\varnothing$

  Practice Exercises
  9. $\varnothing$; inconsistent system
  11. $\{(x, y, z) \mid -x + 5y - 2z = 3\}$; dependent equations

Answers

7.7 Applications of Systems of Linear Equations

Key Terms
1. elimination method 2. substitution

Objective 1
Now Try
1. length: 17 ft; width: 10 ft

Practice Exercises
1. square: 8 cm; triangle: 13 cm   3. 21 cm

Objective 2
Now Try
2. marigold: $12.29; carnation: $17.60

Practice Exercises
5. large: $6; small: $3

Objective 3
Now Try
3. water: 6 liters; 25% solution: 24 liters

Practice Exercises
7. $6 coffee: 100 lbs; $12 coffee: 50 lb

9. water: 9 oz; 80% solution: 3 oz

Objective 4
Now Try
4. Ashley: 5 mph; Taylor: 3 mph

Practice Exercises
11. distance from school: $1\frac{1}{8}$ mi; jogging time: $\frac{1}{8}$ hr or $7\frac{1}{2}$ min

Objective 5
Now Try
6. 20 lb of $4 candy; 30 lb of $6 candy; 50 lb of $10 candy

Practice Exercises
13. $8 coffee: 15 lb; $10 coffee: 12 lb; $15 coffee: 23 lb

Answers

Chapter 8   INEQUALITIES AND ABSOLUTE VALUE

8.1   Review of Linear Inequalities in One Variable

Key Terms
1. interval notation
2. linear inequality in one variable
3. inequality
4. interval

Objective 2
Now Try

1. $(-\infty, -4)$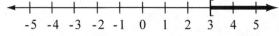

Practice Exercises

1. $[3, \infty)$;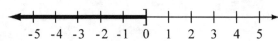

3. $(-\infty, 0]$;

Objective 3
Now Try

2. $(-\infty, -5]$

3. $(-\infty, -8]$

4. $(7, \infty)$

Practice Exercises

5. $[-4, \infty)$;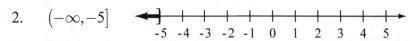

Objective 4
Now Try

5. $(-4, -2]$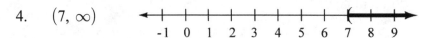

Practice Exercises

7. $(2, 5]$;

9. $(-1, 5)$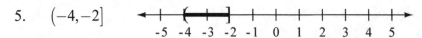

Answers

8.2 Set Operations and Compound Inequalities

Key Terms
1. union
2. compound inequality
3. intersection

Objective 2
Now Try
1. {30, 50}

Practice Exercises
1. {0, 2, 4}
3. {1, 3, 5}

Objective 3
Now Try
2. [12, 16]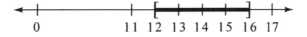

3. $(-\infty, -3)$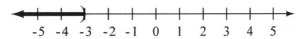

Practice Exercises
5. $[-4, 3)$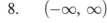

Objective 4
Now Try
5. {20, 30, 40, 50, 70}

Practice Exercises
7. {0, 1, 2, 3, 4, 5}
9. {0, 1, 2, 3, 4, 5}

Objective 5
Now Try
6. $(-\infty, 2) \cup [5, \infty)$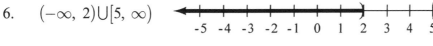

8. $(-\infty, \infty)$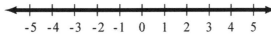

Practice Exercises
11. $(-\infty, \infty)$

8.3 Absolute Value Equations and Inequalities

Key Terms
1. absolute value equation
2. absolute value inequality

Answers

Objective 2
Now Try
1. $\left\{-3, \dfrac{7}{5}\right\}$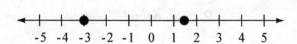

Practice Exercises
1. $\left\{-\dfrac{13}{2}, \dfrac{7}{2}\right\}$ 3. $\{-2, 14\}$

Objective 3
Now Try
2. $(-\infty, -2) \cup (1, \infty)$

3. $(-3, 2)$

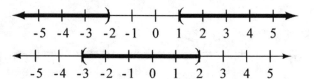

Practice Exercises
5. $(-\infty, -7] \cup [16, \infty)$

Objective 4
Now Try
5. $\{-4, 10\}$

Practice Exercises
7. $\{-2, 3\}$ 9. $\left\{-\dfrac{5}{4}, -\dfrac{1}{4}\right\}$

Objective 5
Now Try
7. $\{-5, -1\}$

Practice Exercises
11. $\left\{-3, \dfrac{11}{3}\right\}$

Objective 6
Now Try
8a. $\varnothing$ 8b. $\{-5\}$ 9a. $(-\infty, \infty)$

9b. $\varnothing$ 9c. $\{4\}$

Practice Exercises
13. $\{-14\}$ 15. $\varnothing$

Objective 7
Now Try
10. between 16.055 and 17.745 oz, inclusive

Answers

Practice Exercises
17. between 16.6465 and 17.1535 oz, inclusive

8.4 Linear Inequalities and Systems in Two Variables

Key Terms

1. boundary line
2. solution set of a system of linear inequalities
3. system of linear inequalities
4. linear inequality in two variables

Objective 1

Now Try

1.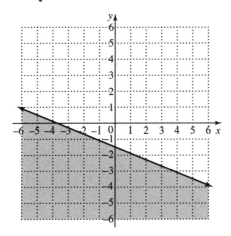
2.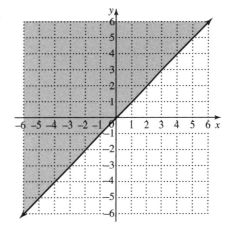
3.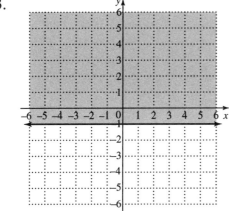

Practice Exercises

1.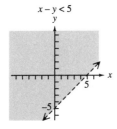
3.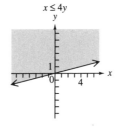

Answers

Objective 2
Now Try
4.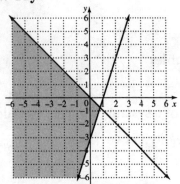

Practice Exercises
5.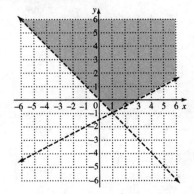

Answers

Chapter 9 RELATIONS AND FUNCTIONS

9.1 Introduction to Relations and Functions

Key Terms
1. range
2. relation
3. dependent variable
4. domain
5. function
6. independent variable

Objective 1
Now Try
1. {(1, 3), (2, 4), (3, 6), (5, 7)}
2a. function
2b. not a function

Practice Exercises
1. {(1, 3), (1, 4), (2, −1), (3, 7)}
3. not a function

Objective 2
Now Try
3. domain: {13}; range: {−2, −1, 4}; not a function
4a. domain: $(-\infty, \infty)$; range: $(-\infty, \infty)$
4b. domain: $[-4, \infty)$; range: $(-\infty, \infty)$

Practice Exercises
5. function; domain: {A, B, C, D, E}; range: {V, W, X, Z}

Objective 3
Now Try
5. function
6. function; domain: $(-\infty, \infty)$

Practice Exercises
7. function
9. function; $(-\infty, -6) \cup (-6, \infty)$

Answers

9.2 Function Notation and Linear Functions

Key Terms
1. linear function
2. function notation
3. constant function

Objective 1
Now Try
1. $f(3) = 17$
3. $g(a-1) = 4a - 11$
4. $f(-6) = 21$
6. $f(x) = -\frac{2}{3}x + \frac{7}{3}, f(-3) = \frac{13}{3}$

Practice Exercises
1. (a) -13; (b) -7; (c) $-3x - 7$
3. (a) 9; (b) 9; (c) 9

Objective 2
Now Try
7. domain: $(-\infty, \infty)$; range: $(-\infty, \infty)$

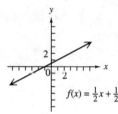

Practice Exercises
5.

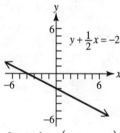

domain: $(-\infty, \infty)$
range: $(-\infty, \infty)$

9.3 Polynomial Functions, Operations, and Composition

Key Terms
1. composite function
2. polynomial function of degree n

Objective 1
Now Try
1. 1

Practice Exercises
1. (a) -7; (b) -17
3. (a) 28; (b) 198

Answers

Objective 2
Now Try
2a. $3x^2 - 6x + 21$ 2b. $9x^2 - 8x + 3$ 3. $42x^3 + 47x^2 + 10x$; -741
5. $2x + 5$ $x \neq 4$; 1

Practice Exercises
5. (a) $-15x^3 - 10x$; (b) 25

Objective 3
Now Try
6. -15 7. -15

Practice Exercises
7. a. 25; b. 97; c. $8x^2 - 7$

9.4 Variation

Key Terms
1. constant of variation
2. varies inversely
3. varies directly

Objective 1; Objective 2
Now Try
1. $25; $P = 25x$ 2. 125 psi 3. 153.86 sq cm

Practice Exercises
1. $k = 0.25$; $y = 0.25x$ 3. 100 newtons

Objective 3
Now Try
4. 640 cycles per second 5. $\dfrac{9}{8}$

Practice Exercises
5. 40 lb

Objective 4
Now Try
6. 1280 psi

Practice Exercises
7. 96 9. 750°

Objective 5
Now Try
7. About 63.5 cm^3

Practice Exercises
11. 9 hr

Answers

Chapter 10 ROOTS, RADICALS, AND ROOT FUNCTIONS

10.1 Radical Expressions and Graphs

Key Terms
1. fourth root
2. radical
3. principal square root
4. radicand
5. cube root function
6. perfect square
7. negative square root
8. index (order)
9. cube root
10. square root function
11. square root
12. radical expression

Objective 1
Now Try

2a. 13 2b. 41 2c. $\dfrac{3}{7}$

3a. 19 3b. 37 3c. $n^2 + 5$

Practice Exercises

1. 25, −25 3. $\dfrac{30}{7}$

Objective 2
Now Try

4a. irrational 4b. rational 4c. not a real number

Practice Exercises

5. not a real number

Objective 3
Now Try

5a. 5 5b. −5 5c. 7
6a. 5 6b. −5 6c. none
6d. −5 6e. −5

Practice Exercises

7. −4 9. −1

Objective 4
Now Try

7a.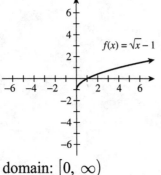

domain: $[0, \infty)$
range: $[-1, \infty)$

7b.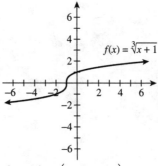

domain: $(-\infty, \infty)$
range: $(-\infty, \infty)$

Answers

Practice Exercises

11.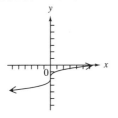

domain: $(-\infty, \infty)$

range: $(-\infty, \infty)$

Objective 5
Now Try
8a. 73   8b. 37   8c. $|n|$
9a. 4   9b. -5 9c. -2
9d. $-x^4$ 9e. w^{10} 9f. $|x^5|$

Practice Exercises
13. 9   15. $-x^4$

Objective 6
Now Try
10a. 3.742   10b. -26.038 10c. 7.554

Practice Exercises
17. -31.464

10.2 Rational Exponents

Key Terms
1. power rule for exponents
2. product rule for exponents
3. quotient rule for exponents

Objective 1
Now Try
1a. 11   1b. -3 1c. not a real number

Practice Exercises
1. -4 3. -15

Objective 2
Now Try
2a. 9   2b. -216 3. $\dfrac{1}{4}$

Practice Exercises
5. 7776

Answers

Objective 3
Now Try
4a. $\sqrt[3]{23}$ 4b. $\left(\sqrt[3]{2x}\right)^4 - 3\left(\sqrt[5]{x}\right)^2$ 4c. $10^2 = 100$

Practice Exercises
7. $4\left(\sqrt[5]{y}\right)^2 + \sqrt[5]{5x}$ 9. $a^{1/4}$, or $\sqrt[4]{a}$

Objective 4
Now Try
5a. $5^{5/2}$ 5b. $\dfrac{x^2}{y^{1/4}}$ 6a. $x^{7/6}$

6b. $x^{1/4}$

Practice Exercises
11. $a^{5/6}$

10.3 Simplifying Radicals, the Distance Formula, and Circles

Key Terms
1. hypotenuse   2. index; radicand   3. legs
4. circle   5. center   6. radius

Objective 1
Now Try
1a. $\sqrt{14}$ 1b. $\sqrt{33mn}$ 2a. $\sqrt[3]{21}$
2b. $\sqrt[3]{35xy}$ 2c. $\sqrt[5]{8w^4}$

2d. cannot be simplified using the product rule

Practice Exercises
1. $\sqrt{42tx}$ 3. cannot be simplified using the product rule

Objective 2
Now Try
3a. $\dfrac{6}{7}$ 3b. $\dfrac{\sqrt{13}}{9}$ 3c. $-\dfrac{7}{5}$

3d. $-\dfrac{a^2}{5}$ 3e. $\dfrac{\sqrt[4]{m}}{3}$

Practice Exercises
5. $\dfrac{\sqrt[5]{7x}}{2}$

Answers

Objective 3
Now Try

4a. $2\sqrt{21}$ 4b. $9\sqrt{2}$ 4c. cannot be simplified

4d. $4\sqrt[3]{4}$ 4e. $-2\sqrt[5]{16}$ 5a. $10y\sqrt{y}$

5b. $4m^2r^4\sqrt{3mr}$ 5c. $-2n^2t\sqrt[3]{4nt^2}$ 5d. $-3y^2\sqrt[4]{5x^3y}$

6a. $\sqrt[4]{11^3}$, or $\sqrt[4]{1331}$ 6b. $\sqrt[5]{z^4}$

Practice Exercises

7. $\sqrt[3]{x^2}$ 9. $5ab^2\sqrt[3]{10a^2b}$

Objective 4
Now Try

7. $\sqrt[6]{63}$

Practice Exercises

11. $\sqrt[8]{28}$

Objective 5
Now Try

8. $12\sqrt{2}$

Practice Exercises

13. 26 15. $6\sqrt{2}$

Objective 6
Now Try

9. $\sqrt{34}$

Practice Exercises

17. 5

Objective 7
Now Try

10. $x^2+y^2=25$

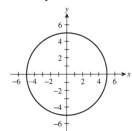

11. $(x+5)^2+(y-4)^2=16$

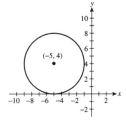

12. $(x-3)^2+(y+1)^2=6$

Answers

Practice Exercises

19. $(x-3)^2 + (y+4)^2 = 25$
21. $x^2 + (y-3)^2 = 2$

10.4 Adding and Subtracting Radical Expressions

Key Terms
1. unlike radicals
2. like radicals

Objective 1
Now Try

1a. $-\sqrt{6}$
1b. $13\sqrt{2z}$
1c. cannot be simplified

2a. $-3\sqrt[3]{2}$
2b. $5z\sqrt[4]{2y^2z}$
2c. $6x^2\sqrt[3]{2x} + 6x^3\sqrt{3x}$

3a. $\dfrac{2\sqrt{5}}{3}$
3b. $\dfrac{17}{w^2}$

Practice Exercises

1. $12\sqrt{x}$
3. $\dfrac{8y^2\sqrt[3]{y}}{15}$

10.5 Multiplying and Dividing Radical Expressions

Key Terms
1. conjugate
2. rationalizing the denominator

Objective 1
Now Try

2a. 12
2b. $6 + 3\sqrt{7} - 2\sqrt{2} - \sqrt{14}$
2c. $11 - 6\sqrt{2}$

Practice Exercises

1. $\sqrt{10} - 4\sqrt{5} + 2\sqrt{3} - 4\sqrt{6}$
3. $4 - \sqrt[3]{25}$

Objective 2
Now Try

3. $\dfrac{2\sqrt{15}}{15}$
4. $-\dfrac{3\sqrt{10}}{8}$
5. $\dfrac{\sqrt[3]{10}}{5}$

Practice Exercises

5. $\dfrac{y\sqrt{21b}}{6b}$

Objective 3
Now Try

6. $-2(\sqrt{3}+2)$

Answers

Practice Exercises

7. $-4\sqrt{3}+8$

9. $\dfrac{\sqrt{3}+2\sqrt{6}+\sqrt{2}+4}{-7}$

Objective 4
Now Try

7. $\dfrac{3+2\sqrt{15}}{4}$

Practice Exercises

11. $\dfrac{2-9\sqrt{2}}{3}$

10.6 Solving Equations with Radicals

Key Terms
1. extraneous solution   2. radical equation

Objective 1
Now Try
1. {10}   2. ∅   3. {5}
5. {4}

Practice Exercises
1. {11}   3. {−9}

Objective 2
Now Try
6. {2}

Practice Exercises
5. {−4}

Objective 3
Now Try

7. $h = \dfrac{3v}{\pi r^2}$

Practice Exercises

7. $L = Z^2 C$

9. $r = \dfrac{a}{4\pi^2 N^2}$

Answers

10.7 Complex Numbers

Key Terms
1. complex number
2. complex conjugate
3. imaginary part
4. real part
5. standard form (of a complex number)
6. pure imaginary number
7. nonreal complex number

Objective 1
  Now Try
  1. $4i$
  2. $-\sqrt{30}$
  3. 5

  Practice Exercises
  1. $-9i\sqrt{2}$
  3. $6i$

Objective 2
  Practice Exercises
  5. imaginary

Objective 3
  Now Try
  4. $10 - 9i$
  5. $24 + 4i$

  Practice Exercises
  7. $-4 + i$
  9. $2 - 3i$

Objective 4
  Now Try
  6. $-14 + 8i$

  Practice Exercises
  11. $-8 + 6i$

Objective 5
  Now Try
  7. $\dfrac{18}{29} + \dfrac{13i}{29}$

  Practice Exercises
  13. $\dfrac{4}{5} - \dfrac{7}{5}i$
  15. $\dfrac{15}{13} + \dfrac{16}{13}i$

Objective 6
  Now Try
  8. 1

  Practice Exercises
  17. i

Chapter 11 QUADRATIC EQUATIONS, INEQUALITIES, AND FUNCTIONS

11.1 Solving Quadratic Equations by the Square Root Property

Key Terms
1. quadratic equation
2. zero-factor property

Objective 1
  Now Try
  1a. $\{-7, -1\}$
  1b. $\{-10, 10\}$

  Practice Exercises
  1. $\{-4, -2\}$
  3. $\{-7, 5\}$

Objective 2
  Now Try
  2a. $\{-9, 9\}$
  2b. $\{-\sqrt{23}, \sqrt{23}\}$
  2c. $\{3\sqrt{2}, -3\sqrt{2}\}$, or $\{\pm 3\sqrt{2}\}$
  2d. $\{2\sqrt{3}, -2\sqrt{3}\}$, or $\{\pm 2\sqrt{3}\}$
  3. About 2.7 seconds

  Practice Exercises
  5. $\{-7\sqrt{2}, 7\sqrt{2}\}$

Objective 3
  Now Try
  5. $\left\{\dfrac{3 \pm 4\sqrt{2}}{7}\right\}$

  Practice Exercises
  7. $\{-6, 2\}$
  9. $\left\{\dfrac{1}{5}, \dfrac{4}{5}\right\}$

Objective 4
  Now Try
  6a. $\{-4i\sqrt{2}, 4i\sqrt{2}\}$
  6b. $\{-2-7i, -2+7i\}$

  Practice Exercises
  11. $\{-1-6i, -1+6i\}$

Answers

11.2 Solving Quadratic Equations by Completing the Square

Key Terms
1. perfect square trinomial
2. square root property
3. completing the square

Objective 1
Now Try

3. $\left\{\dfrac{11-\sqrt{89}}{2}, \dfrac{11+\sqrt{89}}{2}\right\}$

Practice Exercises

1. $\{-4-2\sqrt{3},\ -4+2\sqrt{3}\}$
3. $\{-9, 7\}$

Objective 2
Now Try

4. $\left\{\dfrac{5}{4}, \dfrac{11}{4}\right\}$
5. $\left\{\dfrac{-3-\sqrt{11}}{2}, \dfrac{-3+\sqrt{11}}{2}\right\}$

Practice Exercises

5. $\left\{\dfrac{-2-\sqrt{10}}{6}, \dfrac{-2+\sqrt{10}}{6}\right\}$

Objective 3
Now Try

7. $\left\{\dfrac{-3\pm\sqrt{33}}{2}\right\}$

Practice Exercises

7. $\left\{\dfrac{-1-\sqrt{13}}{2}, \dfrac{-1+\sqrt{13}}{2}\right\}$
9. $\{-2-\sqrt{2},\ -2+\sqrt{2}\}$

11.3 Solving Quadratic Equations by the Quadratic Formula

Key Terms
1. discriminant
2. quadratic formula

Objective 1
Objective 2
Now Try

1. $\left\{\dfrac{4}{3}, \dfrac{3}{2}\right\}$
2. $\left\{\dfrac{1-\sqrt{7}}{2}, \dfrac{1+\sqrt{7}}{2}\right\}$
3. $\{1-2i,\ 1+2i\}$

Practice Exercises

1. $\{2, 4\}$
3. $\{5-3i,\ 5+3i\}$

Answers

Objective 3
 Now Try
 4a. 81; two rational solutions; factoring

 4b. 0; one rational solution; factoring

 4c. –48; two nonreal complex solutions; quadratic formula

 Practice Exercises
 5. C

11.4 Solving Equations Quadratic in Form

Key Terms
 1. standard form   2. quadratic in form

Objective 1
 Now Try
 1. $\left\{-7, \dfrac{5}{4}\right\}$

 Practice Exercises
 1. $\left\{-\dfrac{35}{4}, -3\right\}$ 3. $\left\{-7, -\dfrac{7}{2}\right\}$

Objective 2
 Now Try
 2. 2 mph

 Practice Exercises
 5. 550 mph

Objective 3
 Now Try
 4a. {2}   4b. {8}

 Practice Exercises
 7. {2, 5}   9. {9}

Objective 4
 Now Try
 6. {–2, –1, 1, 2}   7a. {–2, 10}   7b. {–1, 27}

 Practice Exercises
 11. $\{-27, -3\sqrt{3}, 3\sqrt{3}, 27\}$

Answers

11.5 Formulas and Further Applications

Key Terms
1. quadratic function   2. Pythagorean theorem

Objective 1
Now Try
1. $t = \pm \dfrac{\sqrt{mxF}}{F}$ 2. $q = \dfrac{-k \pm k\sqrt{5}}{2p}$

Practice Exercises
1. $d = \dfrac{k^2 l^2}{F^2}$ 3. $a = \dfrac{-c \pm c\sqrt{2}}{b}$

Objective 2
Now Try
3. south: 72 mi; east 54 mi

Practice Exercises
5. 10 ft

Objective 3
Now Try
4. 2.5 in.

Practice Exercises
7. 2.5 ft   9. 4 ft

Objective 4
Now Try
5. 4.1 sec

Practice Exercises
11. 1.2 sec

Answers

11.6 Graphs of Quadratic Functions

Key Terms
1. axis
2. vertex
3. quadratic function
4. parabola

Objective 1; Objective 2
Now Try
1. vertex: (0,−1), axis: $x = 0$; domain: $(-\infty, \infty)$; range: $[-1, \infty)$

2. Vertex: (−3, 0); axis: $x = -3$ domain: $(-\infty, \infty)$; range: $[0, \infty)$

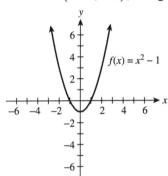
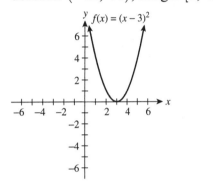

3. vertex: (−2,−1); axis: $x = -2$ domain: $(-\infty, \infty)$; range: $[-1, \infty)$

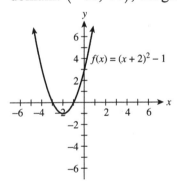

Practice Exercises
1.

Vertex: (0, −4)
Axis: $x = 0$
Domain: $(-\infty, \infty)$
Range: $[-4, \infty)$

3.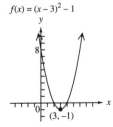

Vertex: (3, −1)
Axis: $x = 3$
Domain: $(-\infty, \infty)$
Range: $[-1, \infty)$

Answers

Objective 3
Now Try
4. vertex: $(0, 0)$; axis: $x = 0$
domain: $(-\infty, \infty)$; range: $(-\infty, 0]$

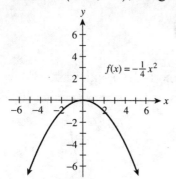

5. vertex: $(1, 1)$; axis: $x = 1$
domain: $(-\infty, \infty)$; range: $[1, \infty)$

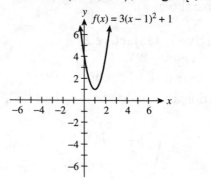

Practice Exercises
5. down; narrower; vertex: $(-1, 0)$; domain: $(-\infty, \infty)$; range: $[0, \infty)$

Objective 4
Now Try
6. $y = 0.01x^2 + 0.77x + 8.36$

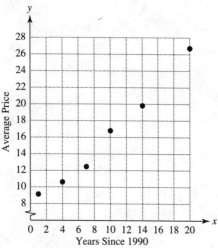

Practice Exercises
7. linear; positive

9.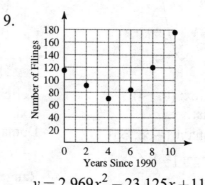

$y = 2.969x^2 - 23.125x + 115$

Answers

11.7 More about Parabolas and Their Applications

Key Terms
1. discriminant
2. vertex

Objective 1
  Now Try
  1. (3, –5)
  2. (1, 1)
  3. $\left(\dfrac{3}{2}, \dfrac{1}{2}\right)$

  Practice Exercises
  1. (1, 3)
  3. (–6, 0)

Objective 2
  Now Try
  4. vertex: (–2, 1), axis: $x = -2$;
     domain: $(-\infty, \infty)$; range: $[1, \infty)$

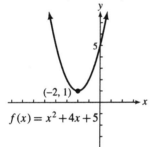

  Practice Exercises
  5.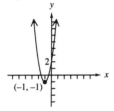

  Vertex: (–1, –1)
  Axis: $x = -1$
  Domain: $(-\infty, \infty)$
  Range: $[-1, \infty)$

Objective 3
  Now Try
  5. 1

  Practice Exercises
  7. 0
  9. 1

Answers

Objective 4
  Now Try
  6. maximum area: 125,000 sq yd; length: 500 yd; width: 250 yd

Practice Exercises
11. 24 (a square)

Objective 5
  Now Try
  9. vertex: (0, 2); axis: $y = 2$
     domain: $(-\infty, 0]$; range: $(-\infty, \infty)$

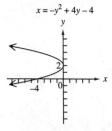

Practice Exercises
13.

Vertex: (2, 0)
Axis: $y = 0$
Domain: $(-\infty, 2]$
Range: $(-\infty, \infty)$

15.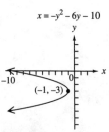

Vertex: $(-1, -3)$
Axis: $y = -3$
Domain: $(-\infty, -1]$
Range: $(-\infty, \infty)$

11.8 Polynomial and Rational Inequalities

Key Terms
  1. rational inequality   2. quadratic inequality

Objective 1
  Now Try
  1a. $(-\infty, 2) \cup (6, \infty)$ 1b. (2, 6)   2. $(-1, 2)$

  4a. $(-\infty, \infty)$ 4b. $\varnothing$

Answers

Practice Exercises
1. $[-1, 2]$
3. $\varnothing$

Objective 2
Now Try
5. $\left(-\infty, -\frac{3}{2}\right] \cup \left[-\frac{1}{3}, \frac{1}{2}\right]$

Practice Exercises
5. $(-\infty, -5] \cup [-3, 1]$

Objective 3
Now Try
6. $\left(-\frac{3}{2}, -1\right)$
7. $(-\infty, 3) \cup [8, \infty)$

Practice Exercises
7. $(-\infty, 1) \cup [8, \infty)$
9. $[-2, 3)$

Answers

Chapter 12 INVERSE, EXPONENTIAL, AND LOGARITHMIC FUNCTIONS

12.1 Inverse Functions

Key Terms
1. one-to-one function
2. inverse of a function

Objective 1
  Now Try
  1a. Not one-to-one
  1b. One-to-one; $G^{-1} = \{(2, 3), (-2, -3), (3, 2), (-3, -2)\}$
  1c. Not one-to-one

  Practice Exercises
  1. One-to-one; $\{(-1,-3), (2,-2), (3,-1), (4, 0)\}$
  3. One-to-one; $\{(0, 0), (1, 1), (-1, -1), (2, 2), (-2, -2)\}$

Objective 2
  Now Try
  2a. One-to-one
  2b. Not one-to-one

  Practice Exercises
  5. Not one-to-one

Objective 3
  Now Try
  3a. $f^{-1}(x) = \dfrac{1}{4}x + \dfrac{1}{4}$
  3b. Not one-to-one
  3c. $f^{-1}(x) = \sqrt[3]{\dfrac{x+3}{2}}$

  Practice Exercises
  7. $f^{-1}(x) = \dfrac{x+5}{2}$
  9. Not one-to-one

Objective 4
  Now Try
  5.
  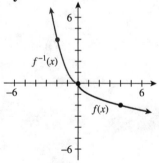

Practice Exercises
  11. Not one-to-one

Answers

12.2 Exponential Functions

Key Terms
1. inverse
2. exponential equation

Objective 1
Now Try
1a. 8.064
1b. 0.172
1c. 1.246

Practice Exercises
1. 3.737
3. 1.442

Objective 2
Now Try
2.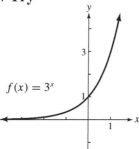
3.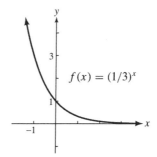
4.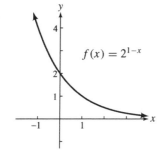

Practice Exercises
5.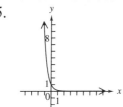

Objective 3
Now Try
5. $\left\{\dfrac{3}{2}\right\}$
6a. $\{-2\}$
6b. $\{-5\}$
6c. $\{-4\}$

Practice Exercises
7. $\left\{\dfrac{1}{2}\right\}$
9. $\{-2\}$

Answers

Objective 4
  Now Try
  7. 256,000   8. about 32,656

Practice Exercises
11. 1 gram

12.3 Logarithmic Functions

Key Terms
  1. logarithm   2. logarithmic equation

Objective 1
Objective 2
  Now Try

1a. $\log_8 64 = 2$ 1b. $16^{-1/2} = \dfrac{1}{4}$ 2a. 2.5850
2b. 2.1962   2c. −2.3219   2d. 1.0792

  Practice Exercises
  1. $10^{-3} = 0.001$ 3. 2.9299

Objective 3
  Now Try
3a. $\left\{\dfrac{1}{64}\right\}$ 3b. {40}   3c. $2\sqrt{3}$
3d. $\left\{\dfrac{1}{6}\right\}$

  Practice Exercises
  5. {81}

Objective 4
  Now Try
4a. 1   4b. 0   4c. 0
4d. 3   4e. 9   4f. 3

  Practice Exercises
  7. 0   9. 5

Objective 5
  Now Try
5. 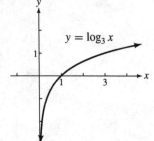   6.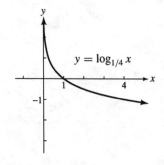

Answers

Practice Exercises

11.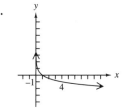

Objective 6
  Now Try
  7. 16 fish

  Practice Exercises
  13. 100 mites

12.4 Properties of Logarithms

Key Terms
  1. special properties
  2. quotient rule for logarithms
  3. product rule for logarithms
  4. power rule for logarithms

Objective 1
  Now Try
  1a. $\log_6 5 + \log_6 3$ 1b. $\log_5 21$ 1c. $1 + \log_4 x$

  1d. $4 \log_5 x$

  Practice Exercises
  1. $\log_7 5 + \log_7 m$ 3. $\log_4 21$

Objective 2
  Now Try
  2a. $\log_5 4 - \log_5 9$ 2b. $\log_6 \frac{x}{3}$ 2c. $2 - \log_4 11$

  Practice Exercises
  5. $\log_6 k - \log_6 3$

Objective 3
  Now Try
  3a. $3 \log_6 4$ 3b. $\frac{1}{2} \log_b 13$ 3c. $\frac{3}{4} \log_3 x$

  3d. $-7 \log_3 x$

  Practice Exercises
  7. $7 \log_m 2$ 9. $\sqrt[3]{7}$

Answers

Objective 4
Now Try
4a. $2 + 3\log_6 x$
4b. $\frac{1}{2}(\log_b x - \log_b 3)$
4c. $\log_b \frac{xy^4}{z}$

4d. $\log_2 \frac{x^2(x-1)}{\sqrt[3]{x^2+1}}$
4e. cannot be rewritten
5a. 7.6293

5b. -3.4594
5c. 9.5097
6a. False

6b. False

Practice Exercises
11. 6.1700

12.5 Common and Natural Logarithms

Key Terms
1. natural logarithm
2. common logarithm

Objective 1
Now Try
1. 2.9928

Practice Exercises
1. 1.7576
3. 4.9401

Objective 2
Now Try
2. pH = 7.2076; rich fen
3. 2.5×10^{-4}
4. 85 dB

Practice Exercises
5. 134 dB

Objective 3
Now Try
5. 4.5850

Practice Exercises
7. 4.3347
9. 3.9120

Objective 4
Now Try
6. 11.9 years

Practice Exercises
11. 3240 years

Answers

Objective 5
Now Try
7. 3.2266

Practice Exercises
13. 1.1887
15. −2.3219

12.6 Exponential and Logarithmic Equations; Further Applications

Key Terms
1. continuous compounding
2. compound interest

Objective 1
Now Try
1. 2.465
2. 6.770

Practice Exercises
1. {1}
3. {−4.324}

Objective 2
Now Try
5. $-1+\sqrt[3]{36}$
7. {1}

Practice Exercises
5. {5}

Objective 3
Now Try
8. $12,201.90
9. 13.89 years
10a. 5309.18
10b. 34.7 years

Practice Exercises
7. $55,200.99
9. 7.7 years

Objective 4
Now Try
11a. 11.3 mg
11b. 30.0 years

Practice Exercises
11. 29 years